canadian garden perennials

A.R. BUCKLEY

Published by
HANCOCK HOUSE PUBLISHERS
in association with
RESEARCH BRANCH, AGRICULTURE CANADA and
PUBLISHING CENTRE, SUPPLY AND SERVICES CANADA

hancock

house

canadian garden perennials

A.R. BUCKLEY

ISBN 0-919654-95-9 Paperback

ISBN 0-919654-78-9 Hard Cover

© Minister of Supply and Services Canada 1977
Government Catalog No. A-22-84-1977

Canadian Cataloguing in Publication Data

Buckley, Arthur R.
    Canadian garden perennials

    DSS cat. no. A-22-84-1977
    Includes index.
    ISBN 0-919654-95-9

    1. Perennials. 2. Perennials — Canada. I. Canada.
Dept. of Agriculture. Research Branch. II. Title.

SB434.B83   635.9'32   C77-002062-3

PRINTED IN CANADA

Published by:

Hancock House Publishers Ltd.
3215 Island View Road
SAANICHTON, B.C. V0S 1M0

Published by Hancock House Publishers in association with
Research Branch, Agriculture Canada, and the
Publishing Centre, Supply and Services Canada.

PREFACE

Herbaceous perennials form the backbone of most gardens; without them no garden is complete. This book is intended to provide the home gardener with practical information on many of these plants.

Most of the perennials mentioned are hardy in Eastern Canada and the milder areas of British Columbia, the Northeastern United States and the Pacific Northwest. Plants suitable for the Prairie regions are noted also.

During the past 10 years, many new varieties of perennials have been developed and are available to the home gardener at most garden centers and nurseries. Wild plants, such as corral bells, loosestrife and black-eyed Susan, have been improved so much that they are now a highlight in many gardens. New hybrids of the old-fashioned, common garden perennials are now available in differing forms and colors.

Dr. William Saunders, the first Director of the Central Experimental Farm, Ottawa, Ontario, started a program of testing all kinds of woody and herbaceous plants "to determine the kinds of plants most desirable to grow"—a venture that has continued every year, increasing in scope as new federal Experimental Stations were established in each province. Testing of perennials started at the Central Experimental Farm in 1887, with the development of a mile-long herbaceous border that stretched the length of the Botanic Garden and Arboretum. This book, like the departmental bulletins prepared by W.T. Macoun, Isabella Preston and R.W. Oliver that preceded it, will serve to bring this information to home gardeners, update old-timers and others interested in growing hardy perennials.

The author, A.R. Buckley, is one of Canada's most widely read garden columnists. He retired in 1973 from the Canada Department of Agriculture where he specialized in horticulture for 35 years. During that period, he evaluated new annuals, perennials, shrubs and trees at the Plant Research Insititute, located on the Central Experimental Farm, Ottawa. He received his formal training at the renowned John Innes Horticultural Institute, London, and the world-famous Royal Botanical Gardens at Kew, England. In 1934, Mr. Buckley came to Canada for further training at the Ontario Agricultural College, Guelph. After a brief return to England, he came back to Canada in 1938 as Curator of the Arboretum at the Central Experimental Farm, Ottawa.

The author's professional accomplishments have earned him the Queen's Coronation Medal, the Silver Medal of the Ontario Horticultural Association, and the Certificate of Merit of the Royal Horticultural Society of England. His weekly column—"Garden Notes"—which was carried by more than 100 daily newspapers across the country has been acclaimed by gardeners, amateur and professional alike. Although he has written numerous horticultural publications, Mr. Buckley regards CANADIAN GARDEN PERENNIALS as one of the most complete compilations of these easy-to-grow plants.

CONTENTS

PART I

Chapter 1. Planning and Planting   10

Chapter 2. Basic Perennials   16

Modern day-lilies   16
Delphiniums   17
Tall bearded irises   19
Oriental poppies   21
Peonies for your garden   22
Summer-flowering perennial phlox   24

Chapter 3. Biennials   28

Canterbury bells   28
Forget-me-not   28
Foxglove   29
Hollyhock   29
Iceland poppies   30
Mulleins   30
Pansies and violas   31
Sweet William   32
Wallflowers   32

Chapter 4. Culture of Perennials   36

Propagation   36
General care   38
Renovating the perennial border   41
Winter protection of perennials   42
Diseases   43
Insect pests   47
Insecticides   49

Chapter 5. Perennials for Special Purposes   52

PART II

Photos and Descriptions of Perennials 65
Color Plates 66
Descriptions of Perennials   92
Plans for Seasonal Borders   184
Index of Common and Scientific Names of Plants   192

part I

⁸ chapter 1

planning
and
planting

planning and planting

In their search for new and interesting plants with color and texture, home gardeners, as well as landscape architects, are turning more and more toward herbaceous perennials.

This group includes most plants with roots that live from year to year and plants with leafy stems that die down each fall and start to grow again in the spring. Also included in this book, however, are other plants such as pinks and moss pinks, both of which are evergreen, and plants like candytuft and alyssum, which, because of their woody stems, are more correctly classed as shrubs. They are placed in the herbaceous perennial category because of their stature and their use.

Early in this century, probably because of efforts to reproduce the famous perennial borders of British gardens, these plants were very widely used. As these attempts to duplicate the British border were constantly thwarted by the vagaries of the Canadian climate, plantings composed solely of herbaceous perennials, except the easily grown and more showy types, almost disappeared from Canadian gardens.

Another reason for disappointment with herbaceous perennial borders is the lack of showy blooms at certain times. Usually it is possible to have some plants in flower every day of the growing season. But at precisely the same time the rest of the border may be bare and ragged. Much disappointment can be avoided by not grouping a large number of plants that are all without flowers at the same time, but rather by mingling such types with species of long-lasting floral value. This is the most popular use of perennials. They are best when grown in a general flower border, an area not restricted to herbaceous perennials but one that includes also annuals, biennials, bulbs, small trees, and shrubs.

Location and Types of Beds

Although most flowering plants grow best in full sun, plants that will tolerate almost any degree of shade can be selected. Most important is to keep the border away from large vigorous trees, which quickly deplete the nutrients in newly developed beds.

Flowering borders with perennials of various heights look best when they have a background. You can provide this by planting a slow-growing hedge such as the arborvitae (white cedar) or the deciduous alpine currant, or by building a fence solid enough to give a good backdrop but having enough space between the boards to allow air to pass through. However, wide beds without a background can be designed successfully if smaller plants are used and the plants are so arranged that the bed can be viewed with satisfaction from either side or from any direction.

Informal island beds are very showy in spacious areas where the terrain is sloping or hilly. Here, the slopes themselves may form the background, or shrubs may be mixed with perennials to create a unified bed.

Planning the Border

When you have chosen the type of border, its location, and aspect, decide how the plants must be arranged. To avoid mistakes make a plan on paper. Base your main planting on six very showy groups. Then fill in the rest of the space with plants that best suit your seasonal needs or special color combinations you wish to create in each section of the border. Make a list of the plants you want to grow, and their heights, type of growth, blooming dates, and colors. Fit these plants into your plan. If you are usually away most of the summer, arrange for a spring planting of bulbs, irises, and peonies, and a fall planting of perennial asters supplemented lavishly with outdoor chrysanthemums. If you want your border to look bright all summer, use large plants of delphiniums and day-lilies, augmented by colorful annuals to fill in gaps that occur after the spring-flowering plants have died down.

Include several herbaceous perennials, such as 'Silver Mound' artemisias, plume-poppies, and plantain-lilies, or hostas, which are useful for their distinctive foliage and because they are interesting all summer.

If the border has a good hedge or a grouping of shrubs as a background, it is, of course, best viewed from the "front." Such a border allows for splendid arrangements of plants. It is usually wise to plant the tallest-growing types of perennials, those 1–1.5 m (3–5 ft.) high, at the back; the medium sizes, 0.5–1 m (2–3 ft.) high, near the middle; and the small subjects, under 0.5 m (2 ft.), in the front. However, to avoid a stiff formal composition, merge the tall groupings into the medium-sized ones and the medium ones partly into the spaces given to dwarf plants. By so doing, the border will be much more interesting, because the appearance will be one of modulating bays and rises.

Take care not to place a taller-growing plant in front of a shorter one, unless the tall one has the sole purpose of replacing or screening an early flowering kind. For example, after a June-flowering perennial such as the oriental poppy has finished blooming it mars the tidy appearance of the border, so it needs to become overshadowed by taller vegetation.

1 Spring border (photo by Malak)
2 A summer border
3, 4 Fall borders

Make your border as wide as your garden area permits. A width of at least 2.5 m (8 ft.) accommodates reasonably large groups of plants, which are needed to provide a succession of bloom. If your garden is small and your flower border necessarily narrow, do not use tall-growing rudbeckias, helianthuses, or perennial asters. Choose smaller plants that grow no higher than 1 m (3 ft.).

For best effect, group plants in threes or fives, and repeat the basic perennials in groups throughout the border. The groups do not need to be composed of identical cultivars but of plants similar in shape and manner of growth. In a small border, subjects like peonies or gasplants should be restricted to one plant in a group, because they become so large that several soon dominate the area. Use large numbers of one plant only in a large group. To create interest, vary the shape of each individual group, from circular to triangular or even oblong.

Planting Plans for Seasonal Borders

Seasonal borders provide a much better show of color than those that have to look presentable for the entire growing season. Larger groups of the more floriferous basic perennials can be used, but few flowers may be expected in the off-season. A border with spring- and fall-flowering plants looks drab during the summer, and a summer-flowering border, which may be highly desirable for a resort area, is unattractive at other seasons. See Plans for Seasonal Borders on page 184.

1 English wallflowers predominate a spring garden (p. 32)
2 Summer planting of monardas and Shasta daisies (pp. 76, 113, 114, 155)
3 Pansies at the border of a rock garden (p. 31)

1

2

3

chapter 2

basic perennials

basic perennials

1 Dividing day-lilies
2 Pacific Giant strain delphiniums (pp. 17, 119)

All groups of plants, woody or herbaceous, hardy or tender, indoor or outdoor types, zerophytes or aquatics, contain a few basic subjects that are showier and more versatile than the others. They have been bred and interbred to produce cultivars of the highest caliber that usually need less care and are more adaptable to varying conditions. Such basics are, of course, present in the herbaceous perennials. Without them, our flower borders and beds would be drab indeed. The basic plants described in this section are perennials that must form the basis of any border. They can also be used by themselves to provide a display at one season of the year, but their widely differing cultivars extend their period of usefulness and they contain kinds that may be mixed together to form a balanced perennial border, even if all other herbaceous plants are excluded.

Four of these basic perennials, day-lilies, delphiniums, irises, and peonies, are so popular that each group has a society of admirers that exists solely to further its interest and use. Such societies as the American Iris Society, the British Delphinium Society, and the American Delphinium Society encourage the development of new cultivars and plants of varying colors and forms, investigate newer cultural methods and newer uses, and in general promote interest in the particular plants of their concern. Basic perennials, then, are fundamental, easily cultivated perennials whose perpetuity is ensured by the ever-increasing number of new cultivars introduced each year.

MODERN DAY-LILIES (Color photos p. 66)

Visitors to nurseries and gardens where there are plantings of the new modern day-lilies cannot rest until they acquire these beauties for themselves. It is easy to understand why. Most people associate day-lilies with the yellow and orange species that have escaped to roadsides and neglected areas, and have no idea of what hybridists have accomplished during the past few years.

The newness in day-lilies may consist merely in more beautiful shades and tints. They are as easy to grow as their ubiquitous forebears. Along the moist banks of a stream or around any body of water in a slightly shaded nook or a sun-drenched slope, day-lilies can be relied upon for good bloom. But to excel themselves they must be used with irises, poppies, columbines, light blue lobelias, or white annuals such as petunias.

Newly developed day-lilies have a great diversity of flower, form, and color. Petals and sepals, three of each, may be smooth or ruffled; the veins may be prominent or indistinct; color may be solid or varied in petals and

sepals. Perhaps the most marked development, however, is in the shape and size of the flower. Cultivars more than 200 years old have narrow petals and sepals that combine to form a loose tubular flower. Contemporary types have wide, 4–5-cm (1½–2-in.) petals that curve back and produce a flare up to 20 cm (8 in.) across. The colors of the flowers have now advanced to warm shades of red, pastel pinks, bronze, light cream, and almost white.

An untidy characteristic of the old day-lilies or the near white one, which is present today in even the most modern irises, was the persistence of the dead flowers. In the newer types a flower drops off as soon as it shrivels, the stem is left clean, and a new flower blooms the following day.

Day-lilies are so very easy to grow that it seems hardly necessary to give instructions. Their roots are fleshy and they can be transplanted at almost any time from early spring to late fall. In most of Canada it is perhaps best to plant them in spring, as soon as they can be obtained from the nurseries, because after a full season's growth they are well able to withstand the first and subsequent winters. These plants soon form nice strong clumps and they bloom the first year. In the milder parts of British Columbia it matters little what time of the year they are divided or transplanted, for they will flourish as long as their roots are able to grow and become established.

The new plants should be set out 2.5 cm (1 in.) below the point where roots and stem meet. The plants will eventually have a spread of 1 m (2½–3 ft.). So give them lots of room unless you want to replant them the following year.

After planting, just pull weeds and cultivate shallowly. Thrips may bother day-lilies by feeding on the blooms. Thrips mar the color, especially of the darker flowers. The blemish is displeasing and the flowers may harbor pests that might spread to other types of plants.

Modern cultivars that have performed well in our test garden are:

yellow
'Blithe Spirit', 'Capri', 'Midwest Star', 'Colonel Joe', 'Jewel Russell', 'Swansdown' (nearest to white), 'Moon Ruffles'

yellow and lime
'Painted Lady'

orange
'Cibola', 'Golden Hours', 'Gladys Kendall', 'Cartwheels', 'Springside' (a very early orange yellow of Canadian origin)

red
'Bess Ross', 'Alan', 'War Eagle', 'Crimson Glory', 'Crimson Pirate'

pink
'Dorcas', 'Taylor Russell', 'Lyric', 'Cheery Pink', 'May Hall', 'Francis Fay', 'Neyron Rose', 'Salmon Sheen', 'Evelyn Claar'

lavender
'Luxury Lace'

purple
'Mabel Fuller', 'Potentate', 'Theron'

eyed or banded types
'Colonial Dame', 'Cathedral Towers', 'Burning Daylight', 'Nantahalla', 'Nashville'

DELPHINIUMS (Color photo p. 66)

Of all the flowers in the perennial borders at the Ornamental Gardens of the Plant Research Institute, delphiniums are the most striking. Their stately beauty so enhances the garden that they become the stars of the June picture. The tall-flowering spikes of lavender, blue, and white make them a most satisfying perennial. Delphiniums are not only valuable as garden ornaments, they are useful also in many floral arrangements for large church baskets, narrow vases, and with side shoots, on the dinner table. With very little care they can also be made to bloom in the fall and then they present a very striking contrast to the usual autumn picture.

The massive, showy delphinium of today is one of the masterpieces of plant breeding. American breeders have used some of the best varieties from Europe to produce new hybrids that tower over anything previously developed. These new plants also have individual flowers almost as large as hollyhocks.

The European hybrids are clonal varieties, which are not reproducible from seed. They must be propagated from cuttings, which are taken in spring from young shoots.

The American strains such as the Pacific Giants have been so selected over the years that they come reasonably true from seeds sold by seedsmen, but seeds saved from the home garden seldom produce first-class plants. This strain is, however, much shorter-lived than the European hybrids, and new seeds should be sown when the old plants start to become less productive.

Seeds of the Pacific Giant strain are best sown in the spring if you want plants large enough to set in nursery rows by fall. Fill a flat with sand and soil, and plant the seeds in shallow drills 5 cm (2 in.) apart. Water the flat well after the seeds are sown, and place it in a shaded part of the garden until the seedlings have formed three leaves. When the plants are large enough to handle easily, transplant them either to flats or open ground. If you plant them in the open, space them 15—20 cm (6—8 in.) apart in the row and between rows, and leave them undisturbed until spring, when they should be moved to their permanent location. Plants that have been grown in flats should be set out in the early fall in rows spaced the same way as for seedlings planted directly outside.

The following spring, prepare the ground well where the delphiniums are to grow. Dig it deep and add decayed manure, peat moss, or compost to the soil. Set the plants 0.5 m (2 ft.) apart in groups of three to five, but if you want to devote an entire bed to them, allow 1 in. (3 ft.) between each plant.

When delphiniums start to grow in spring, the young shoots are often damaged by slugs, particularly in damp weather. To prevent this, scatter a mixture of equal parts of copper sulfate and hydrated lime on the soil around them, but do not let it touch the plants.

Old plants of delphinium are apt to produce too many shoots, which, if allowed to mature, will only give poor blooms. So remove all the weakest shoots and leave only about five to seven of the strongest ones on each plant. If you want very good blooms for exhibition, sprinkle around the plants with 20-20-20 fertilizer, at 1 teaspoon to 4.5 litres (1 gal), or you can use any of the special liquid fertilizers recommended for foliage. Before using these, however, wait until the bud shows just above the leaves.

When the warm weather arrives and the plants are growing vigorously, you will need to stake them. For each plant use 3 or 4 stout bamboo canes about 1 m (4 ft.) high. To leave the flower stalks looking as natural as possible, tie three shoots loosely to each bamboo cane. One cane attached with twine around the whole plant completely spoils the beauty of the specimen and makes it look strangled.

When the flowers have faded, cut the stems down to the level of the leaves. This improves appearance and prevents seeds from developing. Occasionally, this pruning brings on a second show of blooms in September.

Cultivars of European Delphiniums   *1.5-1.8 m (5-6 ft.)*

'Anne Page' — light blue, large spikes

'Anona' — semidouble, sky blue flushed pink, white eye

'Arcadia' — single, pale blue and mauve, white eye

'Audrey Mott' — semidouble, sky blue outer petals, soft rosy blue inner petals

'Blackmore's Glorious' — semidouble, mauve and blue, white eye

'Blue Lagoon' — semidouble, brilliant gentian blue flushed with rose, dark eye

'Blue Riband' — intense blue

'Blue Rosette' —double, lilac blue

'Bridesmaid' — semidouble, silvery mauve, white eye

'Cambria' — semidouble, Cambridge blue, large spikes

'Charles F. Langdon' — semidouble, pure medium blue, conspicuous black eye

'C. H. Middleton' — semidouble, medium blue, large florets, sulfur white eye

'Dame Myra Curtis' — semidouble, sky blue, black eye

'Duchess of Portland' — semidouble, ultramarine

'Etonian' — semidouble, cobalt blue

'Eva Gower' — semidouble, bright gentian blue

'Father Thames' — gentian blue, flushed violet rose

'Fleur Celeste' — gentian blue, white eye

'Frederick Grisewood' — semidouble, deep blue, small white eye, well-balanced spike

'F. W. Smith' — bright deep blue, white eye

'George Bishop' — semidouble, mauve and light blue, chocolate-colored eye

'Janice' — semidouble, white, shorter, 1.2 m (4 ft.) high

'Jennifer Langdon' — semidouble, pale blue and mauve, black eye

'Julia Langdon' — semidouble, mauve suffused pale blue, white eye

'Lady Eleanor' — double, sky plue, shaded mauve

'Lady Guinevere' — soft, mauve, large well-balanced flowers

'Mayflower' — single, rich French blue, white eye

'Mrs. Frank Bishop' — semidouble, rich gentian blue, black eye

'Nell Gwynn' — semidouble, mauve flushed rose

'Silver Moon' — large, silvery mauve, white eye

'Sir Neville Pearson' — semidouble, deep bluish purple, black eye

'Swanlake' — semidouble, large, white, black eye

'Sylvia' — semidouble, sky blue, white eye

'Watkin Samuel' — semidouble, rich sky blue

'W. B. Cranfield' — semidouble, deep mauve and blue

'Wrexham Glory' — single, pure sky blue, white eye

TALL BEARDED IRISES (Color photo p. 67)

Tall bearded irises need almost full sun, although they tolerate some shade during part of the day. Even more important, however, is good drainage. This is not a statement to be considered lightly, because, however suitable your soil and other conditions may be, good drainage is absolutely essential for the survival of your plants, particularly where alternate thawing and freezing often occur in winter.

Prepare the soil well, preferably a year before you plant. Dig in a well-balanced complete fertilizer, about a handful to 0.8 m² (a square yard), but do not plant until at least 3 weeks later. You may use lots of manure, but make sure it has had time to decompose and become mellow before you set out the rhizomes.

The best time to divide and plant irises is from the end of July to the end of August. They can be planted successfully later, at the end of October in favorable years, but the plants are not well enough established to survive a difficult winter. The plants may survive, but the flower buds may freeze.

If the irises you are now growing have insipidly colored small flowers with zebra-like stripes on the shaft of the lower petals, do not bother dividing them. Throw them away and get a collection of the more popular, newer kinds.

To ensure large and abundant blooms, divide the plants every 4 or 5 years. You will need a garden hose, digging fork, sharp knife, trowel, shears, and several tie-on labels. Treat each clump as follows. Cut the leaves off 8 cm (3 in.) from the root, then fork the whole plant out of the ground. Turn the clump upside down, wash it thoroughly with a hose, and cut out and discard all decayed roots and shoots. Then divide the clump into small sections, each one containing a piece of rhizome with a fan of leaves attached. Pieces from the outer part of the clump are younger and more vigorous. They are the best ones for replanting.

Make a slanting hole with a trowel, and set near the soil level the part of the rhizome from which the leaf fan arises. Place the roots in the shallow hole and cover them with soil. Leave the surface of the rhizome exposed. Firm the rhizomes well in the ground after planting and water them thoroughly. Deep planting does not kill irises, but it retards their blooming for several years and makes them more susceptible to soft rot disease.

Space the rhizomes 0.6 m (2 ft.) apart. The appearance of the bed is sparse the first year, but in the next year the bare look will have vanished. For an immediate effect, plant in groups of three of a kind, 0.3 m (1 ft.) apart and in groups 0.9 m (3 ft.) apart. In October, in very severe climates, protect newly planted rhizomes by covering them with 2.5 cm (an inch) of soil. Remove this earth in spring.

Irises are almost all colors and all combinations of colors. An iris dealer may sell a little rhizome division for what seems a lot of money. However, the most expensive varieties are not always the best. Prices are usually set in accord with the newness of the variety and its rarity. New kinds are often improvements over the older ones. They have larger blooms with better and purer colors, and differently shaped petals that may be fringed, horned, or lacy.

As hundreds of new cultivars are registered and placed on the market each year, it is impossible for even a specialist to grow them all. By very carefully taking notes of the award-winning seedling irises and, particularly, of the names of the breeders, a fanatic irisarian can often foretell which will, in a few years, top the list of the best 100 in the annual rating given by members of the American Iris Society, based on the number of those previously rated as the best 25 selections. Although the leaders of the best 100 often retain this position for 2 or 3 years, they seldom last for more than this time before other cultivars take their place.

Another very reliable criterion for selecting an iris for your garden is the Dykes Medal Award. This award is given each year to mark an iris that is outstanding for beauty and ruggedness. Available irises that have received this award and the year the award was given follow:

'Allegiance' (1964) — navy blue, overlaid with velvet and with a medium blue beard tipped lemon chrome

'Amethyst Flame' (1963) — very large light violet flowers tinted lavender with light chestnut falls and a white beard

'Argus Pheasant' (1952) — smooth soft brown, with huge flowers having broad flaring falls and wide standards

'Babbling Brook' (1972) — ruffled petals of light blue, with a lemon beard

'Blue Sapphire' (1958) — light blue, ruffled, with a glowing golden beard; leader of the best 100 for 3 consecutive years

'Debbie Rairdon' (1971) — soft yellow, wavy standards, with milk white falls banded with rich yellow

'Eleanor's Pride' (1961) — very large, majestic and beautifully branched powder blue blooms

'First Violet' (1956) — clean light violet self, the standards, falls, and even the beards are the same color

'Mary Randall' (1954) — deep rose pink self, with a bright tangerine beard

'Pacific Panorama' (1965) — spectacular sea-blue self of enormous size with ruffled flowers on 102-cm (40-in.) stems, yellow beard, tipped white

'Rippling Waters' (1966) — pale bluish orchid with a very bright tangerine beard; sprang to popularity very fast

'Skywatch' (1970) — a large-flowered pale blue iris with an orchid undertone

'Stepping Out' (1968) — white, with marginal and pebble markings of pansy purple

'Swan Ballet' (1959) — heavily ruffled pure white with wide flaring falls and yellow beard, or white beard tipped yellow

'Truly Yours' (1953) — very bright yellow, shading to white at the top of the standards and at the base of the falls; edges of the petals ruffled and laced

'Violet Harmony' (1957) — very large light violet flowers

'Whole Cloth' (1962) — new-style amoena type of iris with pure white standards and soft medium blue falls with a white beard tipped yellow

'Winter Olympics' (1967) — ruffle-petalled, dome-shaped white standards and wide large-petaled falls of the same color

In the gardens of the Plant Research Institute many new iris cultivars as well as most of the 100 popular ones are grown every year. In addition to the Dykes medal winners, the cultivars rated highest for performance at Ottawa are the following:

'Annabelle Lee' — orchid pink

'Black Taffeta' — deep black

'Brilliant Star' — red with white beard

'Captain Gallant' — nearest to red

'Celestial Glory' — bright golden orange

'Celestial Snow' — pure white with white beard

'Edenite' — sooty black

'Gracie Pfost' — blend of henna brown and copper rose

'Gypsy Lullaby' — ruffled standards of butterscotch and violet, with red violet falls

'Henry Shaw' — giant white with crimped and ruffled edges

'Milestone' — yellow standards, purple falls overlain brown with blue center

'Nob Hill' — deep golden yellow

'One Desire' — the nearest to a true pink

'Orange Parade' — ruffled orange duotone (tones of the same) color) between orange and yellow

'Rainbow Gold' — buttercup yellow with a red beard

'Rococo' — ruffled and fluted plicata (spotted) bright blue on a clean white ground with a yellow beard

'Spanish Gift' — bright orange with deep orange red beard

'Stepping Out' — plicata similar to Rococo but with margins of pansy violet

'Sterling Silver' — violet duotone edged with silver

'Ultrapoise' — straw yellow with a flush of pink in the standards and vivid reddish tangerine beard

'Wild Ginger' — ruffled plicata or ginger brown and creamy white; quite unique in this color class

'Wine and Roses' — a combination of clear rose pink and deep violet purple; an edging of rose pink around the falls matching the rose pink of the standards

'Winter Olympics' — ruffled rounded white flowers

ORIENTAL POPPIES (Color photo p. 68)

The flamboyant oriental poppy, which for many decades has set afire the perennial border in June, now appears in soft iridescent, gleaming colors with pastel tones of light pink, deep pink, salmon, white, and a more subdued red. The mammoth scarlets, 31 cm (12 in.) glowing plates of fire, so prominent in older gardens, are now available with fully double and single 10–15 cm (4–6 in.) flowers.

If there is a single reason why this spectacular perennial has not been extensively grown in every garden, it is surely failure to observe the critical period for successful transplanting. That period is the month of September. It is then, before the roots have started to produce new winter rosettes of leaves, that poppies must be divided and transplanted. Nursery specialists can keep young plants in storage for several weeks before they make delivery. So, if you order poppies at the end of August and do not receive the plants until 2 weeks later, you do not need to worry. If planted right away, they will grow like weeds and produce a new set of vigorous shoots that will better survive the winter. Unless you can get pot-grown plants, it is useless to buy oriental poppies in spring.

To propagate these popular plants cut the thick, rhubarb-like roots into sections. Make a slanting cut at the base so you can readily tell the top from the bottom. Plant the pieces in drills and let them grow for a year. In the following fall transplant them into permanent quarters. Bought plants will have been grown from rooted cuttings, so handle them in the same way.

Planting poppies is not a hard job, but, like all gardening operations, it should be carried out properly. First of all, see that the soil is very well drained. If you have a clay soil add lots of cinders or small stones so that water will drain away quickly. To plant poppies, set the root straight down in a hole deep enough that 8 cm (3 in.) of soil will later cover the crown.

Mulch the newly planted roots the first winter, just after the first frost. By this time a rosette of leaves will have appeared. Do not cover this cluster, because the crown will decay. Arrange the mulch around the rosette.

Full sun and partial shade are equally suited to the cultivation of these plants. While they seem to thrive with lots of organic matter such as well-decayed leaves and peat moss, they do not need a rich soil.

The flower colors are so rich and vivid that no other plants are needed for border effect. But, perennials such as gypsophila, asters, or day-lilies planted near the clumps hide the decaying foliage and fill in the voids left when the foliage dies.

The following cultivars have grown well under test and they are easily obtained:

'Barr's White' — pure white

'Beauty of Livermore' — bright crimson, black center

'Carousel' — red, edged white

'Cerise Beauty' — cerise pink

'Coral Cup' — six-petaled flowers, pink coral

'Crimson Pompon' — small, 8-cm (3-in.), double, scarlet, less than 0.6 m (2 ft.) high

'Curtis' Giant Flame' — huge, scarlet

'Field Marshall van der Glotz' — white with a purple black iron cross in the center

'Helen Elizabeth' — crinkled petals of shell pink

'Juliette' — bushy, shell pink

'Lavender Glory' — huge, lavender

'Mary Finan' — deep salmon red with fringed and laciniated petals

'Mrs. Perry' — a fine single pink

'Olympia' — double, scarlet overlaid with golden salmon

'Pink Lassie' — exquisite crepe-paper pink

'Pinnacle' — white with an edge of scarlet

'Salmon Glow' — double flowers, bright glowing salmon orange

'Salome' — large pink flowers

'Springtime' — light red

'Valencia' — deep scarlet

PEONIES FOR YOUR GARDEN (Color photo p. 69)

Peonies are in a very special class. They not only withstand the summer's heat and drought and the winter's cold, but they come in a wide range of colors and forms.

Unlike most herbaceous plants, they are very long-lived and their cultivars are not easily superseded. Whereas a gladiolus variety may be surpassed in 3 years, a lily in 4, and a rose, nowadays, in 6, most peonies grown in home gardens today have held their own since their origination before 1900. In a popularity pool sponsored by the American Peony Society a few years ago, the voters, who were the best peony growers in the United States, selected 'Jules Elie', 'Festiva Maxima', and 'Sarah Bernhardt' as among the best 100 cultivars, and placed 'Jules Elie' at the top. Yet these three cultivars were introduced over 60 years ago.

In the last 10 years there has been a gradual change in cultivar varieties and types, and while the above-mentioned ones are still among the very best, they are being replaced by a new race of hybrids that are nothing short of sensational. This does not mean that these new hybrids have been bred in the last decade, but that they are only now becoming well known.

One of the foremost hybridizers was Dr. A. P. Saunders, son of the famous Dr. William Saunders, founder of the Central Experimental Farm in Ottawa, who worked on these new peonies for 30 years until his death a few years ago. He collected a large number of peony species from all over Europe and Asia, and, although these were not outstanding in themselves, when intercrossed they produced showy new doubles, singles, and Japanese types. These plants were the forebears of the new race now known in peony catalogs simply as 'hybrid' peonies. In reality they are interspecific hybrids and can be distinguished from all the others by their forms and shapes and by their color range.

The older peonies are, of course, also hybrids, but nearly all of them originated from the Chinese peony (*Paeonia lactiflora*), chiefly from seeds of attractive flowers that had probably been fertilized by insects. In addition, some plants originated from crosses of different species, but their various origins have been lost in antiquity. Perhaps the simplest way to differentiate is to state that the well-established groups contain plants of unknown origin and, more recently, plants from crosses between varieties, whereas the hybrids are plants that are single or multiple crosses of known species.

Although the strikingly new group of 'bomb' doubles contains the most popular of the new hybrid peonies, there are many singles in colors that were unknown a few years ago. In the test gardens at the Plant Research Institute, for example, are yellows such as the single 'Claire de Lune', the new very expensive double 'Oriental Gold', 'Illini Belle' with nearly black petals and red-tipped green carpels, the intriguing deep salmon 'Laura Magnuson', the light scarlet red 'Crusader', and 'Joyce Ellen', which is dark red with a white center.

The very best of the 'bombs' are 'Red Charm', which has a rich red collar of big petals, a sensation at most modern peony shows and currently the leader of the popularity poll; 'Diana Parks', early vivid scarlet; 'Charles Haines', deep salmon pink with some intermingling stamens; 'Convoy', deep purple; and 'Chocolate Soldier', an almost black double cultivar.

Striking though the new and recent hybrids are, certainly the older hybrids are still impressive, and they contain some excellent varieties. The large double type peony, which everyone knows, has huge blooms and an enchanting scent. The best peonies of this type are 'Jules Elie', pink; 'Festiva Maxima', pure white flecked with crimson; 'Hansina Brand', full of pink blooms; 'Kansas', the best light red peony; 'Snow Mountain', pure white; 'Hattie Lafuze', pink to white; 'Mrs. Livingstone Farrand', salmon pink; and 'Philip Rivoire', dark red.

Single peonies such as 'The Bride', white; 'Sea Shell', pink; 'President Lincoln', red; and 'America', white, have one row of petals and a center cluster of yellow stamens.

An exotic and exquisite class is known as the Japanese single or anemone type. Plants of this type have remarkably graceful flowers, composed of one or more circles of guard petals that form a chalice surrounding a beautiful center mound. The middle of each flower has enlarged, often twisted, staminodes, which are modified stamens. These petal-like staminodes form a center cushion, which makes the flower look much like a chrysanthemum. The best in this group are 'Isani Gidui', pure white with gold center; 'Mrs Wilder Bancroft', red with red-tipped golden center; 'Mrs. G. E. Hemerik', rosy pink outer petals with a huge golden sunburst center; and 'White Cap', deep rose with a large center cushion of white.

Each type of plant has individual peculiarities, so it is impossible to generalize about how peonies should be treated. For example, the worst way to divide a peony is to dig with a spade in spring alongside an old clump

1 New buds on peony roots
2 When planting peonies the eyes should be
5 cm (2 in.) below the surface of the soil.

and to shave a portion of it off for planting. This may work well with some perennials, but not with peonies. If planted in spring such a division seldom makes satisfactory growth and usually takes a long time to become established. Shaving with a spade invites decay of the fleshy roots.

The peony's year begins not in spring, but in late summer or early fall, which is the only time a clump should be divided.

The large fleshy roots extend downward. Each year the stalks die but after the flowers bloom in the spring or early summer, new buds, called eyes, begin to form. In the following year, the eyes develop into stalks.

Commercial nurserymen often dig as early as the first week of August, but home gardeners should wait until later in the month, or September, when the new buds are about 6 mm (¼ in.) long and the roots are almost dormant.

It is better to divide an old clump than to reset the whole plant in another spot. Each division will make a whole new plant, which will last for many years without deteriorating.

To divide a peony, cut off the foliage 15 cm (6 in.) above the ground. Then sever the roots with a spade in a circle of the clump. Cut under the roots as much as possible before you pry the clump loose, then dig up the whole plant. Wash or shake off the soil. Some growers prefer to let the clump stand for several hours so that the roots become less brittle. When you divide the roots, leave at least two eyes, preferably three, a piece of the crown, and some roots in each division. Cut away all decay from the divisions and trim the roots back to a length of 10 or 12 cm (4 or 5 in.).

After dividing each clump it is a good idea to dip your knife in a solution of permanganate of potassium (1 teaspoon to 4.5 litres [1 gal.] of water) to prevent spreading any disease that might be present.

A better and more certain precaution is, just before you set the divisions in the soil, to soak them in a formaldehyde solution (1 tablespoon of commercial formalin in 4.5 litres [1 gal.] of water). This solution will kill injurious fungi, and prevent disease from spreading to new plants.

When you plant the divisions, make each hole large enough to avoid crowding the roots. Make sure the eyes are 5 cm (2 in.) below the surface of the soil. Fill soil in around the roots and work it in with your fingers until no air spaces are left. Then firm the soil well with your foot, but do not injure the roots.

When you have filled in the hole and firmed the soil, water well and let the soil settle, then put more soil around the plant, and mound it up a few inches to give it protection during the winter.

SUMMER-FLOWERING PERENNIAL PHLOX

The herbaceous summer-flowering phlox *(Phlox paniculata)* is a popular perennial for flower borders. It is almost indispensable, because no other perennial presents as colorful an effect during July and August as this one does in its brilliant and diversely colored display. Its warm sweet fragrance is the very essence of summer.

The uses of these plants are almost limitless. They are not only outstanding in perennial flower borders in summer, but they are also brilliant enough to provide a bedding display of their own. The effectiveness of these plants when they are used lavishly is shown in the test planting in the ornamental grounds at the Plant Research Institute. By choosing a few kinds and planting them in irregular drifts an even more magnificent display can be achieved. Many cultivars are in full bloom from mid-July until late August, and a few late-flowering trusses appear until the end of September.

Summer perennial phloxes are quite compatible with other flowers, and many refreshing combinations may be obtained. For example, white cultivars may be placed with lavender and purple Japanese irises, red or purple colors with bright snapdragons or day-lilies, and tall blue veronicas with white phlox and bright red astilbes.

The bright-colored cultivars are quite showy when flanked with a background of white mugwort *(Artemisia lactiflora)* and in front of the double dropwood *(Filipendula vulgaris* 'Flore Pleno'). This planting provides a good contrast of bloom and later the unattractive basal foliage of the phlox is neatly hidden. Soft blue spiky annual salvias *(Salvia farinacea* cultivars) and the perennial 'Icicle' veronica and 'Summer Snow' physostegia blend nicely with the deeper pinks and reddish tones. In a border, purple or magenta phlox cultivars are not always harmonious in color with other flowers. Silver-foliaged artemisia and nepetas with a lacy veil of perennial baby's-breath *(Gypsophila paniculata)* will help to soften the conflict of color.

The cultivation of this splendid perennial is quite simple; it grows best in a good, rich soil and needs plenty of air circulation. It will withstand light shade, but do not plant it in motionless areas overshadowed by large buildings.

Divide phlox every 3 years. Dig the plants out of the ground and break off the outer shoots and replant them. Discard the center of the clump, because it seldom produces healthy flowers. If you do not divide the plants regularly, they will become more susceptible to rusts and mildew diseases. If you do not divide the plants, then thin the shoots on each plant to produce a good show of bloom.

Although summer phlox need a lot of water, be careful not to wet the leaves. Many home gardeners destroy their plants by using overhead irrigation, which encourages and spreads disfiguring mildew and leaf spot diseases.

To maintain a succession of bloom, remove the main truss as it fades. Then the side shoots will produce flower heads that have many smaller florets to a truss. Do not allow seeds to form, because when they drop they are likely to produce plants, which, if not pulled out, will be the objectionable magenta hues of the original wildling.

Although many of the older cultivars are still grown and some such as 'Miss Lingard', an early white, are still indispensable in gardens, the newer ones introduced in the past decade are more vigorous and slightly more disease resistant.

The following cultivars, arranged according to their average height, are rated as excellent at Ottawa.

Dwarf cultivars   *under 53 cm (21 in.)*

'Gnome' — only 46 cm (18 in.) high, purple red flowers with deeper eye

'Marlborough' — very compact cultivar, dark green leaves and deep rose pink flowers with darker eye

'Stirling' — fine, compact cultivar with large trusses of deep rosy-red

'Vintage Wine' — glowing wine red with much darker center

Medium-sized cultivars   *53-76 cm (21-30 in.)*

These cultivars are very much more popular and many splendid ones are available.

'Barnell' — large flowers, deep pink with dark eye

'B. Symons Jeune' — deep pink red center, named after the late Capt. B. Symons Jeune, who was responsible for many superior cultivars in recent years

'Caroline van den Berg' — soft lavender purple, nearest to blue of any

'Elizabeth Arden' — soft pink with darker eye

'Flamboyant' — very large flowers, clean-colored deep maroon

'Gleneagle Glory' — vigorous orange red

'Joan' – vibrant rich scarlet flowers with reddish purple or crimson eye

'Miles Copijn' – almost pure pink with crimson eye

'Pastorale' – very large warm pink with a purple eye

'San Antonio' – dark blood red star-shaped open-petaled cultivar

'Spatrot' – deep crimson red

Taller cultivars   *over 76 cm (30 in.)*

'Annie Laurie' – deep salmon pink

'Brigadier' – very bright orange red with deep crimson eye

'Bruno von Zeppelin' – white with red eye

'Dodo Hanbury Forbes' – very large, pure pink with reddish center

'Fairy's Petticoat' – pale mulberry purple with deeper center

'Firefly' – peach pink

'Gaiety' – very striking cherry red, suffused orange

'Graf Zeppelin' – white with carmine eye

'Hampton Court' – bright carmine with especially good, rich foliage color

'Harewood' – bright red purple

'Henderson's Late White' – reliable white

'Iceberg' – excellent, pure white, under test

'July Glow' – high rating because of its very long period of bloom and its large brilliant deep rose flowers

'Leo Schlageter' – brilliant scarlet crimson with darker eye

'Lord Lambourne' – red with white zone in the middle of the flowers surrounding reddish purple eye

'Olive Symons Jeune' – deep pink with darker center

'Tenor' – outstanding red

'Windsor' – rosy pink with purple center

1

2

3

4

chapter 3

biennials

biennials

Biennials are plants that make most of their growth the first year, then flower and die the following year. Some perennials, such as pansies, are treated as biennials, since they flower the second year from seed, and, although they do not die after flowering, they become so straggly the next year that they have to be removed. Most biennials may be grown outside all the time; some need the protection of cold frames during winter and others need a light covering of straw.

Nearly all biennials are raised from seed sown during the summer months. The time of sowing the seed is important, because the plants must attain almost their ultimate size before winter sets in and yet not become so mature that they tend to produce a flower shoot in the fall, or so leafy and soft that they do not survive the winter.

Prepare the soil well, sow the seed in small drills about 15 cm (6 in.) apart, and place a shade over the bed. Keep the seedbed moist at all times. Never let it become dry. As soon as the young plants show above ground, gradually accustom them to more light. Begin by removing the shading early and late, putting it on only during the bright part of the day. After a few days remove it altogether. As soon as the plants are large enough to handle, transplant them to a nursery bed or cold frame.

Because each kind of biennial requires a special kind of care, in the following list the most popular species are entered separately, and the kinds that have rated highly in trials of biennials carried out at the Plant Research Institute, Ottawa, are noted afterward.

CANTERBURY BELLS (Color photo p. 70)

The Canterbury bell (*Campanula medium*) has large, loose racemes of inflated bell-shaped violet blue, lavender, pink, rose, or white flowers borne in June and July. The cup-and-saucer type, 'Calycanthema', has petaloid calyx lobes, which are somewhat deeply divided and sometimes joined to form a saucer-shaped structure that is much larger than the one in the common type. 'Weigandii' has yellow leaves and blue flowers and 'Imperialis' is much more floriferous.

Almost all Canterbury bells flower the first year after the seed is sown and the seedlings are transplanted. Those that fail to flower produce very large plants that bear a great profusion of bloom the second year.

Sow seeds of this elegant biennial in late May or June into sifted soil in flats or directly into a cold frame. If the seeds are sown outside select a partly shaded area,

sow in drills 1.3 cm (½ in.) deep and 15 cm (6 in.) apart. When the plants are large enough to handle, transplant them 15 cm (6 in.) apart in a nursery border and leave them there until the end of September or October. At this time or in very early spring, set out the plants 38 cm (15 in.) apart, where they are to bloom. In the Ottawa area, even though during most winters they survive unscathed, cover them with a light layer of straw. In colder areas place them in a cold frame covered with mats or straw for the winter.

Canterbury bells may be grown in patio containers or in pots for moving later to the terrace. To do this, take plants from the nursery in September or October and set them singly in 20-cm (8-in.) pots containing good sandy compost. Keep them in the cold frame all winter, plunging the pots in sandy soil and covering in the same way as for plants to be set out in beds.

Besides the cultivars just mentioned, a group of hybrids named Deans Hybrids rated very highly in our trials in 1964 and 1971. This strain showed a good assortment of clean colors, pink, blue, and white, and abundant large single blooms on strong uniform plants. All the plants of this strain survived the winter, so possibly they are hardier than the others. The white and rose double cultivars were 60% true from seed and were quite hardy.

In the garden these plants are best planted in front of delphiniums or hollyhocks or similar tall plants that are compatible in colors. A big advantage over other biennials is that these plants may be moved to any location when in full bloom, and they will not wilt or become unsightly.

FORGET-ME-NOT

Quite a lot of confusion exists regarding the correct botanical name for the forget-me-not. The true biennial species is *Myosotis sylvatica*, but it often is referred to in lists as *M. alpestris*, a very fine little alpine. The confusion is a result mainly of the common name, forget-me-not, which probably more correctly belongs only to *M. scorpioides*, a common European wayside perennial, but which is used for any member of the Myosotis family, all having the same kind of light and medium blue flowers and a distinct yellow eye.

Once it has been set out in a border, the biennial forget-me-not will come up every year as a straggly little plant or in a colony in places where it cannot be called a weed, because it is not very persistent and can be easily removed. It often forms very graceful little groups that fit in nicely with all the spring-flowering bulbs.

For the best results treat this plant as a true biennial by sowing the seeds each year at the end of June and transplanting them to nursery beds for the summer. By late September or October transplant the large specimens to the location where you want them to flower. Treated this way, they will make large mounds of bright blue flowers that are extremely effective as a foil for tulips, particularly pink-flowered tulips. They are also useful for planting along the front of the flower border in groups of three or a edgings to small flower beds. Selected cultivars such as 'Blue Bird', 'Blue Ball', and 'Victoria' are available commercially; these are preferable to the type species.

FOXGLOVE (Color photo p. 70)

The common purple foxglove *(Digitalis purpurea)* is a true biennial that is native to Western Europe and Great Britain; in gardens in America and England foxglove and its hybrids have been grown for many years.

They are excellent for planting in semishaded areas such as those bordering woodlands and in wild gardens. They are just as pleasing and will grow as well in sunny perennial borders, where their vertical spike-like effect relieves the horizontal forms of most perennials. The better hybrids form large stems 1–1.5 m (4–5 ft.) high and are suited to the rear of the border, where they combine well with other biennials such as Canterbury bells, and sweet Williams. They produce a good effect with a planting of Madonna lilies.

For best results with foxgloves, buy seeds of the best strains and sow them in June; transplant the seedlings 0.3 m (12 in.) apart into nursery beds. Shade them lightly during the summer and put some boughs over them in late fall to protect them from intermittent freezing and thawing. In the Prairie Provinces they need more protection, sometimes by carrying them over winter in cold frames and transplanting in spring, or by covering them with a thick layer of straw until early spring.

The giant Shirley strains produce stately 0.9–1.2 m (3–4 ft.) steeples of crimson-spotted white or rose blooms in June and July. The Excelsior strain excels the Shirleys, from which it originated, with a better range of colors and greater all-round beauty.

The new 'Foxy' foxglove was introduced and acclaimed for its ability to flower the first year from seed, but it is more like a biennial, because it flowers again the second year. Although the color selection and size of spike and bloom are much inferior to the Shirley strain

and the improved Excelsior strain, it is much hardier and more durable.

HOLLYHOCK (Color photo p. 70)

The hollyhock *(Althaea rosea)* is a stately majestic perennial plant that towers up to 3 m (10 ft.) high in good fertile soil and has large crinkled leaves and single or double flowers in many colors. Although it is a perennial, it is best treated as a biennial or at best a short-term perennial that might flower 2 or 3 years in succession. To get the best results from some of the newer selections, treat them as biennials. The plants start to flower in July and continue to blossom over a long period.

They may be planted as single specimens or in masses against walls or buildings, in groups at the back of the perennial border, interplanted with low shrubbery, or in bold masses along drives or walks. Most plants seeded in June or July give their best efforts the year after seeding, so this fine garden biennial is of special benefit to those who are just starting a garden. When your shrubs are still low and the garden looks bare, they give height to borders and foundation plantings.

Hollyhocks thrive best in a very deeply dug, fertilized, well-drained soil. In May sow the seeds in nursery beds or seed boxes, and, when they are large enough to be moved, transplant the seedlings 25 cm (10 in.) apart. Leave them until fall if the soil is well drained or light, but do not move them until the following spring if the soil is clay or heavy.

If you have not grown hollyhocks in recent years, you will be surprised to find that many new and exciting strains are offered. The Chater's strains have reached almost the summit of perfection with their fine, rich, double flowers in scarlet, yellow, maroon, pure white, and salmon rose. The Imperator strain has large 12–15-cm (5- or 6-in.) flowers of double rosettes surrounded by a broad collar of fringed and frilled petals. Colors in this strain are from cream to deep scarlet. Sometimes the central rosette has a different color from the outside petals.

The Triumph strain is another type similar to Chater, with double flowers in pastel shades.

Mr. Simonet, of Edmonton, has used Chater and another selection to produce a new strain named Powderpuff. This is a delightful strain that has fluffy double flowers in many soft pastel colors on medium-sized stems.

The so-called annual hollyhocks produce a large number of blooms the first year from seed if sown indoors in March, and then they behave as biennials. These are of value only if you want the early blooms, because the flower stalks are not tall and the color selections or form of the flowers are not highly desirable.

ICELAND POPPIES

The Iceland poppy *(Papaver nudicaule)*, a native of subarctic regions, is a very fine garden flower for all Canada, except where the temperature in summer is too hot or humid. Its delicate creped saucer-shaped blooms are orange, yellow, salmon pink, and other tones, and are produced on fairly long stems that are invaluable for cutting. Unlike the perennial poppies, they produce blooms over a long season, because as one bloom is fading another opens out while yet another is in the bud stage.

Sow seeds in early spring to produce flowers in late summer and again the following year, or sow the seeds thinly in a fine, sandy soil or flats in June for transplanting into nursery beds in September. Since they are not always easy to transplant, some gardeners prefer to sow a few seeds in a small pot and then pull out all but one plant after they germinate. These pots are plunged in a frame for the summer and the plants set out in their permanent beds in early fall.

In recent years many good strains of Iceland poppies have been developed. In 1969 in the test gardens at the Plant Research Institute, Ottawa, the following strains rated highly:

Akabana Scarlet   *36 cm (14 in.)*   *Apr.-June*
Extremely vigorous. The deep red orange blooms, darker than those of Red Cardinal, abundant from April to late June.

Coonara   *46 cm (18 in.)*   *Apr.-June*
A fine mixture of pastel shades of salmon pink and rose.

Kelmscott   *56 cm (22 in.)*   *Apr.-July*
Excellent in every respect. Good color range of white, salmon, golden orange, and yellow. The crop of many blooms began in late April and reached a peak during the second week of July.

Pencilstalk Giant   *46 cm (18 in.)*   *Apr.-June*
Produced a few weak plants. The mixed colors included yellow, white, salmon, and orange.

Red Cardinal   *46 cm (18 in.)*   *Apr.-June*
The abundant striking orange red flowers on compact, uniform plants were the most outstanding Iceland poppies in the test.

Sutton's Lemon Yellow   *46 cm (18 in.)*   *Apr.-July*
A very bright yellow, completely true to type.

Sutton's Orange   *41 cm (16 in.)*   *Apr.-June*
Very bright orange flowers gave an excellent bedding effect.

Sutton's Pink Shades   *28 cm (11 in.)*   *Apr.-June*
True to type and color, the uniform compact plants produced a large number of bright pink flowers on strong stems.

Tangerine   *48 cm (19 in.)*   *Apr.-July*
Similar to Red Cardinal, but the color of the bright orange blooms was not quite so intense.

MULLEINS

Many of the 300 or more species of mulleins, *Verbascum*, have contributed to the production of the garden cultivars and strains we have today. Most have descended from Caledonia, a plant with stout bushy stems to 1.5 m (5 ft.) high and an abundance of large yellow orange blooms through June and July. Of the true species that are biennial, the most imposing are *V. olympicum*, which reaches 1.8 m (6 ft.) high and has gray white leaves and yellow flowers, and the common mullein, *V. hapsus*, which also has gray leaves and yellow flowers.

While not truly spectacular, if they are planted in groups of three in the border and combined with lupines and delphiniums, these biennials or short-lived perennials make a striking contrast. When not in flower, the silvery gray leaves of the species and some of the cultivars are quite attractive.

To grow these plants successfully they must be given a sandy soil and a sunny location. Although the easiest way of raising the strains is from seeds sown in May and treated as other biennials, the cultivars named here do not come true from seeds and must be propagated by root cuttings, or obtained by purchasing new plants. Since they often last more than 2 years this is a worthwhile outlay.

The best of the cultivars are:

                         June, July

'C. L. Adams'   *0.6-1.1 m (2-4 ft.)*
Deep yellow flowers, large branching stems, silvery foliage

'Cotswold Beauty'   *0.9-1.2 m (3-4 ft.)*
Biscuit color with lilac anthers

'Cotswold Gem'   *0.9-1.2 m (3-4 ft.)*
Rosy brown with a purple center

'Cotswold Queen'   *0.9-1.2 m (3-4 ft.)*
Buff terra cotta

'Gainsborough'   *1.2 m (4 ft.)*
Pale yellow

'Harkness Hybrid'	1.5 m (5 ft.)
Deep rich yellow	
'Miss Willmott'	1.5 m (5 ft.)
Creamy white, very large flowers	
'Pink Domino'	1.1 m (3½ ft.)
Deep rose pink	

PANSIES AND VIOLAS (Color photo p. 71)

Botanically all pansies are violas, but among gardeners, viola is the name commonly used for a group of plants obtained by crossing the garden pansy with the horned violet *(Viola cornuta)* that bears the group botanical name of *Viola williamsii.* These are better known in British Columbia and the Pacific Northwest states as bedding violas. They differ from the pansy by their unmarked clear flowers, which are slightly smaller than the pansy, but, unlike the flowers of that group, they do not get smaller as the season progresses but stay the same size. In climates with cool summers, they last much longer and are more perennial in nature than pansies. They are usually propagated by cuttings and sold as named cultivars; the best known is 'Maggie Mott'.

In Eastern Canada the name viola is most often linked with violettas, derivatives of the cross between the horned violet *(V. cornuta)* and the graceful violet *(V. gracilis)* whose group name is *V. visseriana.* This group contains hardier cultivars than the ordinary pansy and bedding viola, but they are short-lived and best when treated as biennials. They have much smaller flowers than the pansy and bedding violas, but are extremely useful for planting in rock gardens, for edgings to borders, or for planting among spring-flowering bulbs.

The garden pansy *(V. wittrockiana)* is a group of hybrids derived mainly from the common heartsease *(V. tricolor)* that varies a great deal in habit and flower color and has the characteristics of *V. lutea* and *V. altaica.*

In 1830 about 400 named cultivars of clean-colored round-petaled types were available commercially from selections made originally by T. Thomson, gardener for Mr. Sorel Gambier of Iver, Buckinghamshire, England. At the same time, selections in Belgium and France showed greater range of colors and less formal shade. The modern large-flowered pansy has been derived from intercrossing these two strains.

From these strains three distinct groups are known in gardens today: the bedding kinds, which are bush and compact plants with almost self-colored blooms except for markings in the center of the flowers; the Trimardeau hybrids, the groups under which the number and strains are classified as well as the large-flowered pansy cultivars that come true from seeds such as 'Delft', 'Berna', and 'Ullswater'; and the winter-flowering groups, which consist of cultivars that flower very early in the year in Europe and in British Columbia. Such cultivars as 'Celestial Queen', 'North Pole', and 'Winter Sun' are in this category and should be grown in British Columbia, Niagara Falls, or the Windsor area, where the climate is milder and the beauty of very early pansies can be better appreciated.

If you sow a packet of good pansy seed in May or June, you will likely have enough plants to border all your flower areas and have some to spare. Also, you will have vigorous stocky plants that will flower next year over a longer period than those you are most likely to buy locally.

When you grow your own plants from seed, you can choose the kind of pansy you like best, instead of selecting your plants from what is available the next spring.

Be sure to buy good seed from a reliable firm that specializes in pansy seeds. You may be surprised at the price you have to pay for the very select kinds, but you will receive more enjoyment from the larger blooms and purer colors these seeds produce.

Sow in prepared flats filled with a mixture of peat moss, soil, and sand in equal parts and add 42.5 g (1½ oz.) of superphosphate to each bushel of the mixture. Spread the seeds over the flats and cover them with a fine layer of soil, or plant them in rows 5 cm (2 in.) apart. Water the flats well after sowing and for a few days shade them or place them in a shady part of the garden.

After the seedlings have formed two or three leaves, transplant them to other flats filled with the same prepared soil and mix in 1 cup of complete fertilizer for each bushel of soil. Or, set them in nursery rows in the garden. Shade them for a few days until they have recovered from the shock of transplanting.

About the middle of September transplant the plants to a cold frame for the winter. Leave the frame open until the ground freezes and then cover the frame with boards.

In early spring set out the plants where they will show to good advantage. Remove fading blooms from all pansy plants or they will soon stop flowering.

32

Cultivars and Strains:
Violas, violettas, or *V. cornuta* types
'Blue Carpet' — purple violet, white eye
'Buttercup' — deep yellow
'Chantreyland' — amber apricot
'Duchess' — cream, edged with lavender
'Jackanapes' — yellow, and dark brown
'Jersey Gem' — deep purple blue
'Lorna' — deep lavender blue
'Picotte' — white, edged violet
'Violetta' — white

Bedding Violas
'Archer Grant' — indigo blue
'Better Times' — soft yellow
'Bridal Morn' — light purple blue
'Eileen' — deep-rayed blue
'Enchantress' — bright lavender blue
'Gladys Findlay' — white-banded purple
'Iden Gem' — purplish violet
'Lily' — lemon yellow
'Maggie Mott' — pale blue with creamy center
'Mauve Queen' — mauve
'Mrs. Morrison' — mahogany
'Mt. Spokane' — snow white
'Queen Elizabeth' — ruffled flowers of royal purple
'Windsor' — gray blue

Pansies — bedding types
'Apricot Queen' — apricot with black center
'Coronation Gold' — rich golden yellow
'Gipsy Queen' — terra-cotta and bronze
'Goldie' — large deep golden yellow
'Orange King' — rich orange

Giant or Trimardeau cultivars (Color photo p. 71)
'Alpenglow' — rich orange with black blotch
'Berna' — dark velvety purple
'Delft Blue' — light blue
'Jet Black' — almost black
'Rheingold' — yellow
'Ullswater' — rich blue
'Wine Red' — wine

Strains
Englemann's Giant
Felix
Majestic hybrids
Majestic — white with blotch
Maple Leaf Exhibition
Masterpiece
Oregon Giant
Steel's Jumbo

SWEET WILLIAM (Color photo p. 71)

The sweet William *(Dianthus barbatus)* is a great favorite among gardeners. It brings the delight and freshness of an Old World garden to the most modern settings. It has always been popular in gardens because of its great diversity of color, its long season for flowering, the sweetness of its perfume, and its ease of cultivation. The plant is capable of reproducing itself from its abundant seeds that it self-sows in the soil where it is planted, but unfortunately these seeds seldom produce plants true to type. The species *Dianthus barbatus*, a native of Southern Europe, was introduced into English gardens before 1575.

There are annual forms of this biennial, but most of the better cultivars and strains are biennial types. Sow the seeds in June and transplant the seedlings in the fall. Make the nursery rows 30 cm (12 in.) apart and plant the seedlings 23 cm (9 in.) apart in the row. In September or October move them into the beds or borders where they are to flower.

In 1964 at the Plant Research Institute, Ottawa, an attempt was made to grow all the existing kinds of sweet Williams. Some of the most delightful cultivars were 'Pheasant Eye', with brilliant crimson florets accented by a prominent pure white zonal marking; 'Scarlet Beauty', short, compact plants with uniform scarlet red flowers; 'Harlequin', with two-toned pink and white flowers on very vigorous plants; and the extremely beautiful short, compact strain, 'Indian Carpet', which grew only 15 cm (6 in.) high and had flowers ranging from white to pink and deep red.

WALLFLOWERS

The English wallflower *(Cheiranthus cheiri* and *Erysimum hieracifolium)* is a native of the Canary Islands, Madiera, and the Mediterranean regions, where it is a perennial. In Canada, it is best treated as a biennial. It will not grow in most areas of North America, where the winters are too cold and are followed by very hot summers. The only area that is suitable for these plants in Canada is the milder and more humid parts of British Columbia. The Siberian wallflower, however, is extremely hardy and survives the winters well in Eastern Canada and flowers as an excellent companion to tulips. When it is almost finished flowering, cut back the plants and a second period of bloom will follow.

The English wallflower has many garden strains ranging from the familiar old-fashioned yellow, found on walls and among ruins, to the modern large-flowered forms that vary from ivory through yellow, orange, red, and purple.

Sow seeds of the English wallflower outdoors in June or July in well-prepared soil. Sow them thinly and evenly to 6 mm (¼ in.) deep. When 5–8 cm (2–3 in.) high, plant the seedlings 23 cm (9 in.) apart in nursery rows 38 cm (15 in.) apart. In colder climates transplant them into cold frames or a heated frame and leave them there until spring. In a mild climate transplant them in the fall.

Sow seeds of Siberian wallflower in June and transplant the seedlings to a nursery bed. In the fall plant them in beds or areas where they will flower the following May and June. In harsh climates cover them with a layer of straw for winter protection.

English wallflower cultivars:
'Blood Red' — deep blood crimson
'Carmine King' — rich carmine
'Cloth of Gold' — golden yellow
'Eastern Queen' — chamois, changing to rosy salmon
'Ellen Willmott' — ruby red
'Fire King' — vivid orange-red
'Harper Crew' — double gold
'Vulcan' — velvety crimson
'White Dame' — creamy white

Siberian wallflower cultivars:
'Golden Bedder' — gold
'Lemon Queen' — light yellow

chapter 4

culture of perennials

culture of perennials

PROPAGATION

In general the propagation of perennials presents very few difficulties. Most of them can be propagated by seeds, division, stem cuttings, or root cuttings. A few must be increased by means of grafting. The method adopted depends upon the type of plant to be increased. Most often vegetative propagation offers the only method of acquiring new plants from outstanding cultivars. Seeds offer some very good possibilities of getting new species and interesting forms from other gardens at home or abroad, since seeds usually travel well and if packaged in polyethylene bags will not lose their viability.

Seeds

Seed is the quickest and easiest means of growing a large number of plants of the same species, of strains, and of certain cultivars, but not often of the same cultivar. For example a package of seed of a cultivar of *Phlox paniculata* 'Daily Sketch' or of *Papaver orientale* 'Juliette' will not produce plants true to type, though it will most likely produce a number of interesting perennial phlox or Oriental poppies of various colors and shapes. The phlox will be predominantly mauve and the poppies red, but a few will be salmon, white, or purple. Similarly the seeds of a double crimson hollyhock or pyrethrum will not produce all-double or all-crimson progeny.

Most named horticultural cultivars will not breed true to type from seeds; for this reason gardeners have devised vegetative methods of propagation, such as division, cuttings, and grafts, so that the interesting cultivars will be reproduced exactly like their parents, once they have been created by breeding and selection.

Seeds must not be buried too deeply. Very fine seeds like those of *Campanula* must not be covered with more than 2.5 mm (1/10 in.) of soil or they will most likely fail to germinate.

With most large seeds a good general rule is to cover them with soil no deeper than three or four times their own thickness. If sowing outside, this must be modified according to the types of soil.

Never let sprouting seeds become dry. Give very close attention to watering, particularly during the first 2 or 3 weeks after sowing.

Flat seeds will often give better results if they are carefully sown on edge, because there is less danger of them rotting before they germinate.

Seeds with hard outer coats (such as *Lupinus*, everlasting peas) usually take a long time to germinate, because the water required to activate the embryo within is very slow in penetrating. For this reason soak the seeds of these plants for 6 to 8 hours in lukewarm water before sowing. Chipping seeds of lupine, which means removing a very small part of the outer seed coat with a sharp knife, often gives good results. If this is not possible with the tools at hand, use a small file.

Seeds of plants that by nature inhabit high altitudes and northern climates appear to be so adjusted as to need an after-ripening period at nearly freezing temperatures before they will germinate. Without this chilling period of from a few weeks to several months, many will not germinate. Classical examples among perennials are the woodruff *(Asperula odorata)*, the gasplant *(Dictamnus fraxinella)*, and the fringed gentian *(Gentiana crinita)*. Seeds of many alpine and rock garden plants used for the front of the perennial border need to be left outside to the elements for at least two winters unless they may be stored at freezing temperatures 0°C (32°F) for 2 to 6 months, depending upon the species. Keep them either between two layers of cheesecloth covered with moist sand, or in polyethylene bags. After this period, plant the seeds in pots or boxes, and put them in a cold frame. So treated most of them will germinate the first year.

Seeds of irises, lilies, delphiniums, and many other perennials must be sown as soon as they are ripe, even though they often will not germinate until the following year. If they are kept and stored dry in seed bags, germination will be delayed for a longer time or will be very irregular.

Trillium seeds require two periods of chilling before they can develop both roots and leaves. Sow them late in the summer as soon as they are ripe, but they will not produce any growth above ground until after the second winter. To gain time and to be more sure of success, give two properly timed periods of artificial chilling in a refrigerator, 1 month of chilling interrupted by 3 months at 15.6°C (60°F), which will approximate normal conditions. Other plants that behave similarly are bloodroot *(Sanguinaria canadensis)*, blue cohosh *(Cauliphyllum thalictroides)*, lily-of-the-valley *(Convallaria majalis)*, and Solomon's seal *(Polygonatum multiflorum)*. Seeds of the Christmas rose *(Helleborus niger)* that are picked in June, dried on a dry sunny shelf, cleaned and stratified in sand, sown in October or November in a cold frame will germinate in the spring. If kept dry during the summer the seeds probably will not germinate for 2 years.

Many, if not most, seeds of perennials, unlike those previously described that are inclined to remain dormant though viable for many years, are quite short lived and remain viable only for a short time. These seeds should be sown as soon as they are collected.

Division

Where new plants of selected cultivars that will not come true from seed are required, division of the old clumps is generally recommended. For the average gardener who probably has no propagating equipment, division of old clumps is the most practical method. Most perennials, although not all, may be increased this way.

The objective of propagation by division is to divide the root stock into the largest number of pieces that will grow when planted in nursery rows. To achieve this end, dig out the old plant carefully so as to retain as many roots as possible. Shake off much of the earth and wash the roots thoroughly with a hose. This will make it easier to determine where the divisions are to be made. This work is best done on a dull day or in a shady area.

In general early flowering perennials are best divided in September or October and late flowering kinds in the spring. There are exceptions to this rule as for most rules concerning organized operations for living plants. Irises are best divided in July or August, Oriental poppies in August, and peonies in early September.

The method of division varies with the type of root growth produced by the different groups. *Anthemis, Astilbe, Centaurea, Coreopsis, Doronicum, Gaillardia, Chrysanthemum, Scabiosa, Stachys, Veronica,* and similar genera have heavy woody roots that are joined together in the root ball. Cut these with a sharp knife, dividing the plants into large or small clumps as required, but making sure each division has roots and one or two buds at the base from which new growth can be started.

Anemone, Aquilegia, Hepatica, Heliopsis, Primula, Saxifraga, Trollius, and similar genera are formed of crowns that may be separated easily by pulling them apart, but this must be done carefully to keep the crowns and roots intact.

Species of perennials that bloom late in the season such as *Aster, Achillea, Chrysanthemum, Echinacea, Helianthus,* and *Rudbeckia* should be divided in early spring as soon as new growth starts. Cut or pull the young rooted shoots at the side of the clump either separately or in groups of three or four together.

Some mat-forming plants like *Alyssum, Aubrieta, Arabis, Cerastium, Dianthus,* and *Phlox subulata,* and other alpine sorts like *Sedum* and *Thymus* send out roots whenever their stems touch the ground. Dig these in early fall, make them into small clumps with or without roots attached, and replant them in sandy soil in a shady cold frame.

Stem Cuttings

Many herbaceous perennials that are difficult to divide and are hard to grow from seed or will not reproduce themselves true from seed may often be propagated by means of stem cuttings. This is also a good method to adopt in cases where a larger number is required than can be obtained by division.

In most areas of Canada where the winters are severe, it is best to root cuttings in the spring or early summer, because when cuttings are taken later in the season they will not root in time to make the necessary growth to withstand the winter.

Modern materials and new techniques have made it possible for the average homeowner to propagate some of the best herbaceous perennials from stem cuttings. One of the easiest ways of doing this is by using the 'quonset hut' method, a technique involving the use of a propagating case, covered with polyethylene plastic.

To make such a structure, get a box or flat about 51 cm (20 in.) long, 36 cm (14 in.) wide, and 10 cm (4 in.) deep. Fill the box to within 1.3 cm (½ in.) of the rim with a mixture of equal parts sand and peat moss. Firm the medium into the box with a tamper made from a block of wood. Now get some willow twigs, thin bamboo, or heavy-gauge wire. Bend them like hoops and insert the ends in the inside of the flats making a framework similar to that of a covered wagon. Have at hand some sheets of polyethylene, or a number of laundry or fruit bags made from this material, to be attached to the frame after the cuttings have been inserted.

From mid-May until the end of July is a good time to take cuttings of perennials. When you are ready to start, equip yourself with a good sharp knife, some labels, twine or plastic wire, and a polyethylene bag. Select shoots about 10–15 cm (4–6 in.) long that arise from the base of the plant and cut these just at or below ground level. Remove the leaves from the lower nodes and treat the cuttings by dipping the ends in a mild rooting compound. There are many kinds of these root-inducing hormones on the market. They are available in powder and liquid forms and all of them make rooting faster. Cuttings of some perennials such as phlox and asters may be made from the tips of the plant in early July, but they are more difficult to root. Dianthus and viola cuttings are best taken in July. Shoots of dianthus are pulled from the plants and these serve as excellent cuttings after a few basal leaves have been removed. Viola cuttings are taken by severing a few inches from the ends of tip shoots. Cuttings of the creeping moss pinks (*Phlox subulata*) may be taken from early spring until midsummer. These are made from small shoots not more than 2.5 cm (1 in.) or so long. In general, cuttings of perennials taken from the basal shoots that arise in early spring are the most successful. *Phlox paniculata*, asters, and delphiniums may be made to induce basal shoots later in the season by cutting them down to ground level just before flowering.

After the cuttings are prepared, insert them 5–7 cm (2–3) in. deep in the medium and space them 2.5 cm (1 in.) apart in the box, in rows 5–7 cm (2–3 in.) wide, according to their size and substance. A flat with the dimensions given will hold 75–100 cuttings. When you have inserted all your cuttings, water them well from above with a fine rose watering can. This will set them firmly in the medium. Now attach the polyethylene to the willow or other framework by means of racks or pegs and arrange the whole frame so that it looks like a covered wagon. Make sure it is air tight, but at one end allow for it to be opened so that you may push a watering can spout through.

Leave the frame in a sunny place but keep it shaded by means of a cloth during the hottest part of the day. It should not need watering for a week or 10 days, but it is a good plan to wet the leaves occasionally as they dry. You may need to do this every 4 or 5 days.

In about 3 weeks test the cuttings by pulling one out to see if it has formed roots. Some cuttings root faster than others. When they are rooted, plant the cuttings in the garden in some spot such as a corner of the vegetable area, and sprinkle them daily from above with a fine rose watering can until they no longer wilt. Leave them in the nursery until next spring, then plant them in the flower garden, where they will flower freely the first summer.

Cuttings of dianthus and viola are best taken in July. Shoots of dianthus pulled from the plants make excellent cuttings after a few basal leaves have been removed. Viola cuttings are cut 7–10 cm (3 or 4 in.) from the tips of the shoots. The moss pink cuttings are necessarily small, not more than 2 or 3 cm (an inch or so) long.

Root Cuttings

Many plants that are difficult to root from stem cuttings can be increased readily by means of root cuttings. In general the plants are dug in the fall; roots about 1 cm (½ in.) or more in diameter are selected and cut into sections 2.5–7.5 cm (1–3 in.) long. Place these horizontally in flats filled to within an inch of the top with a mixture of equal parts sand, peat moss, and soil. Place the cuttings one-half inch apart and cover them with sand. Water them thoroughly and store in a cold frame for the winter. In spring, young plants will develop from these root cuttings. When they grow 5–7 cm (2 or 3 in.) high, they should be either planted in pots or set out in nursery beds. Japanese anemone, *Phlox*, *Verbascum*, *Gaillardia*, *Echinacea*, *Rudbeckia* 'The King', *Brunnera*, and *Statice* may be propagated this way.

Oriental poppies are also grown from root cuttings, but the technique differs somewhat in timing and method. The root sections are taken in August by cutting the roots in lengths of 7 cm (3 in.) and inserting them in a trench or a nursery bed, or in 10-cm (4-in.) pots that are plunged to the rim in peatmoss or sand in a cold frame. Potted cuttings will root easily in 18 months and in the spring will be large enough to set out in the border, and because they are in pots they will suffer very little setback. Trench cuttings will remain in the soil until the following September. In most localities some winter protection the first year must be given.

GENERAL CARE

Preparing the Soil

Any good average garden soil can be used to grow most perennials, but the soil must be well drained. Because perennials are usually permanent features of the

Before you do your planting, rake the soil, remove stones and large undecomposed lumps, and add a complete fertilizer such as 5-10-15 or 6-9-6 at 1.4 kg/9 m² (3 lb. for 100 sq. ft.) to the top layer of soil.

Prepare your flower border in August so that you can do your planting in September and the next spring, depending on the type of plants you want and their availability.

Planting

Most perennials can be planted at any time except when the ground is frozen hard, but there are seasons when they can be planted with the least injury to the plants and the most assurance that they will survive for a long time. Early blooming hardy perennials are best when planted in early fall, which enables the roots to get well established before flowering time in spring or early summer. Late flowering perennials are best when planted in spring, when they are just starting their growth, so they have the long summer season to become established. Exceptions are the irises, Siberian irises, Oriental poppies, doronicums, Virginia bluebells, Madonna lilies, Crown Imperial lily, and bleedinghearts, which are better when planted from late July to late August.

If you have made a planting plan to scale on paper, mark out the area to be planted in the same way. Mark the squares with strings, make furrows with the handle of a rake, or use lime to mark the lines.

When the plants arrive from the nurseryman they are usually packaged well with small wooden labels inside. Arrange them alphabetically, or heel them in the soil until you need them. By this means you will be able to plant each one properly without fear of the rest drying in the sun. If possible, choose a dull day for planting, and do not leave bare-rooted plants lying in the sun for too long before planting.

Use a spade to plant very deep-rooted plants, such as peonies, baby's-breath, or anchusa, so that the roots can be set down deep enough. Before planting cut off all the broken or smashed roots to prevent rot. The root systems of most perennials are small enough that they can be planted with a trowel. Follow these good rules for planting.

1. If the roots are packed in nursery packages, spread out the roots when planting and do not plant them all bunched together.

2. Do not lay out too many plants on the soil at one time.

3. Make the holes wide enough and deep enough to hold all the roots.

garden, it is important to prepare the soil carefully so that it will remain in good condition for many years.

The best method of soil preparation for garden flower beds where herbaceous perennials are to be grown is the trenching method. Mark off a strip 0.6 m (2 ft.) wide across the bed, and remove the top 20 cm (8 in.) of soil, or one spade's depth. Pile this soil at the far end of the bed. Lossen the subsoil by turning it over with a digging fork, and mix in a dressing of 8–10 cm (3 or 4 in.) of topsoil, peat moss, decayed leaves, or compost. Take the top 20 cm (8 in.) from the next 0.6 m (2 ft.) width and place it on top of the subsoil just turned, then loosen the bottom of the trench as before. Repeat this process throughout the length of the border, placing the topsoil removed from the first section into the last hole.

The next best method, requiring less manual labor, is to rent a rotary cultivator and to break up the soil to 46 cm (18 in.) deep. To cultivate the soil deep enough, you will probably have to go over the ground several times.

When you know that a soil is badly drained, and water remains standing after a heavy rain, use another method of preparing the soil. Remove the soil completely, as for trenching, so that the bottom hardpan (most likely the reason for the bad drainage) may be broken up. Before you add the topsoil, place an 8-cm (about 3-in.) layer of coarser stones or gravel at the bottom of the trench for extra drainage. This might seem like a lot of hard work, but it is absolutely necessary in order to avoid winter loss and puny garden plants. In extreme cases, you may have to put in drain tiles.

4. Set the plants out at the same depth that they were growing before they were dug from the nursery beds.

5. Spread out the roots in the hole and add soil gradually; step gently on the soil to firm it before you add more soil.

Maintenance

Too many gardeners think that when perennials have been set out they need no further attention. This popular fallacy is based on a comparison with annuals, which must be started afresh each year from seed. Although perennial plants such as peonies, delphiniums, irises, phlox, and veronicas give good displays year after year when left undisturbed except for a little weeding, these plants show a greater and much more prolonged abundance of bloom and larger individual flowers when a few cultural practices are followed during the growing season.

After you have removed the winter coverings from some of the more tender perennials, the next operation in early spring in the border is to reset plants that have been lifted from the ground by the action of alternate thawing and freezing. Also, at this time, remove the decaying leaves that were left on irises and other perennials the previous fall.

Established perennials are likely to become overcrowded with flowering shoots, which, if left to develop will produce an abundance of small weak blossoms susceptible to mildew and other diseases that will eventually destroy their beauty. Therefore, the shoots of perennial phlox, asters, delphiniums, and such plants should be thinned each year or dug up and divided. This operation should be done early in the spring before the shoots are more than 8 cm (3 in.) high. Leave 9 to 15 shoots on each plant, depending on the age of the plant; remove the unwanted shoots by cutting them about 2.5 cm (1 in.) below soil level. These severed shoots of some perennials will root quite readily in clean sand if kept shaded from strong sunlight.

Some perennials such as hardy chrysanthemums and asters will grow much better if lifted and divided every spring. Each single shoot of a hardy chrysanthemum will produce a large plant and a bountiful display of flowers by fall if kept pinched back.

A complete fertilizer such as 6-9-6, 7-7-7, or 10-10-10 at 102 g/m² (3 oz. to a sq. yd.) should be forked into the borders in spring. Sprinkle the fertilizer around the plants but try not to get it on the new shoots because it may burn them. To introduce as much organic matter as possible, mulch the border occasionally with well-decayed leaves or compost.

A necessary operation, often neglected, is supporting the plants by staking. Small perennials such as gypsophilas, pyrethrums, and Shasta daisies can be neatly supported by placing twiggy clippings from shrubs around the plants. Delphiniums, asters, and phlox need strong supports such as bamboo canes. Place the stakes among the plants and tie the shoots to them loosely so that they do not appear crowded. A strong wire hoop supported by a stake is a good support for peonies. Stake the plants before they are in bud. If you have a weeping willow you may use some of the twigs from this tree for making the hoops.

In hot dry summers watering is usually necessary, but do not sprinkle the perennials every time you water the lawn. When the weather has been dry for a couple of weeks, sprinkle the perennial border some evening. A mulch can be very helpful during the hot summer months. A good mulch put on the border during early July after a soaking rain will often eliminate the need for watering. Before you apply mulches such as straw, peat moss, or sawdust, it is necessary to add extra nitrogen to the soil to compensate for the large amount used by bacteria as these materials decompose. The exact amount of nitrogen and the form in which it should be used depends upon the nature of your soil, and the kind of mulch you find most economical to use. For example, with sawdust on a light soil use 170 g/9 m² (6 oz. of sulfate of ammonia for 100 sq. ft.) on alkaline soils, and use 227 g (8 oz.) of nitrate of soda on acid soils just before the mulch is spread. After 6 weeks, apply the same amount of fertilizer on the mulch and water it. Mulches of inert substances such as cocoa shells, buckwheat hulls, gravel, and crushed white stone do not need this extra fertilizer. The depth of the mulch should be about 5 cm (2 in.) and it should be replenished, if necessary, as the season progresses.

Some perennials such as chrysanthemums need to be pinched back to produce a neat bush and more flowers. This should be done soon after planting in June and again in July.

Disbudding is an important operation that is nearly always neglected. Nearly all plants produce superfluous buds, which seem entirely unnecessary and often spoil the appearance of the plant. It is possible that the surplus buds have a natural use in uncultivated plants, but, when they are clustered on perennials such as chrysanthemums or peonies, they do not form perfect flowers and usually detract from the main ones. Therefore, with these plants it is best to allow the main central bud to

remain and to pinch out all the others before they grow too large. By this means all the strength of the shoot goes to the remaining bud, causing it to grow much larger.

Removal of old faded flowers throughout the season serves several useful purposes: the plants look tidier, plant energy is not spent on the development of seeds, unwanted seedlings do not spring up, and the flowering period of the plant is usually prolonged. Just take off the actual flower or flower cluster, because quite often the foliage and stem below will continue to manufacture food that will help produce a healthy plant next year.

RENOVATING THE PERENNIAL BORDER

The perennial border should be remade every 5 or 6 years, regardless of how carefully the border was planned in the first place.

Perennials tend to become too large and so overgrown that the larger plants spread over and smother the smaller ones, and some kinds such as irises and peonies become so crowded within themselves in 4 or 5 years that the flowers become smaller and less abundant. Also, because there are so many new and interesting varieties to add and the older plants have to be re-located, renovation soon becomes inevitable.

New homeowners who have lawns to make, hedges to plant, and many other jobs to do should heel in temporarily the perennials given to them by friends until there is time to plant them in a permanent location. Consequently, tall plants are often placed by mistake at the front and smaller ones are hidden at the back, resulting in complete lack of harmony and a need to remake the border after the second year.

At the Plant Research Institute's Ornamental Grounds we like to do the major part of this operation in September, when so many kinds of plants may be moved and divided. Oriental poppies, for example, are only successfully moved in September; at this time, you may also dig and plant Madonna lilies, although it is better to move most lilies later on. In an emergency they can be successfully transferred from one place to another if you are careful to leave the soil intact around the clumped bulbs.

Peonies are best divided in September. Although irises should be divided at the end of July or in August, this can be done later in the fall if about 2.5 cm (1 in.) of soil is heaped over the exposed rhizomes.

First, dig up and discard all the plants that you want to eliminate. Then select the ones you want and label them with 30-cm (12-in.) stakes. Move these into an unoccupied part of the garden and heel them in temporarily. If you do not have any space in the garden, put them in flats or boxes or on a piece of canvas spread out on the lawn. In any case, cover the roots with moist soil and keep them moist until they are replanted.

Unless it is absolutely necessary, do not move plants that are hard to transplant, such as dictamnus, anchusa, and alyssum. Also, do not move young peonies that flowered for the first time during the summer. Chrysanthemums and asters may be moved successfully when in flower, if this is done carefully and lots of soil is left on the roots.

When the area is almost completely clear of plants, dig it deeply and work in plenty of organic matter. When the soil is dry, tramp it down thoroughly.

Before you start planting, it is an excellent idea to catalog the plants you have available and then work out a plan on paper, so that you know just about where each plant or group of plants is to be placed. It is also advisable to write the name of the plants on stakes and put the stakes where the plants are to be placed: tall ones in groups at the back of the border, medium ones in the middle and occasionally at the front, and dwarf border plants close to the edge.

For some plants you need to dig a hole with a spade; others may be planted with a trowel.

Here are a few rules to follow when planting.

1. Prevent roots from drying out by working quickly; do not place more than six plants on the soil at a time.
2. Spread out the roots of new plants when planting if the roots are bunched from packing.
3. Make the holes large enough to take the roots without crowding.
4. Set the plants at the same depth they were in their previous location.
5. Work the soil among and over the roots and then firm it with your feet or hands.
6. Do the planting when the soil is in good condition and not too wet.

Leave spaces here and there for other plants that will be ordered in the spring: lilies that may not arrive until late October, spring-flowering bulbs that may not yet be available, and plants such as perennial phlox, which should not be divided until the next spring, even though you may transplant them in the fall.

WINTER PROTECTION OF PERENNIALS

Herbaceous perennials are often called hardy perennials. The term hardy may promote the belief that all herbaceous perennials are rugged enough to survive the severest winters under all conditions, but this is true of these plants only in the mildest parts of British Columbia. In most of Canada, proper care and cultivation are needed to bring the plants through the winter.

The first step to ensure winter survival is to select a hardy species or variety. It is not very difficult to choose hardy woody plants, but herbaceous perennials are a different matter. The ability of nonwoody perennials to live through the winter seems to depend more on the particular kind of winter weather than it does on the hardiness of the species.

A serious deterrent to survival is heaving by frost. When alternate freezing and thawing take place, the entire root systems of some plants may be pushed out of the ground, especially if the plants are growing in heavy soil.

When there is plenty of snow, heaving is less of a problem, because snow keeps the ground temperature constant. In most of Eastern Canada heaving occurs in early spring after the snow has melted. When this happens, push the plants back into the soil and cover the roots.

The most important protection against winter can be provided when perennial beds or borders are being made. If the soil is poorly drained, you must find a way to improve the drainage. You may even have to dig out the bed and add a layer of coarse stones or drain tiles to lead to a natural drainway.

Trenching the soil or very deep digging is usually satisfactory in most areas. All that can be done with a badly drained border once it has been planted is to dig a trench in the lowest place and run drain tiles to the surrounding land.

If you have a hedge as a snow fence that collects snow on your border, little extra protection is needed. If you do not have a hedge, you should trap the snow some other way. This may be done by leaving the stems on

Straw placed around perennials for winter protection

some medium-sized perennials, or by cutting tall-growing plants such as asters and helianthus to half their height. This method means more work in spring, but it may save your plants from being killed in winter.

If you have a friend on a farm who can spare a few evergreen boughs, these can be scattered on the border to help collect snow, or you can use your Christmas tree.

Some kind of protection is always needed for a few tender plants, such as Canterbury bells, foxgloves, Madonna lilies, pansies, and Siberian wallflowers, which start the winter with green leaves. Do not use any covering that will mat down and form a compact, wet mass. The best way to protect these plants is to wait until the ground has frozen 4 or 5 cm (1 or 2 in.) deep, then put a layer of partly decayed leaves or peat moss under the lowest leaves of the plants. Next, place evergreen boughs or boxes over them, or put a loose covering of straw, excelsior, or pine needles over the plants. Do not put on any winter covering until the ground has frozen, and do not cover the crowns of peonies and delphiniums.

Perennials newly planted in the fall need special protection against heaving during winter thaws. This is best done with a light mulch, as mentioned previously, or with the use of evergreen or deciduous boughs and prunings from woody plants.

Irises planted in August and September are usually well enough established to escape winter injury, but the safety of new and precious cultivars can be ensured by hilling about 2 or 3 cm (1 in.) of soil over the rhizomes. This earth must, of course, be scraped off in early spring to allow the sun and air to penetrate the rhizome, a necessary precaution for the prevention of the soft-rot diseases.

If you have hardy chrysanthemums in your border, it is best to transplant them to a place near the warm foundation wall of your home. To collect snow, leave the flower stems or cover the plants with boughs. Provide an extra covering of straw if your plants are on the south side of the house. If your plants come through the winter successfully, in spring, divide them into small clumps, each with one or two shoots. These new plants will produce more abundant blooms and larger flowers.

In general, unless you are attempting to grow the more tender biennials and perennials or are in an exposed situation, the border needs no special winter attention and should thrive for years.

The research stations in Western Canada have found that in the Prairie garden the position of a perennial border is very important. It is best facing north. If a border faces south, exposed to the direct rays of sun in summer and the alternate freezing and thawing in winters of light snowfall, it is difficult to grow any perennials except the most drought-enduring ones. A location where snowdrifts collect is desirable, and boughs strewn about the border will help hold the snow and prevent too rapid thawing in spring.

Straw makes an effective mulch, but it should not be applied until the ground is frozen. In November there are frequent warm spells of quite long duration when a straw mulch is not always sufficient to prevent the soil around the roots of the plants from thawing. Therefore do not put down the mulch until winter has definitely set in.

DISEASES

For a gardener to select and apply the proper control measures for a disease occurring in his garden, he must be able to recognize the disease or the group to which it belongs. Also, he must have some knowledge of the causes of plant diseases and the symptoms of a group of diseases or specific ones.

44

A plant disease may be defined as a deviation from the normal condition caused by the continuous irritation of some factor, which may be a living organism or an unfavorable environmental condition or cultural practice. Living organisms that cause plant diseases are fungi, bacteria, and viruses. Also, abnormal conditions or disorders caused by nematodes are generally considered to be diseases. These living organisms that cause disease are called pathogens. Diseases caused by unfavorable environment are called physiological diseases. Plant injury caused by insects and other animals is not considered to be a disease.

Most of the diseases found in the garden are caused by fungi, but diseases caused by viruses are also common. These two groups of diseases occur over a wide range of environmental conditions. Diseases caused by bacteria are not common, because they are found mainly in wet or humid seasons or damp locations in the garden. However, bacterial diseases may be found in dry years on plants that have been watered often by overhead sprinkling.

Diseases that affect all plants may be grouped according to their symptoms or their causes. The symptoms may be local, that is, confined to a small area or to a part of the plant such as a leaf spot or a canker, or they may be general, such as a wilt. In most cases control measures for all the diseases having the same symptoms are the same. The most common diseases of herbaceous perennials are given in the following paragraphs, arranged in symptomatic groups. Because regulations regarding the use of chemicals for disease and insect control differ among provinces and are subject to change, no recommendations for their use are given in this publication. For up-to-date recommendations consult your provincial agricultural representative or faculty of agriculture.

Blights

The areas of diseased tissue attacked by blight are fairly large and often result in the death of the leaves, flowers, or shoots. The symptoms are sudden wilting and browning. The most common and disastrous blight is sweet chestnut blight, which has at one time or another attacked or destroyed all the plants on the American continent. Another well-known disease in this group is bacterial fireblight of fruit trees and of ornamentals in the rose family.

Blight is not only caused by fungi and bacteria, but can be caused by unfavorable environmental or cultural conditions. The most common blight of perennials is the one that defoliates *Phlox paniculata*. It is entirely a physiological condition caused in the spring when the old stems are unable to supply enough water to prevent drying of the lower leaves after the demanding new shoots arise and take up all the moisture.

Other perennials that are seriously affected by this disease are garden chrysanthemums, delphiniums, lupines, peonies, and poppies.

Sanitation is very important in the control of blights. Gather and burn all affected leaves, stems, and flower petals and debris on the ground. Cut out affected parts of the plant, and keep the garden clear of weeds and rubbish. Leaf blight of phlox is difficult to control; the best method is to divide the plants often and to provide ample water in dry weather. Other bacterial and fungus blights are stimulated by overwatering.

Leaf Spots

Diseases that cause leaf spots are quite common among herbaceous perennials; the spots appear as circular dead areas in leaves or fruit and are usually brown in the center with a yellow or reddish border. Some spots, particularly those caused by bacteria, have a water-soaked border that is darker green than the leaf. These leaf spots are common on most perennials.

Leaf spots are also a symptom of a group of diseases called anthracnose, caused by a fungi; these are common on lilies-of-the-valley, foxgloves, violets, and peonies.

Most leaf-spot diseases can be effectively controlled in the early stages by picking off affected leaves and shoots, and by good sanitation practices.

Powdery Mildews

Powdery mildews are the most common diseases in all Canadian gardens, mainly in late summer and early fall when they are apparent on many plants, particularly asters, phlox, delphiniums, erigerons, shrubs such as lilacs, and annuals such as zinnias. These diseases are easily recognized by the white or grayish growth of powdery mildew fungi that grow on the surface of leaves and stems, and are most noticeable after cloudy moist nights in areas such as gardens surrounded by high hedges or walls protected from air currents. Under these conditions the spore-bearing stalks grow out through the stomata (pores) of the leaves and shed their spores into the dew or raindrops that cling to the leaves. The mildews may be seen as distinct spots or they may cover a large area over the whole surface of the leaf and stem. Some spots are brown in the center.

Rots

Any organism that causes extensive disintegration of the plant tissue may cause a rot, which may be dry, wet, soft, or hard. The most common rot diseases are crown and rhizome rot of delphiniums, irises, and aconites, and botrytis rhizome and bacterial soft rot of irises, which is wet and extremely vile smelling.

If you know that crown rot of delphiniums and other plants is present in the soil, treat the soil with a soil fumigant, or sterilize it under steam pressure, or remove the soil and replace it with clean soil. For rhizome rot of iris, destroy the badly infected rhizomes and treat the rest of the plants with a fungicide. Bacterial soft rot, which is very common in iris plantings, usually gains access to the plant through wounds made by borers. The best control is to eliminate the borers by burning old leaves and other rubbish that may have accumulated and to spray with an insecticide. Cultural practices such as providing as much sun as possible help eliminate bacterial soft rot. At the test gardens of the Plant Research Institute, iris leaves are cut back to within 10 cm (4 in.) of soil level in October and the cuttings are carted away. When planting, take care to plant the rhizome very shallowly, with about half of it showing above ground.

Rusts

Although the name rust may be applied to symptoms of other diseases, it is used here to refer to those diseases caused by a specific group of fungi that can only exist on living plants. Many rusts require two different kinds of host plants to complete their life cycle, and they form different types of spores on each host. Other rusts have all their spore forms on one host and so they do not need a second host plant.

The symptoms of rust are yellow, orange, reddish brown, or brown spores in powdery pustules or gelatinous tendrils. The spores may be found on the leaves, particularly on the lower surfaces, and the fruits and stems.

The erigeron, dianthus, gaillardia, helenium, helianthus, liatris, monarda, oenothera, most members of the mallow family, sempervivum, and viola are commonly affected by rusts. Rusts are not as serious a problem on other hosts as they are on the hollyhock.

To control rust on hollyhock pick off and burn infected leaves and parts as soon as you see them. Cut out any stem showing rust canker, galls, or swellings. When the flowering season is over, cut back the plants to the base and burn the old parts. On the other perennials previously mentioned as hosts of rust, pick off and destroy infected leaves.

Wilts

Wilting is caused by a lack of water in the leaves and stems. It may be caused by dry soils, injury to the root system, or the effects of various kinds of parasites. It is often difficult to determine the cause of wilting unless the symptoms of a disease or an injury to the stem are apparent.

Wilt diseases are caused by fungi or bacteria that invade the water-conducting vessels in the stems and roots. The pathogen may plug the vessels of the plant by its growth or by producing a gummy substance, or it may produce toxic substances that cause wilting and are carried in the sap stream.

Other symptoms that accompany wilting are yellowing followed by browning, curling, and drying, and the eventual falling of the leaves. Some herbaceous perennials that are subject to wilts are garden chrysanthemums, foxgloves, helianthuses, and peonies.

Sanitation is most important in controlling wilt diseases. Clean up all plant refuse and destroy affected plants. Most wilt-causing fungi are soil borne, so do not transfer infested soil to other areas. After you have removed infected plants, disinfect your tools and do not plant susceptible plants in that location for several years. Keep plants in good health by regular watering and fertilizing, and providing good drainage.

Nonparasitic Diseases

These maladies or disorders result from unfavorable cultural practices or environmental factors such as drought, mineral deficiencies, overwatering, underwatering, improper soil acidity, extreme temperatures, gases or fumes produced by a leak in gas lines, or by some industrial process, and toxic spray materials applied to the plant by mistake, such as the misuse of weed killers or an overly strong solution of spraying chemical.

All plants must have a balanced diet to grow well. Plants grown in a soil that is deficient in one major element or is lacking completely one or more minor elements will not be normal. A mineral deficiency of the element iron is most common in gardens, and shows up in many perennials, such as *Dicentra*, by turning the foliage yellow. Many problems with plants, especially in gardens near highways, are caused by air pollution, which usually discolors or disfigures flowers.

Leaves and shoots that are disfigured in early spring and appear tattered when they open were most likely injured by frost. Many perennials are killed each year by frost heaving them from the soil and severing their roots. Unless they are tramped down and kept in contact with the soil, they will wilt and die.

High temperatures and strong drying winds may cause leaves to brown at the tips and along the edges. Some leaves will curl and shrivel and then turn brown. This is called leaf scorch and is often mistaken for blight damage.

To control nonparasitic diseases, you must first ascertain the cause of the disorder and then rectify it.

Plants vary in their sensitivity to environmental factors, just as they sometimes differ in susceptibility to a particular disease-causing organism. If the environmental conditions have been changed or improved since the symptoms first appeared, diagnosis can become very difficult. You must know the growing conditions, cultural practices, and treatment before you can diagnose and evaluate the types of plant ailments.

To prevent frost injury, cover tender plants with straw after the soil has frozen. As soon as the soil is dry enough to walk on in spring, tread around all your perennials to prevent them from damage by heaving.

During hot weather, water the garden in early morning to help eliminate bad scorch. To eliminate most if not all nonparasitic ailments, follow carefully the directions in this publication for the correct culture and special requirements of the plants mentioned.

Virus Diseases

Viruses are responsible for initiating many diseases of garden plants. Viruses are ultramicroscopic bodies, too small to be seen with an ordinary microscope. Yellows, stunt, and mosaic of *Chrysanthemum;* mosaic, curly tip, and stunt of *Delphinium*; iris mosaic, and mosaic and leaf curl of peonies are some of the common viruses that affect herbaceous perennials.

The most common symptoms of virus diseases are changes in flower color, often shown by streaking or blotching of the petals, or the whole flower may be greenish; yellowing of the leaves; mottling of the leaves with yellowish or light green areas in the usually green leaves; stunting, when a plant is dwarfed much more than usual and shows malformed leaves and buds; rosetting, a condition where the leaves are crowded because of shortening of the stem; and leaf deformities such as curling, twisting, or crinkling.

To control virus diseases remove and burn the infected plants, and keep the garden free of insects, particularly leafhoppers and aphids. It is best to buy new chry-

santhemum cultivars from a specialist who treats the stock plants by subjecting them to heat and propagates the plants under aseptic conditions.

Nematodes

Plant-feeding nematodes are transparent microscopic worms that move about in the water film of the soil and obtain their nourishment by puncturing plant cells with their hollow needle-like spears. In most of Canada where the winter is severe, only a few nematodes are troublesome plant pests. Of these, the peanut root-rot nematode, which attacks *Paeonia, Chrysanthemum, Delphinium,* and *Papaver,* is the most destructive. The stunts, leaf blotches, and blights caused by nematodes are usually classed as diseases.

It is hard for the average gardener to determine whether his plants are being damaged by nematodes. The symptoms are not always readily apparent. Nematodes seldom kill plants, but usually they cause a decline in vigor.

If you carefully examine the roots at the first sign of abnormality, certain symptoms characteristic of nematodes can be seen: galls or swellings on the root, brownish root lesions, or stubby root systems.

Preplanting soil fumigants are quite effective against nematodes, but they are not practicable in the garden. If you suspect the presence of nematodes, send the affected plant to the Experimental or Research Station nearest you, and you will be given advice on controlling the disease. The chances are that the problem will turn out to be some other disease, and the control will be much simpler.

To obtain a diagnosis of any disease and the recommended control measures, select a specimen that shows typical symptoms. Write a description of the symptoms, the way the disease started, and the conditions under which the plant is growing. Include in your notes any special or unusual treatment the plant may have received. If you are sending the specimens by mail, enclose this information in a separate letter and send it to the same address that you send the parcel.

Pack leaves flat in newspapers or between sheets of blotting paper. Wrap soft stems in newspapers, but place woody stems in a box with rough newspapers around them to keep them from shifting around. Pack the larger parts of the plants, such as fruits or roots, from which all soil has been removed, in a cardboard box with crumpled or shredded newspapers. Allow any moisture on the surface of specimens to dry before you pack them. Specimens dried on the surface will ship well if placed in a plastic bag, provided they are not en route too long or the temperature is not too hot. Do not pack specimens in wax paper. Make sure your name is on the parcel and on the accompanying letter.

INSECT PESTS

Insect pests must be controlled in order to preserve the beauty and health of herbaceous perennials. Insect damage can often be prevented by carrying out good cultural practices such as providing nutrients to the soil and following a good sanitation program. It may, at times, be necessary to use insecticides.

For advice on insecticides consult your local agricultural representative, agricultural college, or the nearest research establishment of the Canada Department of Agriculture.

Common insect pests of all garden ornamentals, including herbaceous perennials, can be divided into two groups, according to the way they feed: those that suck the plant juices and those that chew the leaves or bore into the stems.

The first group consists of aphids, which are easily seen on the leaves or stems of the plant; mealybugs, which are uncommon, but appear as small cottony masses in the leaf axils or on the leaves and stems; plant bugs, which often produce small dead spots in the leaves where they have fed and cause curling and stunting; the leafhoppers, whose feeding produces small white spots in the leaf tissue; and spider mites, which are very hard to see, but whose presence can be detected by a whitish stippling and general lack of green color in the leaves. When spider mites are abundant, a webbing of silk may also be seen. Scale insects and lace bugs, which also belong to this group, are usually found on trees and shrubs and seldom on annuals or perennials in the garden.

The second group consists of insects that chew the leaves or bore in the stems. These are larvae and adults of beetles; larvae of butterflies and moths; and occasionally the larvae of sawflies (primitive wasps). These larvae and adults are sometimes quite hard to find, but you can easily detect their presence by the telltale evidence of their feeding.

Contact insecticides are used to control insects in the first group, which are the piercing and sucking insects that feed on the tissues below the surface of the leaves and fruits and so do not get enough of the surface layers to be affected by stomach poisons. These contact insecticides applied as sprays or dusts must actually touch the insect in order to destroy it.

Stomach poisons are used to control insects in the second group, which are insects with chewing mouthparts. These poisons must be applied evenly and thoroughly on plants in such amounts that the insect gets a fatal dose in taking its usual amount of food. There are two types of stomach poisons: protectants that are put on the plant to protect it against the injury caused by the insects, and poison baits, which are prepared by mixing the insecticide with a substance more attractive to the insect than its usual food.

Ants

Ants do not usually attack ornamental plants directly, but they often act as a means of transportation for aphids, particularly root aphids. Because of this and the fact that they may build their nests near roots of perennials, they may have to be controlled. Ants are often found around peonies, and they are indirectly responsible for the sticky buds of these perennials. However, unless they build their nests around the roots of the plants, they do no harm.

Aphids

Aphids are the most common of all garden insects. They seem to attack nearly all herbaceous perennials to some extent, but *Delphinium* and *Rudbeckia* 'Golden Glow' seem to be affected more than others.

Root aphids attack asters and primroses particularly. These aphids are mostly nurtured by ants, who keep them over the winter and then in the spring take the eggs or young to the stems and roots of the plants. Control of these aphids, therefore, must be mainly concentrated on the destruction of the ants.

Beetles and Weevils

Beetles are chewing insects that have a horny sheath covering membranous hind wings. The larval stage is a white grub, often softish, with a brown head. Weevils are beetles that have a snout and wing covers that are fastened down so they cannot fly. These insects usually feed at night.

The Japanese beetle, the most destructive of all beetles, is not a serious pest in Canadian gardens, except in isolated areas in southern Ontario. The cucumber beetle is a serious pest of *Coreopsis*, *Delphinium*, and *Digitalis*. The strawberry root weevil attacks *Heuchera*. In spring, many plants may be destroyed by groups of weevils that feed on roots and crowns.

Beetles and weevils in the adult stage can be destroyed by handpicking and dropping the insects in a can of kerosene. In the fall, clean the garden of all decayed leaves and debris where the beetles might overwinter.

Borers

Borers are caterpillars, grubs, or sawfly larvae that work in the stems of woody and herbaceous plants and not only cause complete collapse of the stems and sometimes kill the whole plants, but have the deleterious side effect of providing entry for fungus diseases. This is especially true of borers that attack irises; these insects are very damaging to the plants and are responsible for introducing the bacterium that causes the evil-smelling soft rot.

Sanitation is very important in controlling borers, and when carried out properly will often eliminate pests before they do any damage in the spring. Collect and destroy all the aboveground parts of perennials and nearby weeds that die down naturally in the fall.

After the plants have been attacked, it is possible to destroy the borers by inserting a wire in the stem at the point where the borers entered, or by sticking pins through the stalk in many places.

Caterpillars

Many different kinds of caterpillars feed on herbaceous perennials, although they never are as completely devastating as those found on trees and shrubs. Cutworms occasionally will destroy *Dianthus* and *Delphinium*.

Leafhoppers

Leafhoppers are elongated wedge-shaped sucking insects that hold their wings in a roof-like position. They have the peculiar habit of running sideways when disturbed. They feed on the undersides of the leaves, and produce a stippled effect on the upper surfaces.

Leaf Miners

Leaf miners are the larvae of some species of small flies and moths that feed between the two leaf surfaces making disfiguring blotches or, in *Aquilegia* and some other perennials, forming serpentine trails. The herbaceous perennials that are most affected by leaf miners are *Aquilegia, Bergenia, Euphorbia,* and *Delphinium*.

Mealybugs

Mealybugs are mealy, sucking insects that are better known as pests of houseplants. They are small, flattened, elongated, oval insects usually covered with a waxy mealy substance and often forming a series of short filaments around the margin of the body. There is a ground mealybug that attacks the roots of *Delphinium*.

Mites

Mites are minute animals commonly called spider mites or red spiders. They are barely visible to the naked eye, but their effect on the leaves of plants is quite apparent. They turn leaves yellow or gray and, when the attack has taken hold, many species also produce a very fine cobwebby appearance. This web, in fact, protects them from haphazard spraying.

The cyclamen mite, a very active pest of *Delphinium, Aconitum,* and *Antirrhinum,* makes no web, but it deforms and darkens leaves, flower stalks, and buds. In *Delphinium*, leaves become thick and deformed, the flower stalks gnarled, twisted, and darkened, and the buds almost black. The effect of this mite is so severe that it is often called the 'blacks disease' and is thought to be a real disease.

The adult cyclamen mite winters in the crowns of delphiniums and in the spring emerges, crawling from one plant to another on the leaves, or is spread on tools or the hands of gardeners.

Quite often, a good forceful spray with the garden hose will keep these pests under control if the plants affected have leaves that can stand it. This method is especially effective during hot dry weather, when the infestation has just started.

Slugs and Snails

Slugs and snails are soft-bodied mollusks that leave a slimy trail behind them as they crowd around the garden eating large holes in the leaves of almost all perennials that grow close to the ground. They are particularly damaging to *Delphinium*, hollyhocks, violets, and *Hosta*.

Thrips

Thrips are very small slender insects with rasping mouthparts that scrape the plant tissue and then suck the juices. The adults are brown or black and have bristle-like wings. Because they are very small they are seldom noticed, and by the time they are seen they are in such numbers that a lot of damage has already been done. The feeding causes silvering or blanching of affected parts and distortion of leaves and blooms. On blooms of *Gladiolus* and *Hemerocallis* they seem to rasp off the color in patches. Besides *Hemerocallis* they also attack *Iris* and *Lilium*.

Many species of thrips feed on grasses and weeds, and from these they move to cultivated plants when the native vegetation dries up. Because they are constantly moving from one place to another, they are hard to control except by repeated applications of pesticides.

INSECTICIDES

Spraying versus Dusting

There is really no answer to the question of whether it is best to dust or to spray plants for disease or insect control. Both sprays and dusts will likely be used in the garden; the use of one or the other will depend on the plant, the disease or insect, the weather, how much time the gardener has, and even on the fungicide he wants to use. Dusting is usually faster and no water is required. On the other hand, plants are covered better by spraying, especially on the stems and undersurface of the leaves, because a spray often sticks to these surfaces better than a dust. Many gardeners like to dust the plants while they are still wet with rain or dew. However, it is best not to apply a dust to wet ornamentals, because the dust will form unsightly blotches. I feel that regardless of how well and evenly it may be applied, dust on an ornamental is unsightly. Both sprays and dusts must be applied evenly so that both surfaces of the leaves are covered.

All-purpose Sprays and Dusts

Some spray materials or dusts may be safely combined with others to control fungus or bacterial diseases and insects in one operation. These fungicide—insecticide mixtures, called all-purpose sprays or dusts, are very useful to home gardeners. There are a number of these combination sprays or dusts on the market and they can also be prepared by the gardener to suit his specific needs.

Precautions

There are certain precautions that must be taken when preparing, handling, and storing fungicides and insecticides. It is important that the materials used to make a spray be measured exactly according to the directions given on the package and mixed thoroughly if maximum control of the disease or insect is to be obtained and plant injury is to be avoided. If a fungicide—insecticide mixture is prepared by the gardener it is necessary that he know what materials can be safely combined with others. Never breathe dusts or spray mists. Special notice should be taken of any precautions given on the label concerning plants on which the spray can not be used without injury, and conditions under which it may cause injury to other plants. It must be kept in mind that all these materials used in the control of insects and plant diseases are poisonous. Therefore, proper precautions must be taken in preparing and using them, and they must be kept out of the reach of children, animals, and irresponsible people. Also, remember that the deposits left on the plant by these materials are poisonous to people. Do not let pesticides drift onto nearby edible crops or fish ponds.

chapter 5

perennials
for special
purposes

perennials for special purposes

Bulbous and similar plants for the perennial border

Allium
Bulbocodium
Camassia
Chionodoxa
Colchicum
Crocus
Eremurus
Erythronium
Fritillaria
Galanthus
Galtonia
Hyacinthus
Leucojum
Lilium
Lycoris
Muscari
Narcissus
Ornithogalum
Scilla
Tulipa

Plants for the shade

Aconitum species
Actaea species
*Actaea rubra**
Ajuga reptans
Aruncus sylvester
Anemone hepatica
Anemone hupehensis
Anemone sylvestris
Aquilegia species and strains
Astilbe species and selections
Campanula species
Cimicifuga racemosa var. *simplex**
Convallaria majalis
Dicentra species
Digitalis species and strains
Doronicum species
Dracocephalum species
Epimedium species
Eupatorium coelestinum
Filipendula ulmaria
Geranium ibericum
Helleborus species

*Best for shaded areas

Hemerocallis selections
Heuchera selections
Hosta species
Lobelia cardinalis
*Lobelia siphilitica**
Lunaria rediviva
*Mertensia virginica**
Monarda didyma and selections
Myosotis alpestris
Phlox divaricata
Phlox paniculata selections
Phlox stolonifera
Platycodon grandiflorum
Polemonium species
Polygonatum multiflorum
Primula species
Pulmonaria species
Thalictrum species*
Tiarella cordifolia
Trillium species*
Trollius species
Uvularia grandiflora
Vinca minor
Viola cornuta

Plants for dry sunny situations

Achillea ptarmica
Achillea millefolium
Alyssum species
Anaphalis margaritacea
Anthemis species
Arabis species
Artemisia albula
Artemisia stelleriana
Asclepias tuberosa
Baptisia australis
Campanula carpatica
Centaurea macrocephala
Cerastium tomentosum
Chrysanthemum leucanthemum
Coreopsis grandiflora selections
Dianthus plumarius selections
Dianthus deltoides
Dodecatheon media
Echinops ritro
Gaillardia grandiflora selections
Gaura lindheimeri

Geum chiloense selections
Gypsophila repens
Helianthus species and selections
Iris, tall bearded selections
Linum perenne
Lychnis chalcedonica
Lychnis coronaria
Monarda didyma selections
Oenothera species
Papaver species and selections
Penstemon alpinus
Penstemon grandiflorus
Phlox subulata
Rudbeckia speciosa
Salvia pitcheri
Sedum species
Sempervivum species
Thermopsis caroliniana
Thymus species
Verbascum species and selections

Perennials for wet soils

Aruncus sylvester
Astilbe species and selections
Astrantia major
Caltha palustris
Chelone glabra
Chelone lyoni
Chrysanthemum uliginosum
Cimicifuga racemosa
Epilobium angustifolium
Eupatorium purpureum
Filipendula palmata
Gentiana andrewsii
Hemerocallis selections
Hibiscus moscheutos
Hosta species and selections
Iris kaempferi selections
Iris pseudacorus
Iris sibirica
Ligularia dentata
Ligularia wilsoniana
Lobelia cardinalis
Lobelia siphilitica
Lobelia × *speciosa* hybrids
Lysimachia clethroides
Lysimachia nummularia

Lythrum salicaria selections
Monarda didyma selections
Peltiphyllum peltatum
Ranunculus acris 'Flore-pleno'
Trollius species and selections
Vernonia noveboracensis

Plants that tolerate acid soils

Actaea alba
Chelone glabra
Chelone lyoni
Cimicifuga racemosa
Dicentra eximia
Digitalis purpurea
Mertensia virginica
Trillium grandiflorum

Perennials with fragrant flowers

Aquilegia chrysantha
Aquilegia caerulea
Arabis albida
Artemisia lactiflora
Campanula lactiflora
Convallaria majalis
Dianthus caryophyllus selections
Dianthus plumarius selections
Dictamnus albus
Filipendula ulmaria
Hesperis matronalis
Hosta plantaginea
Iris, tall bearded selections
Malva moschata
Oenothera speciosa
Paeonia selections
Phlox divaricata and selections
Phlox stolonifera
Primula elatior
Primula florindae
Primula × *polyantha*
Primula veris
Primula vulgaris
Valeriana officinalis
Viola cornuta
Viola odorata

Perennials with fragrant foliage

Achillea millefolium
Anthemis tinctoria
Artemisia stelleriana
Asperula odorata
Chrysanthemum balsamita
Dictamnus albus
Hyssopus officinalis
Lavandula spp.
Mentha spp.
Monarda didyma
Nepeta faassenii

Perennials flowering over a fairly long period

Achillea ptarmica 'Boule de Neige'
Anthemis sancti-johannis
Anthemis tinctoria selections
Aquilegia chrysantha
Armeria maritima
Aster frikartii
Callirhoe involucrata
Campanula rotundifolia
Centaurea montana
Chrysanthemum maximum selections
Coreopsis grandiflora
Coronilla varia
Dianthus caryophyllus
Dianthus latifolius
Dianthus plumarius selections
Dicentra eximia
Gaillardia selections
Galega officinalis
Gaura lindheimeri
Geranium ibericum
Geranium sanguineum
Heliopsis helianthoides
Heliopsis helianthoides var. *scabra*
Heuchera sanguinea selections
Lamium maculatum
Linum perenne
Lythrum salicaria selections
Malva moschata
Monarda didyma selections
Nepeta faassenii
Oenothera fruticosa

Oenothera missouriensis
Papaver nudicaule selections
Phlox paniculata selections
Physostegia selections
Platycodon selections
Potentilla pyrenaica
Rudbeckia speciosa
Salvia pitcheri
Scabiosa caucasica
Sidalcea selections
Tradescantia virginiana selections
Veronica spicata selections

Perennials with blue flowers

Aconitum carmichaelii
Aconitum henryi
Aconitum napellus 'Blue Sceptre'
Aconitum napellus 'Newry Blue'
Ajuga genevensis
Ajuga reptans
Amsonia tabernaemontana
Anchusa azurea selections
Aster selections
Baptisia australis
Boltonia latisquama
Campanula lactiflora 'Pouffe'
Campanula lactiflora 'Pritchard's Variety'
Campanula persicifolia 'Blue Bells'
Campanula persicifolia 'Mrs. H. Harrison'
Campanula persicifolia 'Telham Beauty'
Catananche caerulea 'Major'
Centaurea montana
Delphinium × *Belladonna* selections
Delphinium elatum selections
Echinops exaltatus
Echinops humilis 'Blue Cloud'
Echinops humilis 'Taplow Blue'
Erigeron speciosus 'Double Beauty'
Eryngium planum
Hosta ventricosa
Iris, dwarf 'Atomic Blue'
Iris, dwarf 'Dear Love'
Iris, intermediate 'Lilli Hoog'
Iris sibirica 'Blue Herald'

Iris sibirica 'Gatineau'
Iris sibirica 'Mountain Lake'
Iris, tall bearded 'Blue Sapphire'
Iris, tall bearded 'Eleanor's Pride'
Lavandula spica 'Gruppenhall'
Lavandula spica 'Munstead'
Limonium latifolium 'Blue Cloud'
Limonium latifolium 'Elegance'
Linum narbonense 'Heavenly Blue'
Lobelia siphilitica
Meconopsis betonicifolia
Mertensis virginica
Omphalodes cappadocica
Platycodon grandiflorum
'Bristol Bluebird'
Polemonium boreale
Polemonium caeruleum
Salvia azurea
Salvia uliginosa
Scabiosa caucasica 'Clive Greaves'
Scabiosa caucasica 'Challenger'
Tradescantia andersoniana
'Blue Stone'
Tradescantia andersoniana 'Iris'
Veronica latifolia 'Amethystina'
Veronica latifolia 'Royal Blue'

Perennials with orange flowers

Alstroemeria aurantiaca
Asclepias tuberosa
Belamcanda chinensis
Geum × *borisii*
Hemerocallis selections
Inula royleana
Kniphofia selections
Papaver orientale
Trollius asiaticus

Perennials with pink or rose
flowers

Ajuga genevensis 'Pink Spires'
Anemone × *hybrida* 'Montrose'
Anemone × *hybrida*
'Queen Charlotte'
Aquilegia 'Rose Queen'
Arabis × *arendsii* 'Rosabella'
Armeria maritima 'Laucheana'

Armeria maritima
'Laucheana Six Hills'
Armeria maritima 'Vindictive'
Armeria pseud-armeria
'Glory of Holland'
Aster selections
Astilbe 'Erica'
Bergenia cordifolia
Chrysanthemum coccineum
'Jubilee Gem'
Chrysanthemum coccineum 'Helen'
Chrysanthemum coccineum 'Senator'
Chrysanthemum morifolium
selections
Dianthus deltoides 'Caprice'
Dianthus plumarius 'Excelsior'
Dianthus plumarius 'Pink Princess'
Dicentra formosa 'Bountiful'
Dicentra spectabilis
Dictamnus albus 'Purpureus'
Echinacea 'The King'
Erigeron 'Felicity'
Filipendula rubra 'Venusta'
Gypsophila paniculata 'Flamingo'
Gypsophila paniculata
'Rosenschleier'
Hemerocallis 'Neyron Rose'
Hemerocallis 'Cheery Pink'
Hemerocallis 'Silver Sheen'
Hemerocallis 'Evelyn Claar'
Heuchera 'Freedom'
Heuchera 'Scintillation'
Iris, intermediate 'Pink Debut'
Iris, intermediate 'Pink Fancy'
Iris, tall bearded 'Mary Randall'
Iris, tall bearded 'Happy Birthday'
Iris, tall bearded 'June Meredith'
Kniphofia 'Coral Sea'
Lavatera cachemirica
Lupinus polyphyllus 'Betty Astrid'
Lupinus polyphyllus 'Guardsman'
Lychnis coronaria
'Abbotswood Rose'
Lythrum salicaria 'Robert'
Lythrum virgatum 'Morden Pink'
Monarda 'Croftway Pink'
Monarda 'Granite Pink'
Monarda 'Melissa'
Paeonia 'Sea Shell'

Papaver orientale 'Juliet'
Papaver orientale 'Helen Elizabeth'
Papaver orientale 'Salome'
Phlox paniculata 'Elizabeth Arden'
Phlox paniculata 'Firefly'
Phlox subulata 'Betty'
Physostegia 'Summer Glow'
Physostegia 'Summer Spire'
Platycodon grandiflorum
'Bristol Blush'
Primula japonica 'Rose Dubarry'
Primula juliae 'Kinlough Beauty'
Saponaria ocymoides
Thalictrum aquilegifolium
'Thundercloud'
Veronica spicata 'Barcarolle'
Veronica spicata 'Minuet'

Perennials with reddish purple
flowers

Anemone pulsatilla
Aster selections
Dierama pulcherrimum
Echinacea purpureum
'Robert Bloom'
Eupatorium purpureum
Geranium sanguineum
Geranium wallichianum
Hesperis matronalis 'Purpurea'
Iris kaempferi selections
Iris sibirica 'Eric the Red'
Lavandula spica 'Hidcote Blue'
Liatris pycnostachya
Liatris scariosa 'September Glory'
Liatris spicata 'Kobold'
Lychnis viscaria
'Splendens Flore-pleno'
Lythrum 'Dropmore Purple'
Lythrum 'Granite Purple'
Phlox paniculata 'Vintage Wine'
Phlox paniculata 'Flamboyant'
Phlox paniculata 'Tenor'

Perennials with red flowers

Achillea millefolium 'Fire King'
Astilbe 'Red Sentinel'
Chrysanthemum coccineum 'Comet'

Chrysanthemum coccineum 'Harold Robinson'
Chrysanthemum coccineum 'Kelway Glorious'
Chrysanthemum coccineum 'J. M. Tweedy'
Dianthus × *allwoodii* 'Little Joe'
Dianthus deltoides 'Brilliant'
Dianthus plumarius 'Cyclops'
Dianthus plumarius 'Emperor'
Dianthus plumarius 'Highland Queen'
Dianthus plumarius 'Shadow Valley'
Gaillardia × *grandiflora* 'Burgundy'
Geum chiloense 'Mrs. Bradshaw'
Helenium autumnale 'Spatrot'
Hemerocallis 'Bess Ross'
Hemerocallis 'Warrior'
Hemerocallis 'Crimson Glory'
Heuchera 'Bressingham Blaze'
Heuchera 'Scarlet Sentinel'
Heuchera 'Firebird'
Iris, tall bearded 'Captain Gallant'
Kniphofia 'Summer Sunshine'
Lychnis × *arkwrightii*
Lychnis chalcedonica
Mimulus cardinalis
Monarda didyma
Monarda 'Cambridge Scarlet'
Paeonia 'Philip Rivoire'
Paeonia 'Red Charm'
Paeonia 'Diana Parks'
Potentilla nepalensis
Potentilla 'Gibson's Scarlet'
Primula pulverulenta 'Red Hugh'
Pulmonaria saccharata 'Bowles Red'

Perennials with violet flowers

Aconitum carmichaelii 'Wilsonii'
Aster selections
Erigeron macranthus
Erigeron speciosus 'Azure Beauty'
Erigeron speciosus 'Mrs. E. H. Beale'
Eupatorium coelestinum
Hosta fortunei
Hosta lancifolia
Iris, intermediate 'First Lilac'
Iris sibirica 'Tunkhannock'

Limonium latifolium 'Chilwell Beauty'
Mimulus ringens
Monarda 'Blue Stocking'
Salvia superba

Perennials with white flowers

Achillea ptarmica 'Angels' Breath'
Achillea ptarmica 'Boule de Neige'
Anemone × *hybrida* 'Louise Uhink'
Anemone × *hybrida* 'Marie Manchart'
Anemone sylvestris
Arabis caucasica
Arabis caucasica 'Flore Pleno'
Aster selections
Astilbe 'Deutschland'
Astilbe 'W. D. Willen'
Boltonia asteroides 'Snowbank'
Campanula persicifolia 'Mount Hood'
Chrysanthemum coccineum 'Avalanche'
Chrysanthemum coccineum 'Carl Vogt'
Chrysanthemum maximum 'Majestic'
Chrysanthemum maximum 'Esther Read'
Chrysanthemum maximum 'Thomas E. Killeen'
Chrysanthemum maximum 'Wirral Supreme'
Chrysanthemum morifolium 'Powder River'
Chrysanthemum morifolium 'Christopher Columbus'
Chrysanthemum uliginosum
Cimicifuga americana
Cimicifuga racemosa
Cimicifuga simplex
Convallaria majalis
Dianthus plumarius 'Her Majesty'
Dictamus albus
Echinacea 'White Lustre'
Epimedium × *youngianum* 'Niveum'
Filipendula ulmaria
Filipendula vulgaris

Filipendula vulgaris 'Flore Pleno'
Gypsophila paniculata 'Bristol Fairy'
Helleborus niger
Heuchera 'Pearl Drops'
Heuchera 'Snowflakes'
Hosta plantaginea
Iris, dwarf 'Bright White'
Iris, intermediate 'Cloud Fluff'
Iris, intermediate 'Easter Bunny'
Iris sibirica 'Snow Crest'
Iris sibirica 'White Swirl'
Iris, tall bearded 'Brilliant Star'
Iris, tall bearded 'Celestial Snow'
Iris, tall bearded 'Frost and Flame'
Liatris scariosa 'White Spire'
Lysimachia clethroides
Monarda 'Snow Maiden'
Paeonia 'Le Cygne'
Paeonia 'Snow Mountain'
Paeonia 'The Bride'
Papaver orientale 'Barr's White'
Papaver orientale 'Field Marshall van der Glotz'
Phlox paniculata 'Henderson's Late White'
Phlox paniculata 'Iceberg'
Phlox suffruticosa 'Miss Lingard'
Phlox subulata 'The Bride'
Physostegia 'Summer Snow'
Primula denticulata 'Alba'
Primula japonica 'Postford White'
Primula juliae 'Alba'
Scabiosa caucasica 'Mrs. Willmott'
Valeriana officinalis
Verbascum 'Bridal Bouquet'
Veronica spicata 'Icicle'

Perennials with yellow flowers

Achillea filipendulina 'Gold Plate'
Achillea filipendulina 'Parker's Variety'
Achillea taygetea
Achillea tomentosa
Adonis amurensis
Alyssum saxatile
Anthemis tinctoria selections
Aster ericoides 'Golden Spray'

Aster linosyris
Baptisia tinctoria
Buphthalmum salicifolium
Buphthalmum speciosum
Caltha palustris
Cephalaria gigantea
Chrysogonum virginianum
Chrysopsis villosa
Coreopsis grandiflora selections
Corydalis cheilanthifolia
Doronicum caucasicum
‘Madame Mason’
Doronicum plantagineum ‘Excelsum’
Euphorbia epithymoides
Gaillardia × *grandiflora* ‘Goblin’
Geum chiloense ‘Lady Stratheden’
Helenium autumnale ‘Allgold’
Helenium hoopesii ‘Butterpat’
Helenium hoopesii ‘Golden Youth’
Helianthus decapetalus
‘Multiflorus’ selections
Helianthus laetiflorus ‘Mrs. Mellish’
Heliopsis helianthoides selections
Hemerocallis ‘Midwest Star’
Hemerocallis ‘Colonel Joe’
Hemerocallis ‘Jane Russell’
Iris, dwarf ‘Brassie’
Iris, dwarf ‘Golden Fair’
Iris, intermediate ‘Yellow Dresden’
Iris, tall bearded ‘Truly Yours’
Iris, tall bearded ‘Rainbow Gold’
Kniphofia ‘Buttercup’
Kniphofia ‘Bees’ Lemon’
Ligularia dentata ‘Gregynog Gold’
Linum flavum
Lupinus polyphyllus ‘Celandine’
Lysimachia ciliata
Lysimachia punctata
Oenothera tetragona ssp. *glauca*
Oenothera tetragona ‘Highlight’
Oenothera tetragona
‘W. Cuthberston’
Potentilla recta
Ranunculus acris ‘Flore-pleno’
Rudbeckia fulgida var. *speciosa*
Rudbeckia fulgida ‘Goldquelle’
Thermopsis caroliniana

*Ground covers

Thermopsis montana
Trollius europaeus
Trollius asiaticus ‘Byrne’s Giant’
Trollius europaeus ‘Golden Monarch’

Perennials under 0.3 m (1 ft.) high

*Achillea ageratifolia**
Achillea ageratifolia ‘Aizoon’*
*Achillea clavennae**
Achillea filipendulina × *clypeolata*
‘Coronation Gold’
Achillea filipendulina ‘Gold Plate’
Achillea × *lewisii* ‘King Edward’*
*Achillea tomentosa**
Achillea tomentosa ‘Moonlight’*
Adonis amurensis ‘Plena’
Adonis amurensis var. *vernalis*
*Ajuga genevensis**
Ajuga genevensis ‘Pink Spires’*
Ajuga genevensis ‘Brocklebankii’*
*Ajuga pyramidalis**
Ajuga pyramidalis
‘Metallica-crispa’*
Ajuga pyramidalis
‘Tottenham Blue’*
*Ajuga reptans**
Ajuga reptans ‘Alba’*
Ajuga reptans ‘Atro Purpurea’*
Ajuga reptans ‘Variegata’*
Ajuga reptans ‘Gaiety’*
Alyssum saxatile
Alyssum saxatile ‘Citrinum’
Alyssum saxatile ‘Compactum’
Alyssum saxatile ‘Dudley Neville’
Alyssum saxatile ‘Flore Pleno’
Alyssum saxatile ‘Silver Queen’
Anchusa caespitosa
Anemone × *lesseri*
Anemone nemorosa
Anemone patens
Anemone pulsatilla
Anemone pulsatilla ‘Alba’
Anemone pulsatilla ‘Mallandieri’
Anemone pulsatilla ‘Red Clock’
Anemone pulsatilla
‘Mrs. Van der Elst’
Anemone sylvestris
Anemonella thalictroides
*Anthemis biebersteiniana**

Anthemis cupaniana
Anthemis nobilis ‘Chamomile’
Anthemis nobilis ‘Treneague’
Anthemis rudolphiana
Aquilegia alpina
Aquilegia discolor
Aquilegia ecalcarata
Aquilegia einseleana
Aquilegia flabellata ‘Nana’
Aquilegia flabellata ‘Nana Alba’
Aquilegia glandulosa
Aquilegia jonesii
Arabis blepharophylla and selections
Arabis caucasica and selections*
*Arabis kellereri**
Arabis procurrens
Armeria juniperifolia selections
Armeria maritima and selections
Artemisia frigida
Aster acris ‘Nanus’
Aster × *alpellus* selections
Aster alpinus selections
Aubrieta deltoidea selections
Bellis perennis selections
Bergenia cordifolia
Bergenia stracheyi
Boykinia jamesii
Boykinia rotundifolia
Brunnera macrophylla
Calamintha alpina
*Calamintha grandiflora**
*Calamintha nepeta**
Caltha leptosepala
Caltha palustris selections
Campanula carpatica var. *turbinata*
Campanula carpatica var. *turbinata*
‘Pallida’
Campanula cochlearifolia selections
Campanula elatines var. *garganica*
selections
Campanula portenschlagiana
Campanula poscharskyana selections
Celsia acaulis
Cerastium alpinum ‘Lanatum’
Cerastium biebersteinii
Cerastium tomentosum
*Ceratostigma plumbaginoides**
Chrysanthemum arcticum

Chrysanthemum weyitchii
Chrysogonum virginianum
Chrysopis falcata
Codonopsis clematidea
Codonopsis ovata
Convallaria majalis selections*
Coreopsis auriculata 'Nana'
Coreopsis grandiflora
and dwarf selections
Coreopsis rosea selections
*Cornus canadensis**
Corydalis cava (C. bulbosa)
Corydalis cheilanthifolia
Corydalis lutea
Corydalis nobilis
Corydalis wilsoni
Delphinium menziesii
Delphinium nudicaule
Dianthus × *allwoodii* 'Little Joe'*
Dianthus arenarius selections*
Dianthus deltoides selections*
Dianthus gratianopolitanus
selections*
Dianthus plumarius selections
Dicentra cucullaria
Dicentra oregana
Dodecatheon pauciflorum
Dracocephalum grandiflorum
Dracocephalum hemsleyanum
Dracocephalum ruyschiana
Dracocephalum ruyschiana
'Speciosum'
Dryas octopetala selections*
Dryas × *suendermannii**
*Epimedium coccineum**
Epimedium grandiflorum selections*
*Epimedium perralderianum**
Epimedium pinnatum selections*
Epimedium × *rubrum**
Epimedium × *versicolor* selections*
Epimedium × *youngianum**
Epimedium × *warleyense**
*Erigeron aurantiacus**
*Erigeron leiomerus**
*Erigeron mucronatus**
Euphorbia cyparissias
Filipendula multijuga
Gaillardia × *grandiflora* 'Baby Cole'
Gentiana septemfida

*Ground covers

Geranium cinereum selections
Geranium dalmaticum
Geranium endressi selections
Geranium napuligerum
Geranium nodosum
Geranium renardii
Geranium sanguineum
'Lancastriense' selections
Geranium wallichianum selections
Geum × *borisii*
Geum × *heldreichii* 'Superbum'
Geum rossii
Glaucium flavum
Haplopappus coronopifolius
Haplopappus lyalli
Hemerocallis minor (H. gracilis)
Hepatica acutiloba
Hepatica angulosa
Hepatica triloba
Hepatica transilvanica
Heracleum mantegazzianum
Hieracium bombycinum
Hieracium villosum
*Iberis gibraltarica**
Iberis saxatilis selections*
Iberis sempervirens selections*
Iris arenaria (I. flavissima
'Arenaria')
Iris chamaeiris
Iris cristata
Iris douglasiana selections
Iris graminea
Iris innominata
Iris lacustris
Iris mellita
Iris minuta
Iris orientalis 'Nana'
Iris pumila selections*
Iris tectorum
Lamium maculatum selections*
Limonium bellidifolium
Limonium cosyrense
Limonium minimum
Linum alpinum
Linum salsoloides selections
Liriope spicata 'Majestic'*
Lithospermum canescens
Lithospermum intermedium

Lychnis alpina
Lychnis × *arkwrightii*
Lychnis
Lysimachia nummularia selections*
Meconopsis cambrica selections
Mertensia echioides
Mertensia lanceolata
Mertensia longiflora
Mertensia sibirica
*Mertensia virginica**
*Mimulus moschatus**
*Mimulus primuloides**
Myosotis alpestris selections*
Myosotis scorpioides selections*
Myosotis sylvatica
Nepeta × *faassenii**
Oenothera missouriensis
Omphalodes cappadocica selections
Omphalodes verna selections
*Pachysandra procumbens**
*Pachysandra terminalis**
Papaver nudicaule
Penstemon crandallii selections
Phlox adsurgens
Phlox amoena selections*
Phlox divaricata selections*
Phlox nivalis selections
Phlox stolonifera selections*
Phlox subulata selections*
*Polemonium confertum**
Polemonium lanatum selections*
*Polemonium pulcherrimum**
Polygonum affine selections
Polygonum vaccinifolium
Potentilla concolor
Potentilla fragiformis
Primula aurantiaca
Primula auricula selections
Primula denticulata selections
Primula elatior
Primula juliae selections
Primula sieboldii
Primula veris
Primula vulgaris
Pulmonaria angustifolia selections*
*Pulmonaria officinalis**
*Pulmonaria rubra**
*Ranunculus gramineus**

Roscoea cautleoides
Roscoea humeana
Salvia jurisicii
Sanguinaria canadensis selections
Saponaria × *boisseri*
Saponaria caespitosa
Saponaria lutea
Saponaria ocymoides selections
Saxifraga aizoon
Scabiosa lucida
*Sedum acre**
*Sedum ewersii**
Sedum kamtschaticum selections*
Sedum sieboldii selections*
Sedum spurium selections*
Sempervivum tectorum selections*
Solidago brachystachys
Symphyandra wanneri
Thalictrum kiusianum
*Thymus lanuginosus**
*Thymus nitidus**
Thymus serpyllum selections*
Tiarella cordifolia
Tiarella polyphylla
Tiarella trifoliata
Tradescantia virginiana selections
Trollius pumilus selections
Verbena peruviana
Veronica cinerea
Veronica pectinata selections
Veronica prostrata selections
*Veronica repens**
Veronica teucrium selections
*Vinca difformis**
Vinca major selections*
Vinca minor selections*
Viola cornuta selections
*Viola cucullata**
Viola × *florariensis**
Viola gracilis selections
Viola jooi
Viola labradorica
Viola odorata selections
Viola papilionacea selections
Viola pedata

*Ground covers

Perennials 0.3-0.6 m (1-2 ft.) high

Achillea millefolium 'Roseum'
Achillea ptarmica 'Boule de Neige'
Adonis amurensis
Amsonia montana
Anemone hupehensis
Anemone × *hybrida* 'September Sprite'
Anthemis sancti-johannis
Aquilegia alpina
Aquilegia flabellata
Armeria pseud-armeria selections
Artemisia discolor
Artemisia lanata
Artemisia nutans
Artemisia palmeri
Asclepias tuberosa
Aster acris
Aster novi-belgii, dwarf mound type and selections
Aster yunnanensis 'Napsbury'
Baptisia bracteata
Bergenia crassifolia
Bergenia ligulata
Buphthalmum salicifolium
Campanula alliarifolia
Campanula carpatica and selections
Campanula glomerata and selections
Campanula punctata
Campanula rotundifolia and selections
Celmisia spectabilis
Celsia arcturus
Centaurea dealbata and selections
Centaurea montana and selections
Centaurea ruber and selections
Chelone glabra
Chrysanthemum coccineum
Chrysopsis mariana
Chrysopsis villosa
Coreopsis auriculata 'Superba'
Coreopsis verticillata and selections
Delphinium grandiflorum and selections
Delphinium tatsienense
Dianthus caryophyllus
Dianthus knappi
Dianthus superbus and selections

Dicentra eximia and selections
Dicentra formosa and selections
Dodecatheon hendersonii
Dodecatheon meadia
Doronicum caucasicum and selections
Doronicum pardalianches and selections
Dracocephalum forrestii
Echinacea angustifolia
Erigeron macranthus
Erigeron speciosus and selections
Eryngium amethystinum
Eryngium bourgatii
Eryngium dichotomum
Euphorbia epithymoides
Filipendula hexapetala 'Flore Pleno'
Gentiana asclepiadea and selections
Geranium anemonifolium
Geranium armenum
Geranium grandiflorum and selections
Geranium ibericum and selections
Geranium macrorrhizum
Geranium phaeum
Geranium platypetalum
Geranium pratense and selections
Geranium psilostemon
Geranium sanguineum and selections
Geranium sylvaticum and selections
Geum bulgaricum
Geum chiloense and selections
Geum × *heldreichii*
Gillenia trifoliata
Glaucidium palmatum
Helleborus viridis
Hosta crispula
Hosta decorata and selections
Hosta elata and selections
Hosta lancifolia
Hosta plantaginea and selections
Hosta sieboldiana
Hosta tardiflora
Hosta undulata and selections
Hosta ventricosa
Incarvillea compacta
Incarvillea delavayi and selections
Incarvillea grandiflora and selections

Inula ensifolia
Inula grandiflora
Inula hookeri
Inula Oculus-Christii
Iris, intermediate bearded and selections
Iris, Japanese and selections
Iris bulleyana
Iris dichotoma
Iris flavescens
Iris forrestii
Iris japonica
Iris versicolor
Iris wilsonii
Kniphofia galpinii
Lamium galeobdolon
Lamium orvala
Lamium veronicaefolium
Liatris callilepis
Liatris punctata
Liatris scariosa and selection
Ligularia hodgsonii
Ligularia intermedia
Ligularia tussilaginea and selections
Limonium elatum
Limonium exinium and selection
Limonium felicularis
Limonium gmelini
Linaria dalmatica
Lindelofia longiflora
Linum austriacum 'Album'
Linum flavum selections
Linum narbonense selections
Linum perenne selections
Liriope muscari
Liriope spicata
Lychnis coronaria and selections
Lychnis viscaria and selections
Malva moschata and selections
Meconopsis quintuplinervia
Mertensia ciliata
Mimulus cardinalis and selections
Mimulus × *burnetii*
Mimulus cupreus and selections
Mimulus lewisii and selections
Mimulus luteus and selections
Oenothera caespitosa
Oenothera caespitosa 'Eximea'
Oenothera speciosa and selections

Oenothera tetragona and selections
Paeonia anomala
Paeonia emodi
Paeonia mlokosewitschi
Paeonia russii
Paeonia × *smouthii*
Paeonia tenuifolia and selections
Paeonia veitchii and selections
Penstemon alpinus and selections
Penstemon cobaea 'Ozark'
Penstemon diffusus
Phlox ovata and selections
Phlox pilosa
Physalis alkekengi
Physalis franchetii and selections
Platycodon grandiflorum and selections
Podophyllum emodi
Polemonium caeruleum and selections
Polemonium carneum
Polemonium pauciflorum
Polemonium reptans
Polemonium richardsonii
Potentilla × *hopwoodiana*
Potentilla nepalensis and selections
Potentilla recta and selections
Primula alpicola and selections
Primula anisodora
Primula beesiana
Primula chionantha
Primula cockburniana
Primula cortusoides
Primula denticulata and selections
Pulmonaria saccharata and selections
Ranunculus aconitifolius and selections
Ranunculus acris 'Flore-pleno'
Salvia patens and selections
Salvia pratensis and selections
Salvia × *superba* 'Lubeca' selections
Salvia virgata
Saxifraga umbrosa and selections
Scabiosa graminifolia
Scabiosa ochroleuca
Sedum maximum and selections
Sedum spectabile and selections
Sidalcea and selections

Stachys grandiflora and selections
Stachys olympica and selections
Stokesia laevis and selections
Symphyandra pendula
Teucrium chamaedrys
Thermopsis rhombifolia
Tiarella wherryi
Trillium cernuum
Trillium erectum and selections
Trillium grandiflorum
Trillium sessile
Trollius altaicus
Trollius anemonifolius
Trollius asiaticus and selections
Trollius chinensis
Trollius europaeus and cultivars
Trollius laxus
Trollius ledebouri and selections
Trollius yunnanensis
Uvularia grandiflora
Veronica incana and selections
Veronica spicata and selections

Perennials 0.6-0.9 m (2-3 ft.) high

Achillea millefolium
Achillea ptarmica 'The Pearl'
Achillea ptarmica 'Angels' Breath'
Achillea ptarmica 'Lilac Queen'
Achillea ptarmica 'Old Ivory'
Achillea taygetea
Aconitum carmichaelii
Amsonia tabernaemontana
Anchusa azurea 'Feltham Pride'
Anchusa azurea 'Loddon Royalist'
Anchusa azurea 'Pride of Dover'
Anemone hupehensis var. *japonica*
Anemone hupehensis var. *japonica* 'Alba'
Anemone × *hybrida* 'Montrose'
Anemone × *hybrida* 'Margarita'
Anemone × *hybrida* 'Queen Charlotte'
Anemone × *hybrida* 'September Charm'
Anemone × *hybrida* 'Whirlwind'
Anemone vitifolia
Anthemis tinctoria selections
Aquilegia alpina 'Hensol Harebell'

Aquilegia caerulea 'Rocky'
Aquilegia canadensis
Aquilegia chrysantha
Aquilegia longissima
Aquilegia vulgaris
Aquilegia hybrids and selections
Artemisia albula and selections
Artemisia purshiana
Artemisia stelleriana
Aster acris 'Roseus'
Aster amellus and selections
Aster × cordi-belgii 'Pioneer'
Aster × frikartii
Aster linosyris
Aster novi-belgii
mound type and selections
Astible and selections
Astrantia carniolica
'Major' and selections
Belamcanda chinensis
Belamcanda flabellata
Boykinia aconitifolia
Campanula latiloba and selections
Campanula medium
Campanula persicifolia
and selections
Campanula trachelium
and selections
Catananche caerulea and selections
Chelone lyonii
Chrysanthemum balsamita
Chrysanthemum coccineum
and selections
Chrysanthemum maximum
and selections
Chrysanthemum nipponicum
Chrysanthemum rubellum
Chrysanthemum sibiricum
Clematis integrifolia and selections
Codonopsis convolvulacea
Coreopsis grandiflora and selections
Delphinium formosum
Dicentra spectabilis and selections
Dictamnus albus and selections
Dierama pendulum
Doronicum plantagineum
Dracocephalum wilsonii
Dracunculus vulgaris
Eryngium alpinum
Eryngium bromeliifolium

Eryngium oliverianum
Eryngium pandanifolium
Eryngium planum
Eryngium tripartitum
Eryngium 'Violetta'
Euphorbia griffithi
Euphorbia sikkimensis
Filipendula hexapetala
Filipendula palmata
Filipendula purpurea and selections
Filipendula ulmaria and selections
Gaillardia aristata
Gaillardia × grandiflora
and selections
Geranium wlassovianum
Gypsophila manginii
Gypsophila monstrosa
Gypsophila paniculata and selections
Helenium hoopesii
Helleborus corsicus
Helleborus foetidus
Helleborus niger and selections
Helleborus orientalis and selections
Hemerocallis lilioasphodelus
Hemerocallis fulva and selections
Hemerocallis multiflora
Hesperis matronalis and selections
Inula orientalis and selections
Inula royleana
Iris, Siberian and selections
Kniphofia foliosa
Kniphofia tubergeni
Kniphofia uvaria and selections
Liatris graminifolia
Liatris spicata and selections
Limonium latifolium and selections
Limonium tataricum and selections
Linaria purpurea and selections
Linaria triornithophora
Lobelia cardinalis
Lobelia fulgens
Lobelia siphilitica
Lychnis chalcedonica and selections
Lysimachia clethroides
Lysimachia ephemerum
Lysimachia fortunei
Lysimachia punctata
Lysimachia vulgaris

Lythrum salicaria, 0.6-1.5 m (2-5 ft.) and selections
Lythrum virgatum and selections
Meconopsis betonicifolia and selections
Mimulus ringens
Monarda didyma
Monarda fistulosa and selections
Morina longifolia
Nepeta cataria
Paeonia officinalis and selections
Paeonia wittmanniana
Papaver orientale
Peltiphyllum peltatum
Penstemon acuminatus
Penstemon barbatus and selections
Penstemon cobaea
Penstemon glaber and selections
Penstemon × gloxinoides and selections
Penstemon grandiflorus and selections
Penstemon ovatus
Penstemon palmeri
Penstemon utahensis
Penstemon unilateralis
Phlomis cashmiriana
Phlomis fruticosa, 0.6-1.2 m (2-4 ft.)
Phlomis samia
Phlomis tuberosa
Phlomis viscosa
Phlox glaberrima and selections
Phlox hoodii
Polygonatum multiflorum and selections
Polygonum bistorta
Polygonum campanulatum and selections
Polygonum cuspidatum 'Compactum'
Potentilla argyrophylla
Potentilla atrosanguinea
Primula bulleyana
Primula burmanica
Primula florindae
Primula helodoxa
Primula japonica and selections
Primula pulverulenta and selections
Rudbeckia speciosa and selections
Rudbeckia triloba

Salvia nutans
Salvia × *superba*
Sanguisorba canadensis
Sanguisorba obtusa
Saponaria officinalis and selections
Scabiosa caucasica and selections
Smilacina racemosa
Smilacina stellata
Solidago caesia
Solidago missouriensis
Solidago uliginosa
Solidago virgaurea and selections
Thalictrum aquilegifolium and selections
Thalictrum delavayi
Thalictrum diffusiflorum
Thalictrum dipterocarpum and selections
Thalictrum flavum
Thalictrum minus
Thalictrum rugosum
Thermopsis caroliniana
Thermopsis montana
Verbascum bombyciferum and selections
Verbascum chaixii
Verbascum nigrum and selections
Verbena corymbosa
Verbena hastata and selections
Veronica longifolia and selections

Perennials 0.9-1.2 m (3-4 ft.) high

Achillea
Achillea filipendulina
Achillea filipendulina 'Parker's Variety'
Aconitum napellus 'Newry Blue'
Aconitum vulparia
Anemone × *hybrida* 'Kriemhilde'
Anemone × *hybrida* 'Lorelei'
Anemone × *hybrida* 'Louise Uhink'
Anemone × *hybrida* 'Marie Manchart'
Anthemis macedonica
Artemisia schmidtiana and selections
Aruncus sylvester and selections
Asphodeline lutea

Aster cordifolius and selections
Aster ericoides and selections
Aster laevis
Aster novi-belgii and selections
Baptisia australis and selections
Baptisia leucantha
Baptisia tinctoria
Boltonia latisquama 'Nana'
Buphthalmum speciosum
Campanula lactiflora and selections
Centaurea glastifolia
Centaurea macrocephala
Cimicifuga foetida var. *intermedia* and selections
Cimicifuga japonica var. *acerina*
Clematis heracleifolia var. *davidiana* and selections
Clematis recta and selections
Delphinium cheilanthum var. *formosum* and selections
Delphinium elatum
Dicentra chrysantha
Dierama pulcherrima and selections
Echinacea purpurea and selections
Echinops exaltatus and selections
Echinops humilis and selections
Eremurus bungei and selections
Eupatorium coelestinum
Euphorbia corollata
Euphorbia wulfenii
Galega officinalis and selections
Galega orientalis
Gaura lindheimeri
Gentiana lutea
Inula helenium
Iris aurea
Iris delavayi
Iris ochroleuca and selections
Iris orientalis
Iris pseudacorus
Kirengsshoma palmata
Kniphofia
Kniphofia caulescens
Lathyrus rotundifolia
Lavandula officinalis and selections
Liatris pycnostachya
Ligularia clivorum and selections
Ligularia veitchiana
Lobelia × *speciosa*

Lobelia fulgens × *L. cardinalis* and selections
Lunaria annua (*L. biennis*)
Lupinus polyphyllus and selections
Lysimachia brachystachys
Lysimachia ciliata
Malva alcea and selections
Mertensia paniculata
Nepeta tartarica and selections
Penstemon digitalis
Penstemon heterophyllus
Penstemon murrayanus
Phlox maculata
Phlox paniculata
Phygelius
Physostegia virginiana
Polygonatum commutatum
Polygonum amplexicaule and selections
Rodgersia pinnata and selections
Rodgersia podophylla
Rodgersia tabularis
Salvia azurea and selections
Salvia beckeri
Salvia glutinosa
Salvia haematodes
Salvia sclarea and selections
Solidago petiolaris
Symphytum peregrinum
Thalictrum flavum 'Illuminator'
Verbascum phoeniceum
Verbascum thapsiforme
Veronicastrum virginicum

Perennials 1.2-1.5 m (4-5 ft.) high

Aconitum henryi
Aconitum napellus
Anchusa azurea
Anemone × *hybrida*
Asphodelus cerasiferus
Aster novae-angliae and selections
Aster novi-belgii and selections
Boltonia asteroides and selections
Boltonia latisquama
Cautleya robusta
Celsia cretica
Chrysanthemum uliginosum
Cimicifuga americana
Cimicifuga dahurica

Dianella intermedia
Dianella tasmanica
Doronicum plantagineum
'Excelsum'
Eremurus elwesii
Eremurus himalaicus
Eremurus olgae
Eremurus robustus
Eupatorium ageratoides
Eupatorium cannabinum
Helianthus and selections
Heliopsis helianthoides
and selections
Lavatera cachemirica
Lavatera olbia and selections
Lavatera thuringiaca
Ostrowskia magnifica, 1.2-1.8 m
(4-6 ft.)
Rheum palmatum and selections
Romneya coulteri, 1.2-1.8 m (4-6 ft.)
Rudbeckia subtomentosa
Salvia uliginosa
Symphytum asperum
Thalictrum rochebrunianum, 1.2-
1.8 m (4-6 ft.)
Valeriana officinalis
Verbena bonariensis

Perennials 1.5-1.8 m (5-6 ft.) high

Aconitum henryi 'Spark's Variety'
Anchusa azurea 'Dropmore'
Anchusa azurea 'Opal'
Artemisia lactiflora
Cephalaria tatarica
Cimicifuga racemosa
Eupatorium purpureum
Filipendula rubra and selections
Helenium autumnale and selections
Helianthus scaberrimus and
selections
Inula magnifica
Lathyrus grandiflora
Ligularia × *hessei*
Ligularia japonica
Ligularia stenocephala
Ligularia wilsoniana
Malvastrum coccineum
Rudbeckia nitida and selections
Thalictrum flavum 'Glaucum'
Verbascum olympicum

Perennials 1.8-2.4 m (6-8 ft.) high

Aconitum carmichaelii 'Wilsonii'
Aconitum napellus 'Blue Sceptre'
Eryngium agavifolium
Filipendula camtschatica and
selections
Helianthus salicifolius
Helianthus sparsifolius
Lathyrus latifolius and selections
Macleaya (Bocconia)
Macleaya cordata
Macleaya microcarpa and selections
Polygonum cuspidatum
Polygonum sachalinense
Rheum emodi
Romneya trichocalyx
Rudbeckia laciniata

part II

In this book herbaceous perennials are dealt with in two sections: Part I, which includes a special section on basic perennials, those universally known large groups of plants so essential for the herbaceous border, and many cultivars that are constantly changing; and Part II, which comprises perennials that are in smaller groups or do not quite fit into the category of basic or essential perennials. These are no less beautiful or showy, but do not have such an intensive horticultural following as the others.

The herbaceous perennials described here have been selected for their value as garden plants. A few will grow only in the balmy climate of coastal British Columbia, some others may be planted in the Niagara—Hamilton region as well, but most should succeed in Eastern Canada and many in the Prairie Provinces.

Although the ability to withstand cold temperatures is just as important for herbaceous perennials as for other plants, they seem to possess a great deal more and wider tolerance if given the right soil and good drainage. On the other hand, under certain conditions such as bad drainage, coupled with ice storms over a long period even at fairly reasonable temperatures, the complete destruction of fairly hardy groups can occur. Tall bearded irises have been winter-killed at Hamilton, Ontario, and the complete destruction of summer-flowering perennial phlox occurred at Ottawa in 1966.

Snowfall plays a very important part in the survival of herbaceous perennials as well as local climatic conditions that occur in small home gardens or that may be created by the use of windbreaks or fences, or places near the foundation walls of the house.

In the list of recommended plants that follows, particular mention is made of plants hardy only in the milder parts of British Columbia or in the Niagara region and the Hamilton area. Those considered very hardy can be expected to be reasonably hardy in the Prairie Provinces. Those for whom the actual areas of hardiness are not mentioned can be expected to survive in all areas of Canada except the colder Prairie Provinces and northern Ontario, or Zone 2, or colder zones shown on the Canada Department of Agriculture's Plant Hardiness maps.

Extensive botanical descriptions have not been attempted, but in many cases some differences between species serve as a means of identification. Brief horticultural accounts point out the special merits, attractions, or uses.

Also included are the scientific name, the common name if the plant has one, the natural height of the plant, and the season of flowering. Some scientific names may not agree with ones the reader knows. Names are constantly being revised as a result of study by botanists, and the most accurate ones are used in this book. Scientific names are in accordance with the rules and recommendations in the *International code of botanical nomenclature* (1966) and the *International code of nomenclature of cultivated plants* (1969). Many botanical names have been checked with the draft copy of the new *Hortus third*, through the cooperation of the staff of the Bailey Hortorum, Cornell University, Ithaca, New York.

Many cultivar names appear in the text of this book. A cultivar is a group of clearly distinguished cultivated plants. When reproduced, the group retains these characters. Cultivars were formerly called varieties, forms, and kinds. In the following section, cultivars are printed in single quotation marks, for example, 'Aizoon', which is a cultivar of *Achillea ageratifolia*, and 'Gold Plate', a cultivar of *Achillea filipendulina*.

Where possible, cultivar names are listed under the species from which they originate and to which they have the greatest affinity. There are many cases, particularly with herbaceous perennials, where breeding has involved so many species that it is not possible to assign them to any particular one. In some instances a group botanical name such as *Chrysanthemum morifolium* refers to all hybrids known as garden chrysanthemums. In others, the cultivars are listed at the end of the genus under discussion. For examples, see the list following the species of *Eremurus* and *Erigeron*.

The term variety is now usually reserved for a geographical division of the species. In this book, varieties are set in the same type as the specific name. For example in *Anemone hupehensis* var. *japonica*, *japonica* is a variety of *Anemone hupehensis*.

Interspecific hybrids, those obtained by crossing two distinct species, are treated as species and are designated by the sign $\times$ between the generic name and the specific epithet as in *Achillea* $\times$ *lewisii*, an interspecific hybrid between *Achillea tomentosa* and *Achillea argentia*.

The heights given are the average heights under normal growing conditions, and in normal geographical zones. Unless grown in very poor soil or in the shade of tall buildings, herbaceous perennials do not deviate a great deal from the heights given, although some horticultural practices such as pinching and staking can influence the height.

photos and descriptions of perennials

COLOR PLATES

1 A group of English delphiniums (pp. 17, 119)
Day-lilies (pp. 17, 135) :   2 'Cheery Pink'   3 'Francis Fay'   4 'Capri'   5 'Nantahalla'   6 'Bess Ross'   7 'Taylor Russell'

66

Oriental poppies (pp. 21, 158): 1 'Carousel' 2 'Springtime' 3 'Pinnacle' 4 'Mary Finan' 5 'Field Marshall van der Glotz' 6 'Valencia'

1 Peony trials at the Central Experimental Farm, Ottawa
Peonies (pp. 22, 23, 158) : 2 'Red Charm' 3 'White Cap' 4 'Isani Gidui' 5 'Diana Parks' 6 'Laura Magnuson'
7 'Mrs. Wilder Bancroft'

1 Common hollyhock (pp. 29, 30, 94, 137)
2 Cup-and-saucer type Canterbury bells (p. 28)
3 Foxglove 'Foxy' (p. 29, 122)
4 Siberian wallflower (p. 32)

70

1 Pansy (p. 32) : 'Goldie'
Trimardeau hybrid pansies (p. 32) : 2 'Wine Red' 3 'Jet Black' 4 'Berna' 5 'Rheingold'
Sweet Williams (pp. 32, 120) : 6 double 7 single

1 *Alyssum saxatile,* golden tuft (p. 94)
2 *Aconitum napellus* 'Kelmscot', common monk's-hood (p. 93)
3 *Alyssum repens,* madwort (p. 94)
4 *Achillea filipendulina* × *clypeolata* 'Coronation Gold', yarrow (p. 92)

1 *Anemone japonica,* windflower (p. 95)
2 *Arabis alpina* 'Rosea', rock-cress (p. 98)
3 *Anemone hupehensis,* Chinese anemone (p. 95)
4 *Anemone sylvestris,* snowdrop anemone (p. 96)
5 *Anemone pulsatilla,* pasqueflower (p. 96)
6 *Arabis* × *arendsii* 'Pink Charm', rock-cress (p. 99)

1 *Aruncus sylvester,* goat's-beard (p. 101)
2 *Aquilegia* long-spurred hybrid, columbine (p. 97)
3 Aster, dwarf hybrids being tested at the Central Experimental Farm, Ottawa (p. 101)
4 *Armeria maritima,* common thrift (p. 100)
5 *Aster* 'Goldflame' (p. 103)
6 *Artemisia schmidtiana* 'Silver Mound', angel's-hair (p. 100)
7 *Artemisia palmeri,* wormwood (p. 100)

74

1 *Campanula lactiflora,* milky bellflower (p. 109)
2 A collection of astilbes (p. 104)
3 *Bergenia cordifolia,* giant rockfoil (p. 106)
4 *Bellis perennis* hybrid, English daisy (p. 106)
5 *Aubrieta deltoidea* 'Gurgedyke', false rock-cress (p. 105)
6 *Centaurea montana,* mountain bluet (p. 111)

1 Pyrethrum 'White Madeleine' (p. 113)
Shasta daisies (pp. 13, 113, 114):   2 'Wirral Supreme'   3 'Cobham Gold'   4 'Esther Read'   5 'Thomas E. Killeen'
6 *Cephalaria gigantea,* giant scabious (p. 111)

76

1 Trials of hardy chrysanthemums
Hardy chrysanthemums (pp. 114, 115) :   2 'Golden Carpet' 3 'Ruby Mound'   4 'Coquette'   5 'Roll Call'   6 'Pink Haze'
7 'Redskin'

1 *Dianthus* 'Caprice', grass pink, cottage pink   (p. 121)
2 *Coreopsis verticillata* 'Golden Showers', thread-leaved coreopsis   (p. 118)
3 *Coreopsis grandiflora*, tickseed   (p. 117)
4 *Coreopsis grandiflora* 'Golden Plume', tickseed   (p. 117)
5 *Cimicifuga racemosa*, black snakeroot   (p. 116)
6 *Cornus canadensis*, bunchberry or pigeonberry   (p. 118)
7 *Dianthus deltoides* 'Wisley', maiden pink   (p. 120)

78

1 *Dicentra spectabilis,* bleedingheart (p. 122)
2 *Dictamnus albus,* gasplant (p. 122)
3 *Echinacea purpurea,* purple coneflower (p. 124)
4 *Echinops exaltatus,* small globe thistle (p. 125)
5 *Dryas drummondii* (p. 124)
6 *Dicentra formosa* 'Bountiful', western bleedingheart (p. 121)
7 *Dicentra eximia,* plume bleedingheart (p. 121)

Erigeron, fleabanes   (p. 127) :   1 'Quakeress'   2 'Lilofee'   3 'Amity'   4 *E. caespitosus*
5 *Euphorbia epithymoides,* cushion spurge   (p. 128)
6 *Euphorbia griffithi,* spurge   (p. 128)
7 *Eryngium* 'Donard Variety', sea holly   (p. 127)
8 *Gaillardia aristata,* blanketflower   (p. 129)

80

1 *Geranium dalmaticum,* cranesbill (p. 130)
2 *Helianthus decapetalus* 'Multiflorus', thinleaf sunflower (p. 134)
3 *Helianthus decapetalus* 'Badirector Linne', thinleaf sunflower (p. 134)
4 *Helenium autumnale* 'Fountain', sneezeweed (p. 133)
5 *Helenium autumnale* 'Riverton Gem', sneezeweed (p. 134)
6 *Gyposphila paniculata* 'Rosenschleier', rosy veil (p. 133)
7 *Heuchera* 'Gaiety', alumroot, coral bells (p. 136)
8 *Helleborus niger,* Christmas rose, black hellebore (p. 135)

1 *Hosta lancifolia,* lance-leaved plantain-lily (p. 138)
2 *Hibiscus moscheutos* 'Southern Belle' (p. 137)
3 Siberian iris 'Mandy Morse' (p. 143)
4 Japanese Iris 'Pink Frost' (p. 142)
5 *Hibiscus moscheutos* 'Super Rose' (p. 137)
6 Siberian iris 'Snow Crest' (p. 143)
7 *Iris innominata* seedling (p. 144)
8 *Incarvillea delavayi,* Chinese trumpetflower (p. 139)

82

1 *Liatris scariosa* 'Silver Tips', blazingstar, gayfeather (p. 147)
2 *Lavatera thuringiaca*, tree-mallow (p. 146)
3 *Linum flavum,* yellow flax (p. 149)
4 *Ligularia wilsoniana* (p. 148)
5 *Ligularia dentata* 'Desdemona', golden groundsel (p. 147)
6 Russell lupines (p. 151)
7 *Lobelia* × *speciosa* (p. 150)

83

1 *Lythrum virgatum* 'Morden Pink', wand loosestrife (p. 152)
2 *Lysimachia ciliata*, loosestrife (p. 152)
3 *Lysimachia nummularia*, moneywort, creeping Jenny (p. 152)
4 *Lychnis viscaria* 'Splendens Flore-Pleno', German catchfly (p. 152)
5 *Lychnis* × *haageana*, campion (p. 152)
6 *Lychnis chalcedonica*, Maltese cross (p. 151)
7 *Lythrum salicaria* 'Morden's Gleam', purple loosestrife (p. 152)

1 *Mertensia virginica*, Virginian bluebells   (p. 154)
Monarda, wild bergamot   (p. 155) :   2 'Prairie Glow'   3 'Blue Stocking'   4 'Cambridge Scarlet'   5 'Souris'
6 *Macleaya cordata*, plume-poppy   (p. 153)
7 *Meconopsis betonicifolia*, Himalayan blue poppy   (p. 153)

1 *Pachysandra terminalis*, Japanese spurge   (p. 157)
2 *Penstemon barbatus*   'Rosy Elf', beardtongue   (p. 159)
3 *Oenothera tetragona* ssp. *glauca*, evening-primrose   (p. 157)
Phlox subulata   (pp. 24, 161, 162):   4 'Star of Heaven'   5 'Brightness'   6 'Autumn Rose'

86

1 *Potentilla* 'M. Rouillard', cinquefoil   (p. 165)
2 *Platycodon grandiflorum* var. *mariesii*, balloonflower, Chinese bellflower   (p. 163)
3 *Physostegia virginiana*, false dragonhead, obedientplant   (p. 162)
4 *Physostegia virginiana* 'Bouquet Rose'   (p. 162)
5 *Polygonatum multiflorum*, Solomon's seal   (p. 163)
6 *Polygonum cuspidatum* 'Compactum', Japanese knotweed   (p. 164)

87

1 *Primula denticulata* var. *cachemiriana* 'Alba', primrose   (p. 166)
2 *Rudbeckia fulgida* var. *speciosa* 'Goldsturm', coneflower   (p. 169)
3 *Rudbeckia triloba*, gloriosa daisy   (p. 169)
4 *Pulmonaria saccharata*, Bethlehem sage   (p. 167)
5 *Primula × polyantha* Pacific Giant strain, polyanthus primrose   (p. 167)
6 *Primula × juliae* 'Gold Jewel', Caucasian primrose   (p. 166)
7 *Primula vulgaris*, common primrose   (p. 167)

88

1 *Thalictrum aquilegifolium* 'Thundercloud', meadow rue (p. 176)
2 *Sanguisorba obtusa*, Japanese burnet (p. 170)
3 *Sidalcea* 'Rose Queen', prairie mallow (p. 173)
4 *Solidago virgaurea* 'Golden Gates', goldenrod (p. 174)
5 *Thalictrum aquilegifolium,* columbine meadow rue (p. 175)
6 *Symphyandra hofmannii* (p. 175)
7 *Sedum spectabile* 'Brilliant', showy stonecrop (p. 172)
8 *Solidago virgaurea* 'Golden Wings', goldenrod (p. 174)

89

1 *Tradescantia andersoniana* 'Iris Pritchard' (p. 177)
2 *Tradescantia virginiana,* spiderwort (p. 177)
3 *Thermopsis caroliniana,* false lupine (p. 176)
4 *Thymus serpyllum,* mother-of-thyme (p. 176)
5 *Tradescantia andersoniana* cultivars *(left to right)* : 'Red Cloud', 'James C. Weguelin', 'Snow Cap', 'Blue Stone' (p. 177)

90

1 *Veronica longifolia*, beach speedwell (p. 180)
2 *Trollius laxus*, globeflower (p. 178)
3 *Trollius ledebouri* 'Golden Queen', globeflower (p. 179)
4 *Valeriana officinalis*, valerian (p. 179)
5 *Veronica incana*, speedwell (p. 180)

ACHILLEA yarrow (Color photo p. 72)

Plants vary in height and form from creeping rock plants to large showy perennials with small flowers on large flat heads, mostly with leaves dissected into several leaflets, the genus typified by the common yarrow. All species and cultivars withstand hot, dry, sandy soils, and absolute neglect, which is their most important attribute, and all are hardy on the prairies.

A. ageratifolia   *15 cm (6 in.)   May, June*
Dwarf; silvery gray, finely divided leaves; bright white flowers 1.3 cm (½ in.) across, growing well in hot sandy soils.

'Aizoon'   *10 cm (4 in.)   June*
Superior to the type, silvery tufts more compact. Commonly, but incorrectly, called *Anthemis aizoon*.

A. clavennae   *20 cm (8 in.)   May, June*
White flowers in rather loose heads; grayish leaves.

A. filipendulina × clypeolata
'Coronation Gold'   *0.9 m (3 ft.) June-Aug.*
Large masses of 7.6-cm (3-in.) yellow flower clusters; in small gardens better than the taller cultivars of the fernleaf yarrow; new, from England.

A. filipendulina
fernleaf yarrow   *0.9-1.5 m (3-5 ft.)   July-Sept.*
Upright; large flat heads of yellow flowers on leafy stems.

'Gold Plate'   *1.2-1.5 m (4-5 ft.)   July-Sept.*
Golden flowers larger than those of the species, in heads over 15 cm (6 in.) wide.

'Parker's Variety'   *7.6-10 cm (3-4 in.)   June-Sept.*
Flowers not quite as large as those of other *A. filipendulina* cultivars.

A. × lewisii
'King Edward'   *20 cm (8 in.)   June-Aug.*
Small clusters of light yellow flowers arising from 20-cm (8-in.) stems above woolly, gray green leaves. A hybrid of *A. tomentosum × A. argentia*.

A. millefolium
milfoil, common yarrow   *0.8 m (2½ ft.)   June-Aug.*
Finely cut leaves; rather pretty plant; clusters of small daisy-like flowers. A rampant weed, however; treat as such in confined garden areas.

'Cerise Queen'   *0.4 m (1½ ft.)   June-Sept.*
Light rose flowers.

'Fire King'   *0.4 m (1½ ft.)   June-Sept.*
Deep carmine flowers.

'Roseum'   *0.4 m (1½ ft.)   June-Sept.*
Exquisite pink flowers; thin plants each spring; not as vigorous as the species.

A. ptarmica
'The Pearl' sneezewort   *0.8 m (2½ ft.)   June-Aug.*
The form of *A. ptarmica* usually grown in borders; flowers like double daisies. Divide every year; reduce spread every 2 or 3 years.

'Angels' Breath'   *0.6 m (2 ft.)   June-Aug.*
A newer cultivar with a greater profusion of white bloom.

'Boule de Neige' snowball   *36 cm (14 in.)   June-Aug.*
More compact; more fully double flowers than any other sneezewort; the best.

'Lilac Queen'   *46 cm (18 in.)   June-Aug.*
Mauve flowers.

'Old Ivory'   *46 cm (18 in.)   June-Aug.*
Creamy white flowers.

A. × taygetea   *46 cm (18 in.)   June-Aug.*
Pyramidal shape, pale yellow flowers in clusters.

A. tomentosa woolly yarrow   *20 cm (8 in.)   June-July*
Small, showy, canary yellow flowers arising from ferny foliage. Likely to spread rapidly; therefore, thin out each spring. A ground cover, for use between flagstones, in crevices, in rock gardens, for edging.

'Moonlight'   *15 cm (6 in.)   June, July*
More refined than the species; less rampant.

ACONITUM monk's-hood (Color photo p. 72)

An upright plant, this perennial has tall flower spikes of little florets that look like monk's hoods. It is a very useful late-blooming plant; though it grows best in damp shady places, it will stand some sun.

Because all the aconitums have very poisonous roots, take care to keep children away when you are dividing them. Never let any leaf or root sap get in your mouth.

A. autumnale see *Aconitum henryi*

A. bicolor   1.1-1.2 m (3½-4 ft.)   *July, Aug.*
The showiest monk's-hood, white flowers edged with light blue.

A. carmichaelii   0.9 m (3 ft.)   *Aug.-Oct.*
Densely compacted pale blue spikes. Synonym: *A. fischeri*

'Wilsonii'   1.5-2.4 m (6-8 ft.)   *Sept.*
A towering giant, violet blue flowers, goes well with outdoor chrysanthemums. Synonym: *A. wilsoni*

A. fischeri see *Aconitum carmichaelii*

A. henryi   1.2-1.5 m (4-5 ft.)   *July-Sept.*
Dark blue flowers, more open-bracted than those of other species. Synonym: *A. autumnale*

'Spark's Variety'   1.5-1.8 m (5-6 ft.)   *July-Sept.*
Robust flowers of a deeper hue than in the type.

A. lycoctonum see *Aconitum vulparia*

A. napellus
common monk's-hood   1.2-1.5 m (4-6 ft.)   *July, Aug.*
The most poisonous of all the types. Deep blue flowers in tall racemes; leaves most finely divided of all aconitums.

'Blue Sceptre'   2.1-2.4 m (7-8 ft.)   *July*
Erect, tapering, very shapely; flowers blue and white, like *A. bicolor*, but more compact on the stem.

'Bressingham Spire'   1.2-1.5 m (4-5 ft.)   *July*
Erect, very neat, spiked.

'Kelmscot'   1.2-1.5 m (4-5 ft.)   *July*
Deep blue flowers.

'Newry Blue'   0.9 m (3 ft.)   *July*
Tall stiff spikes of deep blue flowers.

A. vulparia wolf's-bane   0.9 m (3 ft.)   *July*
Quite distinct from the other species; small yellow flowers. Synonym: *A. lycoctonum*

A. wilsoni see *Aconitum carmichaelii* 'Wilsonii'

ADONIS

Very hardy dwarf, annual and perennial, early flowering.

A. amurensis   20-46 cm (8-18 in.)   *Apr., May*
Clear yellow flowers.

'Plena'   15-20 cm (6-8 in.)   *Apr., May*
Much dwarfer than the species; double flowers.

A. amurensis var. *vernalis*
spring adonis   20-30 cm (8-12 in.)   *Apr., May*
The best of all; earlier than the other species and with larger flowers.

AGROSTEMMA see LYCHNIS

Most plants of the *Agrostemma* genus are now classified in *Lychnis*. The only species included in this publication is *Lychnis coronaria*.

A. coronaria see *Lychnis coronaria*

AJUGA bugle

Ajugas are useful for the front of perennial borders and for edging. They are excellent as ground cover plants for the shade or sun, and splendid in rock gardens. Flowers are produced in erect spikes. At Morden, Manitoba, ajugas are usually hardy but are sometimes killed during an unusual winter.

A. genevensis
Geneva bugle   13-30 cm (5-12 in.)   *May, June*
Useful in shade; blue flowers, compact green leaves.

'Brocklebankii'   15-30 cm (6-12 in.)   *May, June*
Deep blue flowers; spreads less rapidly than *A. reptans*.

'Pink Spires'   13-30 cm (5-12 in.)   *May, June*
Pink flowers, evergreen foliage.

A. pyramidalis   15-23 cm (6-9 in.)   *May, June*
Tubular deep blue flowers; increases slowly, needs dividing less often than the other species.

'Metallica-crispa'   8-15 cm (3-6 in.)   *June*
After autumn leaves, often sold as the bronze ajuga and listed as *A. reptans*. Bronzy metallic leaves, reddish tinted in autumn.

'Tottenham Blue'   15-23 cm (6-9 in.)   *May, June*
Unique clear steel blue flowers.

A. reptans carpet bugle   *15-23 cm (6-9 in.)   May, June*
Rapid growth of large clumps from stolons, dull blue flowers in whorls, and shiny green leaves make this species the best ajuga for use as a ground cover. The clumps smother less vigorous plants. To be kept thrifty, they need frequent division.

    'Alba'   *10-15 cm (4-6 in.)   May, June*
    White flowers, deep green leaves, not as vigorous as the species.

    'Atro Purpurea'   *15-23 cm (6-9 in.)   May, June*
    Purple bronze leaves make this cultivar very distinctive as a ground cover.

    'Gaiety'
    Very striking form, bronze, yellow, and red leaves.

    'Variegata'   *10-15 cm (4-6 in.)   May, June*
    Yellow mottled leaves.

ALSTROEMERIA

Although from South America, some alstroemerias, including *aurantiaca*, are hardy in the Niagara Peninsula.

A. aurantiaca
Brown-spotted yellow flowers, and some forms with yellow, orange, and red flowers.

ALTHAEA hollyhock   (Color photo p. 70)

A. rosea hollyhock   *0.6-2.7 m (2-9 ft.)   July, Aug.*
In gardening practice, the hollyhock is treated as a biennial (see page 29). The plants with towering spikes of blooms are basic in the herbaceous border.

ALYSSUM madwort   (Color photo p. 72)

Alyssum is a very large genus that contains annuals and perennials. Though some kinds are mostly weedy, several are excellent early flowering rock garden and edging plants and a few are quite suitable for the border. Most of the types suitable for rock gardens are hardy on the prairies.

A. repens   *20-25 cm (8-10 in.)   Apr., May*
Very hardy, tufted, neat perennial, used for edging flower beds or grouping at the front of flower borders.

A. saxatile golden tuft,
gold-dust, rock madwort   *25-30 cm (10-12 in.)   May*
Barely hardy in the Ottawa district. In milder areas, it is much used for drooping over rocks, rocky slopes, and walls because of the profusely produced, showy, bright golden yellow flowers. Grows best in a well-drained sunny position in soil that is not very rich.

    'Citrinum'   *10-20 cm (4-8 in.)   May*
    Compact; pale primrose flowers.

    'Compactum'   *8-15 cm (3-6 in.)   May, June*
    Neater and more compact than 'Citrinum'; pale yellow flowers.

    'Dudley Neville'   *15-23 cm (6-9 in.)   May, June*
    Biscuit yellow flowers desirable.

    'Flore Pleno'   *20 cm (8 in.)   May, June*
    Not quite as vigorous as the species; showy and distinctive; double, yellow flowers.

    'Silver Queen'   *8-15 cm (3-6 in.)   May, June*
    Very compact, lemon yellow flowers, silvery leaves.

AMSONIA

These dependable, slow-growing plants with clusters of blue periwinkle-like flowers on slender stems need protection in very cold localities.

A. montana   *38 cm (15 in.)   May, June*
Semidwarf, neat spikes of pale blue flowers.

A. tabernaemontana   *0.9 m (3 ft.)   May-July*
Good for the border; terminal cymes of steel blue periwinkle-like flowers. In late April, to make a more compact panicled plant, pinch off the tips.

ANCHUSA bugloss, alkanet

Plants particularly noted for their true blue flowers, they are quite hardy and will grow in most soils. The cultivars, produced from cuttings, are so much superior in size and abundance of bloom that it is unwise to grow seedlings. For best effect, grow the plants in bold groups, and cut off the faded flowers to keep the plants from setting seed. Anchusas need protection in winter on the prairies.

A. azurea Italian bugloss   *0.9-1.5 m (3-5 ft.)*   *June, Aug.*
Loose racemes of bright blue flowers, large leaves.

'Dropmore'   *1.2-1.8 m (4-6 ft.)*   *June, Sept.*
The oldest variety, at one time very popular because of its large turquoise blue flowers.

'Feltham Pride'   *0.6-0.9 m (2-3 ft.)*   *June, Aug.*
More compact than the type; bright blue, dense flower clusters.

'Loddon Royalist'   *0.9 m (3 ft.)*   *June, Aug.*
Very large gentian blue flowers; comparatively new cultivar, the best for most purposes.

'Morning Glory'   *1.5 m (5 ft.)*   *July, Sept.*
Tall, sturdier than the species; deep blue.

'Opal'   *1.8 m (6 ft.)*   *July, Aug.*
Habit similar to 'Dropmore'; pale sky blue.

'Pride of Dover'   *0.9 m (3 ft.)*   *June, Aug.*
Much dwarfer than other cultivars, medium blue flowers.

A. caespitosa tufted alkanet   *0.3 m (1 ft.)*   *May, July*
This species is very rare and difficult to grow. A cultivar 'Blue Stars', a delightful dwarf compact plant with white blue-eyed flowers, has almost superseded it, because it will grow in a well-drained soil, in full sun.

A. myosotidiflora see *Brunnera macrophylla*

ANEMONE windflower   (Color photo p. 73)

The many species of perennial windflowers differ in height, color, and usefulness as garden plants. Most anemones do well in ordinary garden soil, but they are not hardy in the Prairie Provinces. The well-known St. Brigid anemone belongs to a group of tuberous anemones and is not a herbaceous perennial.

A. acutiloba see *Hepatica acutiloba*

A. hupehensis
Chinese anemone   *0.3-0.6 m (1-2 ft.)*   *Aug., Sept.*
Flowering habit similar to that of the Japanese anemone, but flowering about a week earlier; not as tall as the Japanese species; up to five attractive pink flowers borne on strong stems well above the leaves.

A. hupehensis var. *japonica*   *0.9 m (3 ft.)*   *Sept., Oct.*
Quite similar to the Chinese anemone but taller; flowers more varied, pink, red, purple, white, and with prominent yellow stamens. Japan. Synonym: *A. japonica*

'Alba'   *0.9 m (3 ft.)*   *Aug., Oct.*
White with prominent yellow stamens; usually called Honorine Joubert.

A. × *hybrida*
Japanese anemone   *0.9-1.5 m (3-5 ft.)*   *Sept., Oct.*
This is the group of modern Japanese anemones that arose mainly through the effort of American and European breeders, who combined selections of *A. hupehensis* var. *japonica* to develop strains taller and showier than the type. Easily raised from divisions or root cuttings. After being transplanted, the plants should be protected with a straw mulch during the first year. They thrive in good rich garden soil in full sun, where they will produce an abundance of their graceful flowers borne on long stems. In Eastern Canada, unless protected by covering at night, the later flowering anemones are often killed by the first precocious frost in September.

Cultivars

'Kriemhilde' — semidouble rose pink	*0.9 m (3 ft.)*
'Lorelei' — delicate rose pink	*0.9 m (3 ft.)*
'Louise Uhink' — semidouble, snow white	*0.9 m (3 ft.)*
'Margarita' — deep pink, conspicuous anthers, semidouble	*0.8 m (2½ ft.)*
'Marie Manchart' — semidouble, white, better than 'Whirlwind'	*0.9 m (3 ft.)*
'Montrose' — large semidouble clear rose	*0.6 m (2 ft.)*
'Queen Charlotte' — pink semidouble	*0.6 m (2 ft.)*
'September Charm' — silvery pink shaded rose single	*0.9 m (3 ft.)*
'September Sprite' — rose pink single	*0.3-0.4 m (1-1½ ft.)*
'Whirlwind' — white semidouble	*0.9 m (3 ft.)*

A. japonica see *Anemone hupehensis* var. *japonica*

A. × *lesseri*   *0.3 m (1 ft.)*   *May, June*
Very striking; brilliant carmine hardy.

A. nemorosa wood anemone   *20 cm (8 in.)*   *June*
Flowers white, rose, and purplish; 'Royal Blue' with double blue flowers.

A. patens prairie crocus   *0.3 m (1 ft.)*   *Apr., May*
The provincial flower of Manitoba. Not much more than 15 cm (6 in.) high, flowering stalks to 30 cm (12 in.); flowers bluish purple, 5 cm (2 in.) wide, appearing just before the leaves develop, covered with silky gray hairs; the plumy silky gray seedpods are also attractive in this and the following species.

A. pulsatilla pasqueflower   *0.3 m (1 ft.) Apr., May*
Similar to the prairie crocus, feathery foliage, flowers blue or reddish purple, bell-shaped, outward-facing, 5 cm (2 in.) across. There is a confusion of so-called species and cultivars in this group; flowers are of various colors, but all obviously belong to the same species. The following cultivars are available from nurserymen:

'Alba' — white   *23 cm (9 in.) Apr., May*

'Mallandieri'   *30 cm (12 in.) Apr., May*
Flowers large, upturned, broad-petaled, rich imperial blue; one of the first anemones to bloom.

'Mrs. Van der Elst'   *23 cm (9 in.) Apr., May*
Rose pink; all varieties do well in light sandy or other well-drained soil in a sunny location. They show to best advantage in a rock garden, but give early bloom to the border if grown in clumps of seven or eight plants.

'Red Clock'   *15 cm (6 in.) Apr., May*
A new variety with upturned flowers, deep crimson, offset by golden anthers.

A. sylvestris snowdrop anemone   *20 cm (8 in.) June*
Very showy; scented, white, slightly nodding flowers, 5 cm (2 in.) wide on 20-cm (8-in.) stems; will grow in full sunshine but prefers semishade, where its white flowers show to best advantage, and last longer; the hardiest of all anemones.

A. vitifolia grape-leaved anemone *0.6-0.9 m (2-3 ft.) Aug.*
Large somewhat coarse; large pale pink blooms on stems 0.6-0.9 m (2-3 ft.) high; leaves resembling grape leaves.

ANEMONELLA

Perennial herbs related to the anemone but differing in the arrangement of flower parts. Leaves are more closely related to those of the meadow rue *(Thalictrum)*.

A. thalictroides rue anemone   *10-15 cm (4-6 in.) Apr.*
White anemone-like flowers, long stems; best suited for naturalizing in the woodland.

ANTHEMIS chamomile, golden marguerite

Very rugged long-suffering plants, admirably suited for hot dry borders and sloped areas. Cultivars are much superior to the species from which they have arisen. Care must be taken not to allow seedlings to develop and crowd out the cultivars, because the seedlings are practically weeds.

A. aizoon see *Achillea ageratifolia* 'Aizoon'

A. biebersteiniana   *30 cm (12 in.) June*
A good dwarf plant with large yellow flowers and silvery leaves.

A. cupaniana
Forms aromatic mats of grayish foliage, from which arise large white flowers with a yellow center.

A. macedonica   *0.9 m (3 ft.) Aug., Sept.*
A very floriferous golden yellow daisy with an excellent uniform habit; useful for midsummer bloom.

A. nobilis chamomile   *20 cm (8 in.) July, Aug.*
Useful as a spreading ground cover to grow into ferny aromatic mats of vegetation with inconspicuous white flowers.

'Treneague'
A selection used for lawns in hot dry areas where grass will not flourish. This is a nonflowering clone.

A. sancti-johannis   *0.4 m (1½ ft.) July, Aug.*
Gray, hairy plant, deep orange flowers; best in sun, well-drained soil.

A. tinctoria golden marguerite

This species becomes a pesky weed in very mild climates and is likely to be a problem in Canada if seedlings are allowed to develop. Many good varieties can be perpetuated by division, an operation that should be carried out at least every 3 years.

'Beauty of Grallagh'   *0.8 m (2½ ft.)*   *July, Aug.*
Very large, deep yellow flowers.

'Golden Dawn'   *0.6 m (2 ft.)*   *July, Aug.*
Double-flowered, requires annual division for survival.

'Grallagh Gold'   *0.8 m (2½ ft.)*   *July, Aug.*
A very deservedly popular bright yellow cultivar.

'Kelwayi'   *0.6 m (2 ft.)*   *July, Aug.*
Deep yellow flowers.

'Moonlight'   *0.8 m (2½ ft.)*   *June, Aug.*
Soft primrose yellow; the best and most refined variety in the border at Ottawa.

'Mrs. E. G. Buxton'   *0.8 m (2½ ft.)*   *June, Aug.*
Very robust, erect, light yellow flowers.

'Perry's Variety'   *0.6 m (2 ft.)*   *July*
Pure yellow flowers; one of the oldest varieties.

'Thora Perry'   *0.6 m (2 ft.)*   *July, Aug.*
Probably an interspecific hybrid of *A. sancti-johannis* × *A. tinctoria* with the orange tints of the former but lacking the vigor of *A. tinctoria*.

AQUILEGIA columbine   (Color photo p. 74)

Columbines have long been favorites for the flower border. Their unique and beautiful dancing-fairy flowers combine harmoniously with elegant and graceful leaves. They are easy to grow from seeds, but because the varieties hybridize very freely, seeds from even distantly related species often produce mediocre plants. The best kinds to use in borders are the selected hybrid strains that have large showy flowers and long spur-like appendages. All are very hardy.

The dwarf types are mainly suited to rock gardens and the larger ones for flower borders or for naturalizing in large informal beds in full sun or part shade.

Columbines are often attacked by leafminers and stem borers. The leafminer makes its presence known by winding trails of white inside the leaf. Affected leaves should be removed at the base, and the plant should be sprayed with an insecticide. A borer may destroy the plant before its presence has been noticed. A systemic spray should be applied in early spring.

A. alpina alpine columbine   *0.3 m (1 ft.)*   *July, Aug.*
A small plant with very large flowers, 5 cm (2 in.) across, powdery blue with quite long incurving spurs. A good long-lasting plant.

'Hensol Harebell'   *0.6 (2 ft.)*   *July, Aug.*
Much taller than the species, flowers the same size, but more intense colors.

A. caerulea
Rocky Mountain columbine   *0.8 m (2½ ft.)*   *May, June*
The state flower of Colorado. The true type has lavender blue sepals, white petals, and long 4-cm (1½-in.) spurs. It has been used often in breeding with other species and hybrids to obtain types with long spurs.

A. canadensis
wild columbine   *0.8-0.9 m (2½-3 ft.)*   *June, July*
Native from Nova Scotia to southern Ontario and south to Georgia. In Canada it is usually found in partly shaded woodlots, and in gardens it should preferably have some shade. It is a graceful plant, having strong, straight stems and nodding flowers with brilliant red sepals, yellow petals, and straight red spurs that are knobbed at the end. It prefers woodland soil, which is composed mainly of well-decayed leaves.

A. chrysantha
golden columbine   *0.8-0.9 m (2½-3 ft.)*   *June, July*
Long-spurred; large 8-cm (3-in.) outward-facing, yellow flowers, delicate rich green foliage. This species has been crossed with others to produce large-flowered, long-spurred strains, which are preferred in gardens.

A. discolor   *10 cm (4 in.)   May, June*
A diminutive exquisite alpine plant, blue and white flowers;
would probably be lost in a border but is outstanding in
pockets of a rock garden.

A. ecalcarata   *30 cm (12 in.)   July*
This rather odd little perennial with spurless reddish brown
flowers makes an attractive graceful plant when combined
with yellow or white flowers of contrasting form. It reproduces itself true from seeds, which is also in its favor, especially because it is short lived and needs renewing after 3 or
4 years.

A. einseleana   *20 cm (8 in.)   May, July*
Very charming, graceful alpine, also of color useful in
flower border. Unique mauve flowers.

A. flabellata fan columbine   *46 cm (18 in.)   July, Aug.*
Mauve, nodding flowers, 5 cm (2 in.) wide, rather short incurved spurs; pale fan-shaped leaves; neat, beautiful, appeal-
ing plant. Japan.

   'Nana' dwarf fan columbine   *15 cm (6 in.)   July, Aug.*
   Neater than the species; large blue and white flowers,
   leaves not as finely cut as in fan columbine.

   'Nana Alba' white fan columbine *15 cm (6 in.)   July, Aug.*
   Neat, dwarf, as in 'Nana'; creamy white flowers, grayish
   green leaves.

A. glandulosa
Altai columbine   *20-25 cm (8-10 in.)   May, June*
Dwarf; large nodding blooms, blue, white. It grows best
from seeds and selections made from the seedlings. Sometimes mistakenly called *A. jucunda.*

A. × helenae   *38 cm (15 in.)   June, July*
A remarkable hybrid *A. caerulea* × *A. flabellata;* blue and
white bicolor; reproduces true from seeds.

A. jonesii Jones columbine   *15 cm (6 in.)   June*
This beautiful little plant from the Rockies is really for the
enthusiast. To produce a profusion of large blue flowers on
healthy plants, make sure it grows in alkaline soil, well
drained and moist at all times.

A. jucunda see *Aquilegia glandulosa*

A. longissima
longspur columbine   *0.6-0.9 m (2-3 ft.)   July, Oct.*
Tall, robust; large, yellow, nodding flowers with pendulous
spurs 13 cm (5 in.) long; extremely distinctive because of
silky hairy leaves and unique spurs.

Strains
Many species such as *A. caerulea, A. flabellata, A. longissima,
A. chrysantha,* and *A. canadensis* have been interbred to
give a wide variety of flower colors and longer, more attractive spurs. These are most often introduced as selected
strains, some of which come quite true from seeds, whereas
others produce a similar but wider assortment of colors.

   Clematiflora Hybrida   *0.6-0.9 m (2-3 ft.)   May, July*
   Large, spurless; clematis-shaped blue, pink flowers.

   Copper Queen   *0.6 m (2 ft.)   May, June*
   Reddish brown flowers.

   Crimson Star   *0.6 m (2 ft.)   May, June*
   Large, long-spurred, crimson, white flowers.

   McKana's Giants   *0.8-0.9 m (2½-3 ft.)   June, Aug.*
   A group of hybrids noted for vigorous growth, excellent
   color variation, and very large flowers with spurs to 10
   cm (4 in.) long.

   Mrs. Scott-Elliot's strain   *0.6-0.9 m (2-3 ft.)   June, Aug.*
   An old strain that has endured many years and still produces excellent plants of high value.

   Rose Queen   *0.6 m (2 ft.)   June, July*
   Soft rose pink, long-spurred.

   Snow Queen   *0.8 m (2½ ft.)   June, July*
   Pure white, long-spurred strain.

A. vulgaris
European columbine   *0.6-0.8 m (2-2½ ft.) June, July*
This is a longer-living perennial than the usual long-spurred
strain, but it is not so effective in a border, because the flower
colors are dull and uninteresting.

ARABIS rock-cress   (Color photo p. 73)

A very popular, useful group of plants for rock gardens,
the front of flower borders, and ground covers. Although the
genus contains more than a hundred species, most of the
species are weedy and only a few are worthy of cultivation.
Occasionally winter-killed in the Prairie Provinces.

A. albida see *Arabis caucasica*

A. albida 'Rosea' see *Arabis × arendsii*

A. alpina   *15 cm (6 in.)   May*
The true type is a neat silvery-leaved plant with compact
racemes of white flowers 8 mm (⅓ in.) across. This is often
mistaken for the wall rock-cress, *A. caucasica,* which is much
more rampant, and has larger flowers and leaves coarsely
toothed toward the apex, and not as they are in this plant,
around the edges.

   'Rosea' — soft pink flowers   *15 cm (6 in.)   May*

Arabis caucasica 'Variegata', wall rock-cress

Grows well in good garden soil, but is more apt to last longer and need less attention in well-drained loose, porous soil in full sunlight or part shade. After the plants have flowered, a light shearing will induce new and stockier growth. Every third year the plants should be divided, especially if grown in a perennial border. Synonym: *A. albida*

> 'Coccinea'   *13 cm (5 in.)*   May
> Very bright rose flowers.

> 'Flore Pleno'   *30 cm (12 in.)*   Apr., May
> Much more attractive than the type; pure white racemes of semidouble and double flowers on taller stalks.

> 'Sulphurea'   *15 cm (6 in.)*   May
> Flowers tinged with creamy yellow.

> 'Variegata'   *20 cm (8 in.)*   Apr., May
> The variegations of creamy white flowers and beautiful foliage make this plant quite distinctive. Unattractive, sparsely produced flowers. New green shoots should be cut out, because they are apt to develop and grow more vigorously than the cultivars. If left untouched, the cultivars will revert back to the species.

A. kellereri   *8 cm (3 in.)*   May
A rock garden gem, which might be lost in a border. Very distinctive; pure white flowers, green cushions of leaves.

A. procurrens   *15 cm (6 in.)* May, June
Similar to *A. caucasica*, but neater, more compact in every way; white flowers on 15-cm (6-in.) stems, shiny green leaves. Synonym: *A. sturii*

A. sturii see *Arabis procurrens*

ARMERIA thrift, sea pink   (Color photo p. 74)

The true thrifts, or sea pinks, are low-growing, summer-blooming perennial herbs, sometimes mistaken for the genus *Statice* or *Limonium*. Because of their strawy-textured flowers, their blooming period is longer than that of many other perennials. To produce abundant flowers, they need full sun and light soil.

A. caespitosa see *Armeria juniperifolia*

A. juniperifolia   *10 cm (4 in.)*   May, June
Very neat little tufts of grass-like foliage from which emerge tiny pink flowers. Needs extra good drainage. Synonym: *A. caespitosa*

> 'Alba'   *8 cm (3 in.)*   May
> Pearl white flowers on tufted mounds.

> 'Beechwood'   *10 cm (4 in.)*   May
> Much larger pink flowers; more vigorous plants.

A. × arendsii   *20 cm (8 in.)*   May
A hybrid of *A. aubrietioides* and *A. caucasica* 'Albida'; rose pink flowers. Synonym: *A. albida* 'Rosea'

> 'Pink Charm'   *20 cm (8 in.)*   May
> Very choice garden form; rose pink blossoms, dark green leaves. Less rampant than *A. caucasica*, but growing fast and requiring similar treatment.

> 'Rosabella'   *13 cm (5 in.)*   Apr., May
> Very neat, sturdy hybrid, deeper rosy pink flowers than 'Pink Charm', but much less rampant, and more suitable for rock gardens.

A. blepharophylla   *15 cm (6 in.)*   Apr., May
Neat mounds of foliage; deep rose, almost red flowers. California.

> 'Spring Charm'   *15 cm (6 in.)*   Apr., May
> Deep crimson or carmine; must be propagated asexually because, unlike some of the previously mentioned hybrids, does not reproduce true from seed.

A. caucasica wall rock-cress   *20 cm (8 in.)*   Apr., May
Very fast-growing, vigorous, trailing; used extensively in front of a flower border, as an edging for beds, for interplanting with tulips, and as a ground cover. In rock gardens, tends to overgrow and smother nearby plants. The profusion of bloom, especially of the double cultivar, makes it well worth planting.

100

'Bevan'   *5 cm (2 in.)   May*
The most compact and choicest form. Tufted; deep rose flowers.

A. maritima common thrift   *23-30 cm (9-12 in.)   July, Aug.*
Probably no other plant has been so confused in the nomenclature as this rugged little perennial. It was once named *Statice*, then *Limonium*, and now *Armeria*, with various specific epithets appended. The true *A. maritima* is seldom, if ever, listed in catalogs. Neat; narrow grass-like ribbed leaves forming dense rosettes; pink globular heads of flowers on strong stiff stems.

'Alba'   *13 cm (5 in.)   May, June*
Very fine compact form, more compact than the species, long-lasting white flowers.

'Alpina'   *20 cm (8 in.)   June*
Light pink flowers on 20-cm (8-in.) stems.

'Laucheana'   *25 cm (10 in.)   June, July*
Large, deep rose flowers.

'Laucheana Six Hills'   *23 cm (9 in.)   June, July*
A selection of 'Laucheana', light pink flowers.

'Vindictive'   *10 cm (4 in.)   June, July*
Dwarf plant; deep pink flowers.

A. pseud-armeria
great thrift   *15-46 cm (6-18 in.)   June, July*
Among the move vigorous cultivars are:

'Bees' Ruby'   *30 cm (12 in.)   June, Aug.*
Large-flowered, glistening cerise pink.

'Bloodstone'   *30 cm (12 in.)   June, Aug.*
Rosy crimson.

'Glory of Holland'   *30 cm (12 in.)   June, Aug.*
Clear pink.

ARTEMISIA wormwood   (Color photo p. 74)

Except for the white mugwort, *A. lactiflora*, artemisias are grown mainly for their grayish or silvery fragrant foliage. Plants are from a few centimetres (inches) to 1.8-2.4 m (6 or 8 ft.) tall, their usefulness depends on their density of growth.

They need a sunny location and well-drained soil. Divide them every 2 or 3 years. Except for *A. lactiflora*, artemisias are very hardy.

A. albula silver king, ghostplant   *0.6-0.9 m (2-3 ft.)*
Silvery leaves; useful for providing mass effects in a border and for mixing with flowers, either fresh or dried for bouquets in winter; needs dividing every year.

'Lambrook Silver'   *0.9 m (3 ft.)*
Slightly more tolerant of cultivation; more robust.

'Silver Queen'   *0.9 m (3 ft.)*
A garden selection with broader more glistening foliage; needs dividing every spring.

A. discolor   *46 cm (18 in.)*
Fairly compact; silvery leaves.

A. frigida fringed wormwood   *20-30 cm (8-10 in.)*
Finely divided, silvery leaves.

A. lactiflora white mugwort   *1.8 m (6 ft.) Aug., Sept.*
Sometimes treated as an overgrown weed, but valuable in a border because of the softening effect of its graceful panicles of creamy white flowers. Aromatic leaves, dark green above, silvery beneath, deeply fissured. Needing a moist rich soil and must be divided every third year.

A. lanata   *46 cm (18 in.)*
Graceful; silver filigree of foliage.

A. maritima 'Nutans' see *Artemisia nutans*

A. nutans   *0.6 m (2 ft.)*
Medium size; feathery, finely divided, silvery, aromatic leaves. Synonym: *A. maritima* 'Nutans'

A. palmeri   *46 cm (18 in.)*
Similar to but smaller and bushier than *A. purshiana*; woolly gray leaves.

A. purshiana   *0.8 m (2½ ft.)*
Textured heavier than most artemisias and, like *A. palmeri*, excellent for contrast in a border reserved mainly for gray-leaved plants; decumbent woolly white stems; leaves silky white on both sides.

A. schmidtiana 'Nana' angel's-hair   *8 cm (3 in.)*
Of particular merit as a low-spreading ground cover or rockery plant where its finely dissected silvery leaves show to great advantage.

'Silver Mound'   *30 cm (12 in.)*
A taller, more compact selection; most effective as a ground cover for dry slopes and rockeries, invaluable as a dot plant in bedding schemes and border foregrounds, or for edging. To keep the plant neat, remove the rather valueless flowers and divide the plant every 3 years.

A. stelleriana old woman, beach wormwood   *0.6 m (2 ft.)*

Densely hairy or woolly white; segmented oblong leaves; yellow racemous flowers. Excellent as a ground cover on sandy soil, particularly in beach areas, or for a flower border where bold silvery foliage is needed as an undercover for pink asters or *Physostegia*. Eastern region of North America from Quebec to New Jersey.

ARUM

A. dracunculus see *Dracunculus vulgaris*

ARUNCUS goat's-beard   (Color photo p. 74)

These are massive plants, best suited to large flower borders or for screening. The large compound leaves are attractive all summer, and the creamy white flowers are arranged in plume-like terminal panicles. The plants need rich moist soil and grow well in part shade.

A. sylvester   *0.9-1.2 m (3-4 ft.)   June, July*

A most remarkable and unmistakable plant with its broad spread; massive plumes of creamy white flowers, large compound leaves. Useful for forming a background for perennials, for shade near patio fences, and as specimens. A very hardy species. Synonym: *Spiraea aruncus*

'Kneiffi'   *0.9 m (3 ft.)   June, July*

Dwarfer, with the same graceful plume of flowers as *A. sylvester,* but finer, with cut leaves. A better choice for the smaller garden.

ASCLEPIAS milkweed, butterflyweed

Most milkweeds are regarded as weedy subjects or as greenhouse semitropical plants, but the species described below is a worthy addition to any perennial or flower border, and is hardy at Ottawa and in similar climates.

A. tuberosa butterflyweed   *46 cm (18 in.)   July, Aug.*

Very showy, particularly adaptable to dry sunny locations, produces abundant umbels of intense orange flowers of striking effect in midsummer. Not rampant like some of the other plants in this genus, spreading and increasing very slowly and, because of its long taproots, not easily transplanted. Usually sold in spring or fall, because the dormant roots are slow to start growth. Seedpods similar to those of the milkweed, good for dried floral arrangements.

ASPERULA

A. odorata   *15 cm (6 in.) June, July*

Very pretty white flowers, trailing, for a semishade.

ASPHODELINE

A. lutea asphodel   *0.9-1.2 m (3-4 ft.)   June*

Forming clumps of narrow leaves; fragrant yellow flowers in finger-shaped clusters 18-38 cm (7-15 in.) long and about 0.6 m (2 ft.) wide on stiff upright stems. Easily grown in ordinary garden soil, either in full sun or part shade.

ASPHODELUS

A. cerasiferus white asphodel   *1.5 m (5 ft.)   June*

White lily-like fragrant flowers on dense branching racemes with buff bracts, clumps of narrow basal leaves. Quite similar to the asphodel described previously, but with buff bracts instead of leafy flower stalks.

ASTER Michaelmas daisy   (Color photo p. 74)

Michaelmas daisies, or perennial asters, are almost indispensable in a flower border, because of their striking appearance in fall. Their soft pastel shades of blue, pink, lavender, rose, and purple provide a unique contrast to the prevailing vivid crimsons, golds, and reds of autumn.

Although most of the 600 species of asters are native to North America, it was English gardeners who, by breeding and selecting the best species and improving them almost beyond recognition, developed a new and invaluable perennial for gardens all over the world. In recent years plant breeders in Canada and the United States have developed an entirely new strain from the English cultivars. This new strain forms mounds of bloom 0.3-1.2 m (1-4 ft.) tall with abundant flowers covering the leaves and stems. (See *Hardy asters for the autumn garden,* Canada Department of Agriculture Publication 1271.)

Most species and garden forms need a well-drained but not very dry soil. Except for species found in woods, these plants do best in full sun and, because of their susceptibility to mildew, in locations where high walls and hedges do not prevent free flow of air.

Once the plants have become established, they should be divided every 3 years to ensure health and vigor. If the plants are not divided they will gradually deteriorate and become more susceptible to mildew. The plants should be sprayed with a fungicide as soon as you notice white spots of mildew. Most species and cultivars are very hardy.

A. acris   *0.6 m (2 ft.)   Sept., Oct.*
Clouds of star-shaped blue flowers.

'Nanus'   *30 cm (12 in.)   Sept., Oct.*
Neater, dwarfer than the species; rosy mauve flowers.

'Roseus'   *0.6 m (2 ft.)   Sept., Oct.*
Light pink flowers.

A. × *alpellus* 'Summer Greeting'   *25 cm (10 in.)   June, July*
A hybrid of *Aster alpinus* and *Aster amellus;* very large light blue flowers on short stems.

'Triumph'   *30 cm (12 in.)   June, Aug.*
Deeper blue flowers than 'Summer Greeting'.

A. alpinus alpine aster   *20 cm (8 in.)   June, Aug.*
A summer-flowering alpine plant with gold-eyed purple daisy-like flowers.

'Beechwood' – mauve flowers   *23 cm (9 in.)   May, June*

'Joy' – lilac blue flowers   *23 cm (9 in.)   May, June*

'Wargrave' – lilac pink flowers   *23 cm (9 in.)   May, June*

A. amellus Italian aster   *0.6 m (2 ft.)   Aug., Sept.*
More dwarfed with larger flowers than the Michaelmas daisy. Usually growing to 46 cm (8 in.); well-branched plants, purple flowers 5 cm (2 in.) across with prominent yellow disk florets in the center. The gray foliage somewhat aromatic, earlier flowers advantageous.

Cultivars   *0.6 m (2 ft.)*
'Brilliant' – bright pink
'King George' – the best known, very large deep blue flowers on compact plants
'Lac de Geneve' – soft blue, long-lasting flowers
'Lady Hindlip' – deep rose pink
'Mauve Beauty' – very large mauve flowers
'Moorheim Gem' – very good companion for 'King George'
'Nocturne' – large rosy lavender flowers
'Perry's Favorite' – deep pink
'Sonia' – very popular, pink
'Sonnewendi' – early flowering, clear pink

A. × *cordi-belgii* 'Pioneer'   *0.8 m (2½ ft.)   Sept.*
A selection from the cross *A. cordifolius* × *A. novi-belgii,* a profusion of rich pink flowers on very distinct neat bushes having the compactness and grace of *A. cordifolius* and the size and freshness of *A. novi-belgii.*

A. cordifolius blue wood aster   *1.2 m (4 ft.)   Sept., Oct.*
Shade-loving; very graceful sprays of tiny blue flowers on 1.2-m (4-ft.) stems, large cordate leaves.

'Sweet Lavender'   *0.9 m (3 ft.)   Sept., Oct.*
A little dwarfer than the type with similar arching sprays of mauve blooms.

'Silver Spray'   *1.2 m (4 ft.)   Sept., Oct.*
Sprays of white flowers.

A. ericoides heath aster   *0.9 m (3 ft.)   Sept., Oct.*
An extremely graceful plant, producing a profusion of very small white flowers, very small leaves.

Cultivars
'Blue Star' – tiny pale blue flowers in elegant sprays
   0.8 m (2½ ft.)   Oct.
'Brimstone' – tiny yellow flowers on bushes
   0.8 m (2½ ft.)   Oct.
'Chastity' – white flowers   *1.1 m (3½ ft.)   Oct.*
'Delight' – white flowers   *0.8 m (2½ ft.)   Oct.*
'Esther' – large sprays of very dainty pink flowers
   0.6 m (2 ft.)   Sept.-Oct.
'Golden Spray' – a color between buff and sulfur
   0.9 m (3 ft.)   Oct.
'Ringdove' – deep mauve flowers   *1.2 m (4 ft.)   Oct.*

A. × *frikartii* wonder of Staffa   *0.8 m (2½ ft.)   July-Sept.*
A very beautiful hybrid of *A. amellus* and *A. thompsonii;* large clear blue flowers and contrasting gold centers. A very long period of bloom. Needs winter protection in areas where the temperature drops to zero or below.

A. laevis smooth aster   *1.2 m (4 ft.)   Sept., Oct.*
The old favorite 'Climax' aster, still grown in England but seldom in Canada, belongs to this species. Very large 0.8-m (2½-ft.) lavender blue flowers on well-branched plants.

A. linosyris goldilocks   *0.8 m (2½ ft.)   Sept., Oct.*
Attractive plants with little yellow daisies. Synonym: *Linosyris vulgaris*

A. novae-angliae
New England aster   *0.9-1.5 m (3-5 ft.)   Sept., Oct.*
This fine towering aster is found along roadsides in Quebec and Ontario. Deep purple flowers on crowded flower heads 4 cm (1½ in.) across, lanceolate leaves 8-10 cm (3-4 in.) long. From this plant many of the more vigorous taller hybrids found in older perennial borders have developed. Because most of these have a rather ragged habit for two-thirds of their height, they should be moved to the back of borders in order to hide their unsightly stems with other plants.

Cultivars   *Sept., Oct.*
'Barr's Pink' – bright rose pink   *1.5 m (5 ft.)*
'Crimson Beauty' – rich rose crimson   *1.2 m (4 ft.)*
'Harrington's Pink' – very beautiful pink   *1.2 m (4 ft.)*
'Incomparabilis – brilliant purple red flowers   *0.9 m (3 ft.)*
'Lil Fardell' – very well-known, deep rose
   1.5-1.8 m (5-6 ft.)
'Lye End Beauty' – soft pink   *1.2 m (4 ft.)*
'Mount Rainier' – white   *1.2 m (4 ft.)*
'Red Star' – deep carmine, rose   *1.2 m (4 ft.)*
'September Ruby' – deep ruby rose red   *1.1 m (3½ ft.)*
'Survivor' – deep pink, outstanding   *1.2 m (4 ft.)*
'Treasure' – lilac, dwarfer than others in this group
   0.9 m (3 ft.)

A. novi-belgii New York aster

A much more refined plant than *A. novae-angliae* with smooth lance-shaped leaves, 10-15 cm (4-6 in.) long, and blue violet flowers, 3 cm (1 in.) across.

The New York aster has been used often as a parent for breeding new and more refined asters. Since all catalogs and most horticultural textbooks list these cultivars as *novi-belgii* hybrids, ease of reference has prompted such a treatment in this publication. Furthermore, because most aster cultivars are interbred, the following entries are arranged by height and form.

Tall cultivars   *1.2-1.5 m (4-5 ft.)*

   'Blondie' — large, double, creamy white
   'Blue Bonnet' — mid-blue semidouble
   'Hilda Ballard' — very large, semidouble, soft lilac pink
   'Picture' — attractive carmine red
   'Queen Mary' — large, light blue, 1.5 m (5 ft.)

Medium cultivars   *0.9-1.2 m (3-4 ft.)*

   'Ada Ballard' — perfectly formed, large, double, blue
   'Appleblossom' — delicate pink
   'Beechwood Challenger' — bright red
   'Chartwell' — rich violet, shaded crimson
   'Crimson Brocade' — abundant, double, red flowers on very well-shaped bushy plants
   'Davey's True Blue' — large, free flowering, vigorous
   'Dawn' — very pleasing, semidouble, pink
   'Destiny' — strong growing, free-flowering, rose violet
   'Elegance' — late-flowering, blue purple, very good for cutting
   'Ernest Ballard' — large, semidouble, deep carmine
   'Fair Lady' — double, lilac mauve
   'Fellowship' — soft pink
   'Flamingo' — very large, double, pink flowers on bushy plants
   'Gayborder Royal' — very bright crimson
   'Gayborder Violet' — rich violet purple
   'Glorious' — long-lasting, deep pink
   'Goldflame' — an unusual light yellow
   'Harrison's Blue' — deep amethyst blue
   'Janet McMullen' — late-flowering, deep pink
   'Mabel Reeves' — large, fully double, deep pink
   'Mauve Ballard' — very large, double, soft light blue
   'Orchid Pink' — light orchid, pink
   'Plenty' — silvery blue with yellow center
   'Red Sunset' — medium size, rich rose red flowers
   'The Archbishop' — semidouble, deep purple blue
   'The Bishop' — semidouble, rosy purple
   'The Cardinal' — neat habit, rich rose red
   'Twinkle' — soft cyclamen purple

Mounds, medium height   *0.5-1 m (1½-3 ft.)*

These and the dwarf mound types produce their blooms in such profusion that they form mounds of flowers and display few leaves and stems.

   'Alpenglow' – rosy red flowers on stiff pyramid growth
   'Beechwood Beacon' – deep rosy crimson
   'Blue Radiance' – large, pale blue
   'Chequers' – rich violet purple
   'Erica' – very strong-growing, stiff-stemmed, compact
   'Eventide' – very large flowers, 5 cm (2 in.) across, purplish blue
   'Fairy' – compact 0.6-m (2-ft.) mounds, pale pink to white
   'Gayborder Charm' – semidouble, rose pink
   'Gayborder Splendour' – semidouble, dark cyclamen rose
   'Guy Ballard' – semidouble, deep pink
   'Lavender Gown' – perfect mounds of lavender blue
   'Little Boy Blue' – mass of luminous lavender blue daisies
   'Little Pink Lady' – similar to 'Little Boy Blue' but with pink flowers
   'Peter Harrison' – mounds of light pink narrow-petaled flowers

The 'Royal Gem' group of cultivars was originated by the Royal Botanical Gardens at Hamilton, Ontario. This group includes:

   'Royal Amethyst' – pale pink flowers   *46 cm (18 in.)*
   'Royal Opal' – pale lilac blue   *46 cm (18 in.)*
   'Royal Pearl' – white with prominent yellow zone
                          46 cm (18 in.)
   'Royal Sapphire' – rounded shape, medium blue
                          51 cm (20 in.)

Mounds, dwarf   *under 46 cm (18 in.)*

   'Audrey' – rosy cerise   *38 cm (15 in.)*
   'Blue Bouquet' – bright blue   *38 cm (15 in.)*
   'Bonanza' – deep pink   *23 cm (9 in.)*
   'Buxton's Blue' – smallest of the dwarfs, blue *10 cm (4 in.)*
   'Canterbury Carpet' – wide-spreading carpet *25 cm (10 in.)*
   'Jenny' – double, violet purple   *30 cm (12 in.)*
   'Lady-in-blue' – semidouble, rich blue   *25 cm (10 in.)*
   'Lavender Midget' – mass of lavender blue *38 cm (15 in.)*
   'Little Blue Baby' – perfect miniature, mid-blue
                          15 cm (6 in.)
   'Little Pink Pyramid' – shapely plants, rose pink
                          46 cm (18 in.)
   'Little Red Boy' – deep rosy red   *30 cm (12 in.)*
   'Margaret Rose' – rosy pink   *30 cm (12 in.)*
   'Midget' – pale blue   *30 cm (12 in.)*
   'Peter Pan' – starry lilac blue   *38 cm (15 in.)*
   'Pink Bouquet' – perfect mounds of luminous starry pink
                          38 cm (15 in.)
   'Pink Lace' – double, pink opening to red   *36 cm (14 in.)*
   'Purple Prelude' – reddish purple, gold center
                          30 cm (12 in.)
   'Purple Feather' – violet blue   *30 cm (12 in.)*
   'Romany' – low, many-stemmed floriferous plant with 4-cm (1½-in.) purple blue flowers   *30 cm (12 in.)*

'Rose Serenade' – rose pink fading to silver pink
                          38 cm (15 in.)
'Snowball' – profusion of shaggy white flowers
                          25 cm (10 in.)

A. yunnanensis 'Napsbury'   *38 cm (15 in.)*   *June, July*
Dwarf, similar to alpine aster but larger, more graceful flowers.

ASTILBE   (Color photo p. 75)

Do not attempt to grow astilbes unless your soil is rich and moist and high in organic matter, and the plants are shaded during the hottest part of the day. The new hybrids are such delightful plants that they are worth the effort of preparing good soil and locating them in semishade. They provide feathery branching spires of bloom in deep and light red, salmon, pink, and white that blend with the mighty spires of delphiniums and form a beautiful picture in June or July. On the prairies all astilbes need very special protection.

Some species are so inferior to the hybrids that have been developed from them that they are not worth growing. Some examples are *A. astilboides*, white flowers; *A. davidii* (synonym: *A. arendsi*) with large compound leaves, elm-like leaflets, and rose pink flowers; *A. japonica* with thrice-compound leaves, lance-oval leaflets and white spire-like flower clusters; and *A. simplicifolia*, a dwarf plant that rarely exceeds 30 cm (12 in.) and has white flowers.

Cultivars

   'Amethyst' – lilac purple spikes, outstanding color
                 0.9 m (3 ft.)   *July, Aug.*
   'Avalanche' – pure white, vigorous, free-flowering
                 0.9 m (3 ft.)   *July, Aug.*
   'Bonn' – deep rose pink spires   *0.8 m (2½ ft.)*   *July*
   'Cattleya' – light pink   *0.9 m (3 ft.)*   *July, Aug.*
   'Coblence' – early, fiery carmine red
                 0.6 m (2 ft.)   *June, July*
   'Cologne' – quite compact, deep carmine rose
                 0.6 m (2 ft.)   *June, July*
   'Deutschland' – very good white   *0.6 m (2 ft.)*   *June, July*
   'Dusseldorf' – bright salmon pink
                 0.6 m (2 ft.)   *June, July*
   'Elna' – dwarf, deep crimson   *0.8 m (2½ ft.)*   *June, July*
   'Erica' – clear pink   *0.6 m (2 ft.)*   *July, Aug.*
   'Fanal' – dark garnet red   *0.8 m (2½ ft.)*   *June, July*
   'Fire' – late, dark red   *0.8 m (2½ ft.)*   *Aug.*
   'Granat' – graceful spikes of deep rose pink
                 0.9 m (3 ft.)   *June, July*
   'Intermezzo' – dwarf, salmon rose
               46 cm (18 in.)   *June, Aug.*
   'Irene Rotsieper' – tall, graceful, buff pink
                 0.9 m (3 ft.)   *July, Aug.*
   'Jo Orphorst' – late-flowering, ruby red
                 0.9 m (3 ft.)   *July, Aug.*
   'Ostrich Plume' – free-flowering, pink
                 0.8 m (2½ ft.)   *July, Aug.*

Astilbe 'Irene Rotsieper'

'Peach Blossom' — pale peach pink

	0.6 m (2 ft)	July, Aug.
'Pink Pearl' — pale pink	0.9 m (3 ft.)	July, Aug.

'Red Sentinel' — spikes of brick red

	0.6 m (2 ft.)	July, Aug.
'Rhineland' — rich rose pink	0.9 m (3 ft.)	June, July
'Tamarix' — late, tall, pink	1.1 m (3½ ft.)	July, Aug.
'Venus' — flesh pink	0.9 m (3 ft.)	July, Aug.

'Vesuvius' — flowering salmon red

	0.6 m (2 ft.)	July, Aug.
'W. D. Willen' — white	0.6 m (2 ft.)	July, Aug.
'White Queen' — fine, white	0.8 m (2½ ft.)	July, Aug.

'William Reeves' — dark crimson, sport of 'Granat'

	0.9 m (3 ft.)	June, July

ASTRANTIA masterwort

This genus represents a small group of Eurasian herbs that belongs to the carrot family. The plants have umbels of small individual flowers that form an attractive flower cluster, which is even more attractive because of the colored bracts beneath.

A. carniolica 'Major'   0.9 m (3 ft.)   June, Aug.
Pale green rose flowers. Synonym: *A. major*

'Maxima'   0.9 m (3 ft.)   June, Aug.
Strong-growing; light rose flowers.

'Rubra'   0.6 m (2 ft.)   June, Aug.
Dwarf; dark red flowers.

'Shaggy'   0.8 m (2½ ft.)   June, Sept.
Very free-flowering white.

A. major see *Astrantia carniolica* 'Major'

AUBRIETA false rock-cress   (Color photo p. 75)

These dwarf plants are usually planted in rock gardens, but are also used in the front of well-drained borders, either as spot plants or as an edging. They have purplish, rose, or lavender flowers, and they may be raised from seeds. The selected cultivars with large flowers and much more brilliant colors must be increased by cuttings or divisions.

A. deltoidea   15 cm (6 in.)   Apr., May
Mat-forming, light purple flowers. Greece.

Cultivars   15 cm (6 in.)   Apr., May
  'Argenteo-variegata' — white variegated leaves
  'Aureo-variegata' — bright yellow and green variegated leaves
  'Bonfire' — rich crimson
  'Borsch's White' — the first white variety
  'Carnival' — large, deep violet
  'Crimson Bedder' — neat-growing, crimson
  'Dr. Mules' — an old favorite, large purple flowers
  'Gloriosa' — rosy pink
  'Gurgedyke' — very good, deep violet
  'Lilac Time' — strong habit, lilac blue
  'Mary Poppins' — new, double, pink, with real merit
  'Maurice Pritchard' — pale rosy pink
  'Oakington Lavender' — pleasing, lavender blue
  'Studland' — compact, light mauve blue
  'Wanda' — double, red

BAPTISIA false indigo

False indigo is a plant that is a little different and yet easy to grow. A very stout plant with compound leaves and pea-like flowers in showy racemes. Thrives in the drier part of the border in full sun and open porous soil. Once they become established, they are best increased by division, but may also be grown from seeds sown in early spring.

B. australis blue false indigo   0.9-1.2 m (3-4 ft.)   May, June
Arching terminal racemes of blue flowers; attractive trifoliate leaflets forming compound leaves.

  'Old Orchard' hybrids   0.9 m (3 ft.)   May, June
  A seedling strain, from tawny violet and buff to yellow and several shades of blue.

B. bracteata   46 cm (18 in.)   June
A rather rare species having axillary racemes of yellow flowers.

B. leucantha   1.2 m (4 ft.)   June

White flowers on 1.2-m (4-ft.) stems; needs partial shade and adequate moisture.

B. tinctoria   1.2 m (4 ft.)   June, July

Easy to grow; racemes of bright yellow pea-shaped flowers. Its leaves are not dense, but rather straggly; a beautiful plant in midsummer.

BELAMCANDA blackberry-lily

An iris-like plant that produces small star-shaped flowers and fruits with blackberry-like seeds.

B. chinensis blackberry-lily,
leopardflower   0.6-0.9 m (2-3 ft.)   July, Aug.

Unique, small, red-spotted orange flowers that wither and die very quickly, but are followed by black fruits that glisten and shine in the autumn sunshine and produce an effect equal to many flowering plants.

B. flabellata   0.6 m (2 ft.)   July, Aug.

Clear yellow flowers, blackberry-like fruits similar to *B. chinensis.* Not very common in gardens.

BELLIS   (Color photo p. 75)

B. perennis English daisy   10 cm (4 in.)   all summer

The common English daisy is a weed that is sometimes a problem in lawns in England and in very temperate climates, such as in parts of British Columbia. The large form 'Monstrosa' is grown and admired in gardens all over the world; it is often called double daisy, from which the following cultivars have been developed:

'Dresden China'   5 cm (2 in.)   Apr., June
Delicate pink, quilled petals.

'Rob Roy'   5 cm (2 in.)   Apr., June
Double; much larger than 'Dresden China'; crimson.

BERGENIA giant rockfoil   (Color photo p. 75)

A genus of plants long known as megaseas or large-leaved saxifrages. The bergenias are rather attractive plants for a border; their large cabbage-like leaves are quite ornamental, green in summer and purple bronze in winter. In early spring they produce nodding clusters of rose, pink, or purple bloom on very thick fleshy stems. They need no special care, but grow well in any soil, in shade or full sun. All are very hardy.

B. ciliata   46 cm (18 in.)   Apr.
Large sprigs of pale pink flowers. Synonym: *B. ligulata*

B. cordifolia   30 cm (12 in.)   Apr., May
Bright rose flowers in large, drooping sprays; round, heart-shaped leaves with wavy or undulated margins.

B. crassifolia   15-41 cm (6-16 in.)   May, June
Pink flowers high above the foliage; leaves similar to, but not as rounded, as *B. cordifolia.*

B. ligulata see *Bergenia ciliata*

B. stracheyi   30 cm (12 in.)   Apr., May
Very large sprays of appleblossom pink flowers.

BOCCONIA

B. cordata see *Macleaya cordata*

BOLTONIA false starwort

These are tall plants that bloom at the same time as perennial asters. They are excellent companions because they have wide spreading flowering branches and are tall enough to use as background plants. They are very easily grown, and need dividing less often than perennial asters.

B. asteroides   1.2-1.5 m (4-5 ft.)   Aug., Sept.
White to pinkish flowers.

'Snowbank'   1.2 m (4 ft.)   Aug., Sept.
As its cultivar name suggests, it forms a bank of snow-white flowers.

Caltha palustris, common marsh marigold

B. latisquama   *1.2-1.5 m (4-5 ft.)*   *Aug., Sept.*
Large light blue flowers.

   'Nana'   *0.9 m (3 ft.)*   *Aug., Sept.*
   Smaller than the species; pale pink flowers.

BOYKINIA

A small genus that is particularly good in a shaded garden. It has creeping rootstocks and lobed or divided leaves. The flowers are not exceptionally showy, but they are dainty.

B. aconitifolia   *0.6-0.9 m (2-3 ft.)*   *July*
A good plant for a woodland garden. Small white flowers in open clusters, rounded five-lobed leaves 10-15 cm (4-6 in.) across.

B. jamesii
Reddish purple flowers. Synonym: *Telesonix jamesii*

B. rotundifolia   *30 cm (12 in.)*   *May, June*
As its specific name suggests, it has rounded leaves, and feathery plumes of white flowers arising from mounds of graceful foliage.

BRUNNERA

B. macrophylla Siberian bugloss   *30 cm (12 in.)*   *May, June*
This species is the only ornamental member of the genus. Some nurserymen also know it by the name *Anchusa myosotidiflora*. Its common name suggests a plant similar to wild blueweed, but it is very much like, though more graceful than, the forget-me-not *(Myosotis)*. Its showy clusters formed by myriads of blue flowers are especially useful for combining with naturalized narcissus, in semishady areas or as a carpet for early flowering perennials such as doronicum, or for shrubs like golden bells *(Forsythia)*. They grow best in partial shade and tolerate most soils, and will survive in the Prairie Provinces most years.

BUPHTHALMUM oxeye

This genus contains two easy-to-grow perennials that have brown-eyed yellow daisies. The species are considered medium hardy at Morden, Manitoba.

B. salicifolium willowleaf oxeye   *0.6 m (2 ft.)*   *Aug.*
Large solitary yellow flowers 5 cm (2 in.) across; narrow willow-like hairy leaves.

B. speciosum heart-leaved oxeye   *1.2 m (4 ft.)*   *Aug.*
A good plant for the back of a border or for growing in a woodland garden because it withstands part shade. Very effective large heart-shaped leaves and deep yellow flowers.

CALAMINTHA calamint

These are small, creeping herbs that have fragrant foliage. They are best used in borders where utilitarian plants are preferable.

C. alpina alpine calamint   *10 cm (4 in.)*   *June, Aug.*
Deep mauve blue flowers.

C. grandiflora   *23 cm (9 in.)*   *July, Sept.*
More erect than alpine calamint; deep rose flowers.

C. nepeta   *30 cm (12 in.)*   *July, Aug.*
Catmint-like foliage; lavender flowers. Synonym: *C. nepetoides*

C. nepetoides see *Calamintha nepeta*

CALTHA marsh marigold

Marsh plants that need moisture; yellow or white flowers, fresh green heart-shaped leaves.

C. leptosepala   *30 cm (12 in.)*   *Apr., May*
A herb that grows in marshes from New Mexico to Alaska; single bloom of white tinged with blue. Not as adaptable to comparative dryland conditions as the common marsh marigold.

C. palustris
common marsh marigold   *30 cm (12 in.)   Apr., May*
One of the earliest spring-flowering plants, it is of tremendous value in gardens for planting alongside pools and streams and for problem areas where drainage is poor and near-bog conditions exist. Deep golden yellow flowers, which, even when growing wild, have great variations of form and intensity of color.

'Alba'   *23 cm (9 in.)   Apr., May*
Neater habit than the other cultivars, but not as floriferous.

'Holubyi'   *23 cm (9 in.)   Apr., May*
Neater habit than the other cultivars.

'Plena'   *30 cm (12 in.)   Apr., May*
Double, yellow, extremely showy flowers; needs rich moist soil to produce good showy clumps.

CAMPANULA bellflower   (Color photo p. 75)

There are so many species and cultivars of bellflowers that one may be selected from the group to serve any purpose in the garden. Some of these plants are extremely dwarf or trailing, useful for wall plantings, others are very tall with long spires of blooms, or showy biennials, useful for larger effects. Most bellflowers are very hardy.

C. alliarifolia spurred bellflower   *0.6 m (2 ft.)   June, July*
Bell-shaped nodding creamy white flowers, hairy oval leaves; suitable for very heavy shade.

C. carpatica
Carpathian bellflower   *30-46 cm (12-18 in.)   June-Sept.*
Very popular for rock gardens and borders. Useful for planting in groups in the front of a border, where its neat tufts of foliage studded with blue, purple, or white flowers are attractive during the summer. Very hardy, adaptable to all kinds of soils and conditions, but best in a sunny well-drained location.

Cultivars
'Alba' – white flowers   *30 cm (12 in.)   June-Sept.*

'Blue Carpet' – clear blue flowers   *30 cm (12 in.)*

'Ditton Blue' – compact dwarf plants;
large bells of deep indigo flowers   *15 cm (6 in.)*

'Harvest Blue' – excellent, violet blue   *23 cm (9 in.)*

'Queen of Somerville'–very fine, pale blue   *38 cm (15 in.)*

'Riverslea' – deep purple cups   *30 cm (12 in.)*

C. carpatica var. *turbinata*   *15 cm (6 in.)   June-Sept.*
Smaller than the species, more compact; deep purple flowers. Transylvania.

'Pallida'   *15 cm (6 in.)   June-Sept.*
A form of the variety *turbinata;* light china blue flowers. Transylvania.

'White Star' – very large white flowers   *30 cm (12 in.)*

C. cochlearifolia   *10 cm (4 in.)   June-Sept.*
This diminutive creeping bellflower is included here specifically because it is useful for trailing over walls, in crevices, or as edgings for patios. Nodding pale blue flowers. Synonym: *C. pusilla*

'Alba' – white flowers   *10-15 cm (4-6 in.)   June-Sept.*

'Miranda'   *10 cm (4 in.)   June-Sept.*
Singularly beautiful little plant producing abundant small silver blue outward-facing bellflowers. Blooms all summer if the flowers are removed as they fade.

'Miss Willmott'   *15 cm (6 in.)   June-Sept.*
Larger bells than the other forms; gray blue flowers.

C. collina   *23 cm (9 in.)   May, June*
Small, compact, tufted, for rock gardens; pendulous purple funnel-shaped flowers.

C. elatines var. *garganica*   *15 cm (6 in.)   July*
Very useful for rock gardens, where its flat, blue star-shaped blossoms forming masses of color are best displayed. Mount Gargano in the Italian Alps.

'Hirsuta'   *30 cm (12 in.)   July*
Light blue flowers, very hairy leaves; grayish silver appearance. Synonym: *C. istriaca*

'W. H. Paine'   *15 cm (6 in.)   July*
White-centered dark blue flowers.

C. glomerata clustered bellflower   *0.6 m (2 ft.)   June, July*
Very hardy; flowers clustered around the top of the stems and in the leaf axis. Because it usually has 12 flowers in each flower cluster, it is known in England as twelve apostles. Its large oval leaves 10-13 cm (4-5 in.) across are a little too coarse for a flower border, but the cultivars are a little showier and more refined.

'Acaulis'   *15 cm 6 in.   June, July*
Attractive, dwarf, compact tufts of much smaller and more graceful leaves than the species. Minute blue violet flowers.

'Dahurica' Dane's-blood   *0.6 m (2 ft.)   June, July*
Much larger deep rich purple flowers to 8 cm (3 in.) across.

'Joan Elliot'   46 cm (18 in.)   June, July
Violet blue flowers, good for cutting.

'Nana Alba'   38 cm (15 in.)   June-Aug.
Dwarf; producing white flowers over a much longer period
than the species.

'Nana Lilacina'   38 cm (15 in.)   June-Aug.
Soft lilac blue flowers.

'Purple Pixie'   38 cm (15 in.)   July, Aug.
New, perfectly formed violet purple flowers on 38-cm
(15-in.) stems.

'Superba'   0.8 m (2½ ft.)   June, July
Clusters of bright violet flowers.

C. istriaca see Campanula elatines var. garganica 'Hirsuta'

C. lactiflora milky bellflower   0.9-1.2 m (3-4 ft.)   June-Aug.
Despite its specific name this species has no more milky
latex than most bellflowers. It is a good, though not too well-
known, perennial. It thrives in part shade and bears an
abundance of milky white or very pale blue flowers in loose
panicles on stiff erect stems. If you char the stems when you
cut them, the flowers will last well in water. Useful for floral
arrangements.

'London Anna'   0.9-1.2 m (3-4 ft.)   June-Aug.
Flesh pink; well worth planting in rich deep soil.

'Pouffe'   23 cm (9 in.)   June-Aug.
An amazing dwarf, in neat mounds of light blue flowers.

'Pritchard's Variety'   0.9 m (3 ft.)   June-Aug.
Stiff, erect; very fine in deep blue flowers.

C. latifolia broad-leaved bellflower   0.9 m (3 ft.)   June, July
Very coarse, broad-leaved. It may become quite weedy because its seeds are likely to scatter throughout the garden.
In a wild garden, where it is not likely to be a pest, its stately
spires of purple flowers are attractive. It lasts well and flourishes in poor soils in semishaded locations. The cultivars
are more attractive and less weedy than the species.

'Alba'   0.9 m (3 ft.)   June, July
Somewhat muddy white.

'Brantwood'   0.8 m (2½ ft.)   June, July
Compact; stately spires of rich violet purple bells.

'Gloaming'   0.8 m (2½ ft.)   June, July
Similar to 'Brantwood,' but having pale mauve flowers.

'Macrantha'   0.9-1.2 m (3-4 ft.)   June, July
Larger purplish flowers than the species.

C. latiloba see Campanula persicifolia
See Chapter 3.

C. muralis see Campanula portenschlagiana

C. persicifolia
peach-leaved bellflower   0.3-0.9 m (1-3 ft.)   June-Aug.
The best campanula for borders; blue or white flowers in
large terminal clusters, narrow finely toothed leaves. Good
border plant may be easily obtained from seeds, but the cultivars obtained by vegetative propagation are much larger, the
colors are much more uniform, and the plants more vigorous.
The basal rosettes should be divided every 2 or 3 years.
Synonym: C. latiloba

Cultivars
                                              June-Aug.
'Beechwood' — vigorous, large, pale blue 0.6-0.9 m (2-3 ft.)

'Blue Bells' — deep blue   0.3-0.6 m (1-2 ft.)

'Blue Gardenia' — double, deep blue   0.6-0.9 m (2-3 ft.)

'Cantab' — very vigorous, clear light blue
                                          0.6-0.9 m (2-3 ft.)

'Fleur de Neige' — large, double, glistening white
                                          0.6-0.9 m (2-3 ft.)

'Moerheimi' or 'Summer Skies' — double, white
flowerhead with blue   0.6-0.9 m (2-3 ft.)

'Mount Hood' — double, white   0.8 m (2½ ft.)

'Mrs. H. Harrison' — Canadian, very large
double blooms, medium blue   0.6-0.9 m (2-3 ft.)

'Pride of Exmouth' — double, dark blue   0.8 m (2½ ft.)

'Snowdrift' — single, pure white   0.8 m (2½ ft.)

'Telham Beauty'   0.6-0.9 m (2-3 ft.)
Introduced from England in 1916, beautiful, 8-cm (3-in.)
bells of rich blue, a great favorite in borders.

'Wirral Belle' — silvery blue violet   0.6-0.9 m (2-3 ft.)

C. portenschlagiana
Dalmation bellflower   10 cm (4 in.)   June-Aug.
Very robust, semiprostrate; grows vigorously in sandy soils
of light purple blue bells on compact clumps of small foliage
all summer. Used mainly in dry walls and the alpine section
of rock gardens. Synonym: C. muralis

C. poscharskyana   *15 cm (6 in.)*   *June-Aug.*

Very robust, semiprostrate; grows vigorously in sandy soils in dry walls, sandy slopes, and large rock gardens. Useful in the front of a border, especially in light shade. Its star-shaped sprays of light blue flowers are abundant all summer.

'Lisduggan'   *23 cm (9 in.)*   *June-Aug.*

Taller than the species; lavender pink flowers.

'Stella'   *15 cm (6 in.)*   *June-Aug.*

Much more compact and having brighter flowers than the species.

C. punctata spotted bellflower   *0.3-0.6 m (1-2 ft.)*   *July, Aug.*

Robust; hairy-stemmed, broad coarsely toothed leaves, the basal leaves on long stalks. Flowers 5 cm (2 in.) long, nodding, bell-shaped, white or mauve, spotted inside. Because this plant spreads from underground roots, it might become weedy. The main reasons for growing it are its large very attractive though dull-colored flowers, and it is very easily obtained from seed.

C. pusilla see *Campanula cochlearifolia*

C. rotundifolia harebell,

blue bells of Scotland   *30-38 cm (12-15 in.)*   *July-Sept.*

Like many Scotsmen this plant is found in many parts of the world. On roadsides in Canada some very fine forms may be found growing wild. At best it is a graceful plant 38 cm (15 in.) high with neat foliage and nodding blue and white flowers that are produced sparingly all summer. The plants do best in sun or part shade.

'Purple Gem'   *38 cm (15 in.)*   *July-Sept.*

Free flowering; deep purple bells.

'Olympica'   *38 cm (15 in.)*   *July-Sept.*

Compact plants; deep lavender blue flowers. Olympia Mountains, Washington, U.S.A.

C. trachelium Coventry bells   *0.6-0.9 m (2-3 ft.)*   *July, Aug.*

Rough hairy perennial; narrow oval leaves 5 cm (2 in.) long, coarsely toothed nodding bluish purple flowers, 3 cm (1 in.) long, in loose clusters. Sometimes weedy.

'Bernice'   *0.6 m (2 ft.)*   *July, Aug.*

Double, powder blue; much superior in every way to the type, showier flower clusters, more compact form, never weedy, and much less rampant.

CATANANCHE

C. caerulea cupid's-dart   *0.6 m (2 ft.)*   *July, Aug.*

Excellent long-lasting cornflower-like cut flowers; should be grown a great deal more than they are. Blue flowers very effective in dried bouquets; also beautiful gray foliage, silvery buds; best in full sun in dry well-drained soil. Usually hardy on the prairies.

'Major'   *0.8 m (2½ ft.)*   *July, Aug.*

Rich lavender blue flowers, larger than the species.

'Perry's White'   *0.8 m (2½ ft.)*   *July, Aug.*

The best white variety.

CAUTLEYA

C. robusta   *1.2 m (4 ft.)*   *Aug., Sept.*

Very stiff spikes of yellow flowers.

CELMISIA

This is a remarkably distinct genus of plants native to New Zealand that belong to the Compositae family. It will not withstand the hot summers or cold winters across most of Canada, but it should be tried in the milder parts of British Columbia. The glistening white daisy-like single flowers usually bloom in April or May. The beautiful rosetted leaves are evergreen and leathery. In New Zealand a large number of species belonging to this genus are available commercially, but many of them are hard to transplant.

C. spectabilis   *38 cm (15 in.)*   *Apr., May*

The only species commercially available, but sold only in England. Leaves 15 cm (6 in.) long, very ridged, covered with silvery white tomentum; glistening white flowers; a most attractive plant all year.

CELSIA

A group of mullein-like perennials that should be treated as biennials, because they are short lived and will not survive cold wet winters. Their flowers are spiky terminal racemes, usually yellow and bracted. It is best to start them indoors and then plant them in the garden during June. They will flower the next year.

C. acaulis   *5 cm (2 in.)*   *May, June*

Crinkled green rosettes, large yellow mullein-like flowers, prominent orange anthers. Ground-hugging plants with spikes 20-30 cm (8-12 in.) high.

C. arcturus Cretan bear's-tail   *46 cm (18 in.)*   *June, July*

Oval leaves either toothed or cut in lyre-like shape. Yellow flowers, prominent purple stamens stalked in a loose raceme.

C. cretica Cretan mullein   *0.9-1.5 m (3-5 ft.)*   *June, July*
Large, mullein-like; spikes of almost stalkless yellow flowers, purple anthers.

CENTAUREA hardy bachelor's button, knapweed   (Color photo p. 75)   *July-Sept.*

Renowned for hardiness, ruggedness, ease of culture, and adaptability. Thistle-like flowers freely produced if faded flowers cut off. Best in poor well-drained soils and full sun. All species drought resistant, most species very hardy except *C. dealbata*, which needs protection on the prairies.

C. dealbata Persian centaurea   *0.6 m (2 ft.)*   *June-Sept.*
Very handsome leaves, dark green above, silvery beneath; large rosy lilac thistle-like flowers, very delicate texture.

    'John Coutts'   *0.6 m (2 ft.)*   *June-Aug.*
    New cultivar; large bright pink flowers.

    'Sternbergii'   *0.8 m (2½ ft.)*   *July, Aug.*
    Very attractive, much more compact than the type; deeply notched rosy red florets surrounding a central disk of white.

C. glastifolia   *1.2 m (4 ft.)*   *June-Aug.*
Fluffy heads of yellow flowers.

C. macrocephala globe knapweed   *1.2 m (4 ft.)*   *June-Aug.*
Large, coarse, showy; yellow thistle-like flowers 10 cm (4 in.) across, large ragged oval leaves; best planted individually, not in massive groups.

C. montana mountain bluet   *0.6 m (2 ft.)*   *May-Aug.*
Large cornflower-like blue flowers, large oblong leaves. Resembles an enlarged annual cornflower but a more rugged vigorous plant.

    'Alba'   *0.6 m (2 ft.)*   *June-Aug.*
    Not as showy as the species; subdued white flowers.

    'Parham's'   *0.6 m (2 ft.)*   *June-Aug.*
    Large lavender purple flowers.

    'Rosea'   *0.4-0.6 m (1½-2 ft.)*   *June-Aug.*
    Pale rose pink flowers.

    'Violetta'   *0.6 m (2 ft.)*   *June-Aug.*
    Deep rich blue flowers.

CENTRANTHUS

C. ruber red valerian   *0.4 m (1½ ft.)*   *June-Sept.*
Useful for hot dry locations; gray green leaves, red flowers, produced intermittently all summer. For compact plants, poor but well-drained soil needed. The generic name refers to the small spurs at the base of the corolla.

'Albus' — muddy white flowers   *0.6 m (2 ft.)*   *June-Sept.*

'Atrococcineus' — bright crimson scarlet flowers   *0.9 m (3 ft.)*   *June-Sept.*

CEPHALARIA   (Color photo p. 76)

C. gigantea giant scabious   *1.2-1.8 m (4-6 ft.)*   *July-Sept.*
Very tall, robust, vigorous; scabrous-like yellow inflorescences borne singly on long stalks, feathery leaves deeply cut into many-toothed segments. Growing in any soil, useful as background specimen in large borders. *C. alpina* is often mistaken for *C. gigantea*. Commercially known as *C. tatarica*.

CERASTIUM chickweed

Of the 100 or more species of chickweeds all but three are weedy herbs, as their name implies, and should be avoided.

C. alpinum 'Lanatum'   *5 cm (2 in.)*   *June, July*
A true alpine; dense woolly tufts of leaves, large white flowers. Difficult to establish except in pockets made up of regular alpine scree.

C. biebersteinii   *15 cm (6 in.)*   *May, June*
Very much like snow-in-summer *(C. tomentosum)* but larger, slightly fragrant flowers, less rampant.

C. tomentosum snow-in-summer   *15 cm (6 in.)*   *June*
Silvery leaved, covered with white flowers. Very useful in front of the border when kept in check by cutting back severely after flowering. Excellent ground cover for sunny dry slopes, ideal for trailing over walls.

CERATOSTIGMA

C. plumbaginoides leadwort   *30 cm (12 in.)*   *Sept., Oct.*
Semiprostrate, shrubby herb; deep blue flowers somewhat resembling the greenhouse plumtago; broad ovate leaves, reddish in autumn. *C. willmottiana*, a woodier species, with light blue flowers on bushes 0.9-1.2 m (3-4 ft.) high, is grown in California. Both species should be adaptable for growing in the milder parts of British Columbia.

CHEIRANTHUS

See Chapter 3, Biennials.

CHELONE turtlehead

This is a North American perennial with irregular 2-lipped flowers on compact terminal spikes, similar to the penstemon. They grow best in moist semishaded woodlands.

C. digitalis see *Penstemon digitalis*

C. glabra turtlehead   *0.6 m (2 ft.)*   *Aug.-Oct.*
Oblong lanceolate leaves, white flowers about 3 cm (1 in.) long. Native from Newfoundland to Georgia and westward.

C. lyonii red turtlehead   *0.9 m (3 ft.)*   *Aug.-Oct.*
Rich dark green leaves, rose purple flowers. Best in shaded, good garden soils.

CHRYSANTHEMUM   (Color photos pp. 13, 76, 77)

This genus covers a very wide range of plants from the huge mop-like Japanese type of greenhouse chrysanthemums to the tall, hardy, moisture-loving moon daisy *(C. uliginosum)* and the dwarf 0.3-m (1-ft.) pink daisy *(C. mawi)*. So many species have been interbred to produce the garden chrysanthemum that the name *C. morifolium,* which has been appended to the group, hardly applies any more, but is still used for both the greenhouse and the hardy plant. Cultivation varies so much from species to species that it is given under each specific group.

C. arcticum Arctic daisy   *30 cm (12 in.)*   *Sept., Oct.*
Very hardy, for borders, useful for fall decoration. Sprawling, prostrate, upright stems to 0.3 m (1 ft.) high, white or lilac aster-like flowers. Very good dwarf for the front of the border, but flowering too late for a good fall effect.

C. balsamita
costmary, mint geranium   *0.6-0.9 m (2-3 ft.)*   *July*
Grown mainly for its fragrant foliage. Insignificant flowers without showy ray florets.

C. coccineum
pyrethrum, painted daisy   *0.3-0.6 m (1-2 ft.)*   *June, July*
Very fine flowers in gardens and excellent cut flower for florists. Dark green fine-textured leaves almost as attractive as the flowers. White, pink, rose, or red flowers; single with prominent yellow dishes, anemone type with tubular central dish, or double in self colors, or occasionally with the central florets and ray florets in contrasting colors. Many cultivars, ranging from small 23-cm (9-in.) plants to tall 0.9-m (3-ft.) specimens. In borders they are best planted in groups of three of the same cultivars 0.3 m (1 ft.) apart.

The plants do best in well-drained enriched soil in full sun or part shade.

After the flower fades the stem should be cut well back, almost to ground level, to induce a second flowering period. Clumps are best divided in late summer every 4 or 5 years. After this operation a protective covering of straw is often necessary.

Single cultivars   *June, July*
except where noted, 0.8 m (2½ ft.)

  'Agnes M. Kelway' — large, bright rose
  'Allurement' — large, light pink
  'Avalanche' — pure white
  'Brenda' — bright cerise, very robust
  'Comet' — early flowering, crimson
  'Crimson Giant' — velvety crimson flowers 10 cm (4 in.) across, 0.9 m (3 ft.)
  'Eileen May Robinson' — light salmon pink, long stem, 0.9 m (3 ft.)
  'Evenglow' — vigorous, salmon red
  'Harold Robinson' — large, rich crimson
  'H. M. Stanley' — blood red
  'Inferno' — intense salmon scarlet
  'Jubilee Gem' — salmon pink
  'Kelway Glorious' — early flowering, rich scarlet
  'Mrs. D. D. Bliss' — coral orange
  'Scarlet Glow' — large, brilliant scarlet, long stems

Double cultivars   *0.8 m (2½ ft.)*   *June, July*
  'Apollo' — large, warm cherry pink
  'Buckeye' — white-speckled deep rose red
  'Carl Vogt' — pure white, used by florists
  'Helen' — soft rose pink
  'J. M. Tweedy' — large, crimson
  'Madeleine' — pale pink
  'Marjorie Deed' — pale reddish purple, crimson center
  'Mrs. E. C. Beckwith' — white
  'Pink Bouquet' — rose pink, silvery center

113

derived from these crosses probably depends upon the inheritance of hardiness given from one or more of these species. In Ottawa the common single-flowered Shasta daisy is hardy, but many of the more refined hybrids are quite short lived.

All the cultivars are useful as cut flowers, and indispensable in flower borders in areas where they are hardy. They do best in good, well-drained, rich, moist soil in part shade, especially the double-flowered specimens. The plants should be divided every 2 or 3 years.

Single cultivars — The flowers, 8-18 cm (3-7 in.) across, may have two rows of ray florets and a prominent yellow dish; 0.8 m (2½ ft.) high except where noted.

'Edgebrook Grant' — flowers 18 cm (7 in.) across	*June, July*
'Everest' — large, single, white, 0.9 m (3 ft.) high	*July, Aug.*
'H. Siebert' — enormous, single, white, frilled petals	*June, July*
'King Edward VII' — very hardy, large blooms	*July, Aug.*
'Majestic' — single bloom, 10 cm (4 in.) across	*June, July*
'Mark Riegel' — flowers 10-13 cm (4-5 in.) across, two rows of petals	*July, Aug.*
'Stone Mountain' — flowers 10 cm (4 in.) across, stocky plants	*June, July*

Double cultivars — The double blooms consist of a central cushion of flowers on modified petals, which may be incurved to form an anemone-centered type or long enough to look like a double garden chrysanthemum; flowering in July and August.

'Aglaia' — large flowers, frilly	*0.8 m (2½ ft.)*
'Beaute Nivelloise' — frilled petals, plum-like appearance	*0.8 m (2½ ft.)*
'Cobham Gold' — creamy white	*0.9 m (3 ft.)*
'Esther Read' — popular, double white	*0.8 m (2½ ft.)*
'Horace Read — double, creamy white	*0.8 m (2½ ft.)*
'Ian Murray' — anemone-centered	*0.8 m (2½ ft.)*
'Jennifer Read' — later, stronger growing than 'Esther Read'	*0.9 m (3 ft.)*
'Marconi' — fluffy flowers 15 cm (6 in.) across	*0.8 m (2½ ft.)*
'Sedgewick' — double, white, very hardy	*0.4 m (1½ ft.)*
'Thomas E. Killeen' — large, white, anemone-centered	*0.9 m (3 ft.)*
'Wirral Pride' — strong stemmed, semidouble, white	*0.9 m (3 ft.)*

'Queen Mary' — salmon-tinted pink
'Senator' — fully double, deep pink
'Vanessa'— fully double, rosy carmine, orange-flushed center
'Venus' — pure white
'White Madeleine' — pure white

C. coreanum see *Chrysanthemum sibiricum*

C. corymbosum
Caucasian daisy, oxeye daisy   *0.6 m (2 ft.)   July*
Very hardy; finely cut foliage, white flowers.

C. maximum Shasta daisy   *0.3-1.1 m (1-3½ ft.)   June-Aug.*
The large, white, single and double daisies classified under this specific epithet are mostly hybrids in which *C. maximum, C. lacustre, C. leucanthemum*, the common oxeye daisy, and *C. nipponicum* were blended. The hardiness of the cultivars

'Wirral Snowball' — fully double, pure white   *0.6 m (2 ft.)*

'Wirral Supreme' — strongest growing of the
double cultivars   *0.8 m (2½ ft.)*

C. morifolium

The plants classified under this specific epithet include inter-specific and intervarietal hybrids and several hybrids of allied species such as *C. arcticum, C. nipponicum, C. rubellum, C. sibiricum, and C. indicum;* the latter is probably a derivative of *C. morifolium* or the converse. Many of the cultivars bloom early enough to be valuable in gardens.

Breeders of chrysanthemums for outdoors have improved the value of the plants as garden flowers for landscape plantings. These very shapely plants produce a massive floral effect and perfectly formed flowers on fairly short stems. They are available in several colors, such as bright glowing fall bronzes or yellows and pastel shades of pink or mauve, and in cultivars that ultimately form spreading mounds for the front of borders and taller mounds for the rear.

When you select varieties for your garden, notice the flowering dates given in the catalog. If you live in Eastern Canada, select those kinds that reach maturity no later than September 20. In southern Ontario this date may be later, but on the prairies and other colder regions the date should be much earlier. Prairie gardeners have little difficulty in selecting outdoor chrysanthemums, because some of the best cultivars were originated in Brandon, Lethbridge, and Morden.

Garden chrysanthemums thrive in well-drained garden loam that has been improved by the addition of lots of organic matter such as peat, leaf mold, well-rotted manure, or compost. They like full sun but will flower well with only 6 hours of sunlight a day.

Get the plants from your garden center or nurseryman by the end of May and plant them by mid-June according to the planting dates for annuals in your area. Most of the plants will arrive in peat pots filled with a mixture of peat and perlite. Soak the pot well and tear away part of the top of the pot, then set the pot and the plant in the soil.

During the summer put on a mulch of peat, leaf soil, buckwheat hulls, or cocoa shells to keep the soil moist and cool and to conserve moisture.

The most important operation in growing outdoor chrysanthemums is to pinch out the terminal shoot when the plant is 13-15 cm (5 or 6 in.) tall. This causes the plant to branch and send up more shoots. Until the middle of July pinch back the new shoots as soon as they reach 13-15 cm (5 or 6 in.). You will then have a bushy plant with lots of blooms.

When the first flower buds appear after this last pinching give the plants a side-dressing of a complete fertilizer at 1 kg/10 m² (2 lb. to 100 sq ft.) of area.

Green and black flies are sometimes pests of chrysanthemums, but the most serious pests are tarnished plant bugs and leafhoppers. These insects attack the buds and cause them to become deformed so that sometimes only half of each flower develops. For up-to-date recommendations on the use of chemicals for controlling insects, ask your provincial agricultural representative.

Virus diseases are harder to control than insects and can be eliminated only by special selection and treatment during the stolon and cutting stages by commercial growers. The stunt disease dwarfs the plant, its leaves, and flowers. All parts become so distorted that the plants are no longer ornamental. Mosaic virus can usually be distinguished by a yellowish green mottling of the leaves. Pull out and destroy all plants that appear to be infected with this disease, and buy new plants each spring.

To overwinter hardy garden chrysanthemums in Eastern Canada, leave the tops on and cover the plants with evergreen boughs after the ground has frozen. Or, dig up the plants and replant them near the warm foundation of your house. The latter method is also successful on the prairies.

More vigorous cultivars in new colors and of neater mound shapes are being developed. The cultivars that rated highest in the test garden of the Plant Research Institute during the past 3 years follow.

bronze 'Brown Eyes', 'Campaigner', Roll Call', and 'Zonta'

pink   'Coquette', very early; 'Knock-out', very uniform; 'Pink Haze', a large-flowered dwarf type; 'Pink Sentinel', silvery pink pompom; 'Salmon Minn Pink', small flower cushion

purple 'Purple Pirate' and 'Purple Star', dark reddish purple, star-shaped flowers; 'Purple Waters', crimson purple, 8-cm (3-in.) blooms

red   'Autumn Song', early crimson; 'Redskin', a large red harvest giant football type; 'Ruby Mound', the best of all mound types; 'Stadium Queen', large incurved blooms

white 'Christopher Columbus' and 'White Cushion', very early; 'Powder River', large-flowered white; and 'Larry', large, white, decorative

yellow 'Astoria', vivid yellow; 'Golden Carpet', cushion type; 'Ontario Nugget', yellow pompom; 'Muted Sunshine', anemone-centered bright yellow, decorative; and 'Prairie Sun', early golden yellow

The best cultivars for the prairies are 'Bonnie Brandon', bright yellow; 'Joan Brandon', bronzy yellow; 'Julie Brandon', reddish orange; 'Acaena', red bronze; 'Ahnasti', purplish pink; 'Beckethon', deep pink; 'Moeis', pale bronzy pink; and 'Nootka', bronzy yellow.

C. nipponicum
Japanese oxeye daisy   0.6 m (2 ft.)   Aug., Sept.
Shrubby, less invasive than *C. leucanthemum,* but showier in areas where it survives the winters. Thick spatulate leaves, white flowers with a greenish yellow zone.

C. rubellum
A single, very hardy chrysanthemum with pink or rosy pink flowers and yellow disks. The triangular leaves are divided into five coarsely toothed parts. The cultivars of the species are hardier, although perhaps not as beautiful as the species.

Cultivars   0.8 m (2½ ft.)   Sept., Oct.
  'Clara Curtis' – the first cultivar of the species
  'Duchess of Edinburgh' – single, fiery red flowers
  'Mary Stoker' – single, soft yellow flowers
  'Paul Boissier' – single, orange bronze flowers

C. sibiricum
Korean chrysanthemum   0.6-0.9 m (2-3 ft.)   Oct.
Although strains of this species have been important in breeding hardier, more colorful, and more rugged outdoor chrysanthemums, this plant is not recommended. Rather straggly; pink or white flowers borne very late in the season. Synonym: *C. coreanum*

C. uliginosum moon daisy   1.7 m (5½ ft.)   Aug.-Oct.
Tall, very bushy; erect stems, broad heads of white flowers 8 cm (3 in.) across. Best in rich, moist soil; tolerant of wet winters. Quite hardy in the prairies.

C. weyitchii   23 cm (9 in.)   July, Aug.
Small compact plants; large lilac pink daisies.

CHRYSOGONUM

C. virginianum goldenstar   23 cm (9 in.)   June-Sept.
Dainty golden yellow star-shaped flowers. Extremely useful as a ground cover, good shade.

CHRYSOPSIS golden aster

This genus contains at least three ornamental species that bear golden yellow daisy-like flowers on compact bushes.

C. falcata   15-23 cm (6-9 in.)   Aug., Sept.
The smallest, most useful for rock gardens; small yellow flowers in corymbs.

C. mariana   46 cm (18 in.)   Aug., Sept.
Not as well known as the other species in cultivation; the most compact and showy of the three species; yellow flowers in corymbs.

C. villosa   30-46 cm (12-18 in.)   Aug., Sept.
The best known species; silvery gray leaves, solitary heads of bright yellow daisies.

CIMICIFUGA snakeroot, bugbane   (Color photo p. 78)

Very hardy plants; double compound leaves, long curving racemes of abundant small white flowers. Best in moist, rich soil and part shade, but will withstand full sun.

C. americana   1.7 m (5½ ft.)   Aug., Sept.
Very graceful creamy white racemes of flowers, rounded 3- to 5-lobed leaves. Synonym: *C. cordifolia*

C. cordifolia see *Cimicifuga americana*

C. dahurica   1.2-1.5 m (4-5 ft.)   Aug., Sept.
Closely packed unopened mauve-tinged flower spikes opening to creamy white on black stems.

C. foetida 'Intermedia' see *Cimicifuga simplex*

C. japonica var. *acerina*   *0.9 m (3 ft.)   Sept., Oct.*
A Japanese variety; maple-like leaves, slender spikes of white.

C. racemosa black snakeroot   *1.8 m (6 ft.)   Aug., Sept.*
Very tall, large deeply cut leaves, small white feathery flowers closely set on long stalks.

C. simplex Kamchatka bugbane   *0.9 m (3 ft.)   Sept., Oct.*
Tall, delicate spires of white; stately for borders but late blooming. Synonym: *C. foetida* 'Intermedia'

'Armleuchter'   *0.9 m (3 ft.)   Sept., Oct.*
Even more robust habit than the species.

'White Pearl'   *1.2 m (4 ft.)   July, Aug.*
Feathery drooping racemes of pearly white flowers; earlier than the species.

CLEMATIS

These are better known as climbing and twining plants, but there are a few herbaceous species that do not climb and can be used in the flower border. Except for *C. heracleifolia* var. *davidiana*, all clematis are perfectly hardy in most parts of Canada.

C. heracleifolia var. *davidiana*
fragrant tube clematis   *0.9-1.2 m (3-4 ft.)   Sept.*
Very vigorous, abundant clusters of pale blue hyacinth-like bell-shaped fragrant flowers in the axils of the leaves. Very striking, spacious plant for borders, flowers excellent for cutting if the leaves below the cluster are removed.

'Côte d'Azure' — deep blue flowers
                              1.2 m (4 ft.)   Sept., Oct.

'Crépuscule' — lavender blue flowers
                              0.9 m (3 ft.) Sept., Oct.

C. integrifolia   *0.6-0.9 m (2-3 ft.)   June, July*
Undivided leaves, nodding white-centered blue flowers. May be staked in the border, or left to sprawl and form a ground cover on a slope or over rocks in the rock garden.

'Coerulea'   *0.6-0.9 m (2-3 ft.)   July, Aug.*
More desirable than the species; larger porcelain blue flowers.

'Hendesoni'   *46 cm (18 in.)   July, Aug.*
Semiprostrate; blue flowers.

C. recta bush clematis   *1.2 m (4 ft.) June, July*
Erect; creamy white, scented flowers in great profusion, followed by a mass of feathery fruits.

'Grandiflora'   *0.9-1.2 m (3-4 ft.)   July, Aug.*
An improved cultivar, more floriferous than the species.

'Mandshurica'   *0.9-1.2 m (3-4 ft.)   June, July*
Very vigorous; more axillary and terminal clusters of flowers than the species.

'Purpurea'   *0.9 m (3 ft.)   June, July*
Very distinctive; purple leaves.

CODONOPSIS

These are Asiatic trailing or decumbent plants that belong to the bellflower family. The flowers are beautiful, particularly when they are examined closely.

C. clematidea   *30 cm (12 in.)   July, Aug.*
Solitary white flowers, so marked with blue and purple that the overall effect is blue.

C. convolvulacea   *0.9 m (3 ft.)   July*
Beautiful, rare; slender twining stems, bell-like blue flowers 4 cm (1½ in.) across.

C. ovata   *30 cm (12 in.)*   *July, Aug.*
Blue-gray bells marked inside with orange and purple. Often mistaken for *C. clematidea*, which lacks orange inside the bells.

CONVALLARIA

C. majalis lily-of-the-valley   *15 cm (6 in.)*   *May, June*
This hardy long-lived harbinger of spring was originally named *Lilium convallium*, which literally means lily of the valley. For large blooms, plant it in shade in good rich soil. Once established it tends to spread, which may be checked by chopping off excess shoots. An excellent ground cover under apple or similar trees, and on the north side of fences or homes.

After a few years, the roots become overcrowded and the flowers noticeably smaller. At this time lift the plant and divide it, and enrich the soil by working in some compost, animal manure, or dried leaves supplemented by a commercial fertilizer.

The pure white fragrant flowers and pale green leaves are so well known that no description of them is needed.

   'Aureo-variegata'   *23 cm (9 in.)*   *May, June*
   Yellow variegated leaves.

   'Flore Plena' see 'Prolificans'

   'Fortin's Giant'   *23 cm (9 in.)*   *May, June*
   Extremely handsome, larger flowers and leaves than the species.

   'Fortunei'   *30 cm (12 in.)*   *May, June*
   Robust; large flowers.

   'Prolificans'   *15 cm (6 in.)*   *May, June*
   Double; just as easy to grow as the species, but harder to obtain. Synonym: 'Flore Plena'

   'Rosea'   *15 cm (6 in.)*   *May, June*
   Rosy-tinted white flowers. At Ottawa just as easy to grow as the other cultivars.

COREOPSIS tickseed   (Color photo p. 78)

A long blooming season of bright yellow flowers. Most species do not live long, therefore, where seeds offer a means of reproduction, start new seedlings each year. Some of the cultivars are much superior to the species. Do not let the cultivars self sow, but keep young plants by frequent division.

In cold climates such as in the Prairie Provinces give them some protection with a light covering of straw. They grow in almost any soil, but moist soil in full sunlight is best.

C. auriculata 'Nana'   *30 cm (12 in.)*   *June-Sept.*
Dwarf, for rock gardens; bright orange-yellow flowers.

   'Superba'   *0.6 m (2 ft.)*   *June-Sept.*
   Showy golden yellow daisy flowers with maroon centers.

C. grandiflora
This and *C. lanceolata* are very often mistaken for one another. Some horticulturists suggest that *C. grandiflora* is more biennial in habit and that *C. lanceolata* is the true perennial. In our tests, however, the cultivar presumed to be a hybrid of the two is long lasting and worthy of cultivation. *C. grandiflora* has narrow, divided leaves and yellow daisy-like flowers.

Cultivars   *June-Sept.*

   'Badengold' — large golden yellow flowers on very strong stems   *0.9 m (3 ft.)*

   'Golden Plume' — double yellow   *0.8 m (2½ ft.)*

   'Goldfink' — new, dwarf cultivar; miniature golden daisies
                                       23 cm (9 in.)

   'Mayfield Giant' — large, rich orange yellow *0.8 m (2½ ft.)*

   'Newgold' — new, improved, double *0.8 m (2½ ft.)*

   'Rubythroat' (Rotkelchem) — dwarf; yellow with a reddish center   *30 cm (12 in.)*

   'Sunburst' — semidouble, golden yellow flowers
                                     0.8 m (2½ ft.)

   'Sunchild' — dwarf cultivar; yellow flowers with a maroon blotch   *30 cm (12 in.)*

C. lanceolata   *0.8 m (2½ ft.)*   *June-Sept.*
Golden yellow flowers, undivided lanceolate leaves.

1 *Corydalis nobilis*, fumitory
2 Close up of *Corydalis nobilis*

C. rosea swamp tickseed   *30 cm (12 in.)*   *Sept., Oct.*
Dwarf, compact plant; tiny pink flowers with yellow disks. Later flowering than other species, aromatic foliage, and creeping stems. It grows wild in many parts of Canada in moist open locations.

    'Nana'   *15 cm (6 in.)*   *June-Sept.*
    Very neat dwarf; grows best in alkaline soils.

C. verticillata
thread-leaved coreopsis   *46 cm (18 in.)*   *July-Sept.*
Pretty "front row" plant; small star-shaped, rich yellow flowers, finely divided leaves.

    'Golden Showers'   *46 cm (18 in.)*   *July-Sept.*
    Very free flowering, brighter yellow flowers than the
    species.

    'Grandiflora'   *0.6 m (2 ft.)*   *July-Sept.*
    Flowers slightly larger on taller plants than the species.

CORNUS   (Color photo p. 78)

C. canadensis bunchberry, pigeonberry   *15 cm (6 in.)*   *May*
A native Canadian woodland plant that is useful for planting in moist shaded areas under trees, where little else will grow. The white dogwood-like bracts form the ornamental effect, because the flower is inconspicuous. In fall its bunches of red fruits are very attractive.

CORYDALIS fumitory

All the species are worth growing for their fern-like foliage alone, but also their unusual spurred flowers, similar to Dutchman's-breeches *(Dicentra cucullaria)*, are attractive in early spring.

C. bulbosa   *15 cm (6 in.)*   *May*
A cormous species similar to *C. cava*.

C. cava   *15 cm (6 in.)*   *May*
Early flowering, light purplish flowers, fern-like foliage.

C. cheilanthifolia
Chinese corydalis   *23 cm (9 in.)*   *Apr.-June*
This species forms tufts of erect, finely divided fern-like leaves from which it gets its specific name *(Cheilanthes* is the lip fern). The yellow flowers are borne in long spikes through spring into summer. The plants are practically stemless, only a few inches high.

C. lutea yellow wall fumitory   *30 cm (12 in.)*   *May-Sept.*
An old favorite plant in Britain, where moist conditions allow it to cling to mossy walls. It is very useful for naturalizing, growing on rock slopes or walls in the milder areas of Canada, where it will give a long display of golden yellow flowers arising from lacy fern-like foliage.

C. nobilis	*30 cm (12 in.)*	*May*

A very interesting plant having dense terminal clusters of white flowers tipped yellow and spotted purple with spurs an inch long, and finely cut leaves.

C. wilsoni	*23 cm (9 in.)*	*Apr.*

Flowers deep canary yellow, leaves gray green, tufted, very delicately cut.

DELPHINIUM perennial larkspur (Color photo p. 66)

Mostly tall, spired blue perennials and some annuals that flower in June or July and intermittently through August and September. The genus contains species with dwarf habits, yellowish and reddish flowers, and loosely open-branched stems. One annual member of the genus is the popular larkspur, *Delphinium ajacis*.

D. × Belladonna
belladona delphinium   *0.9-1.2 m (3-4 ft.)   June*
Unlike the usual tightly compacted flower spikes of the garden delphinium, this type has many open-branched short-stemmed blue flowers that are excellent for cutting. In the border this species provides an excellent foil for peonies. A few good cultivars merit attention.

Cultivars

'Blue Bees' — clear light blue, early, and free flowering
1.1 m (3½ ft.)

'Capri' — rich sky blue   *0.8 m (2½ ft.)*

'Cliveden Beauty' — deep turquoise blue   *0.9 m (3 ft.)*

'Coelestinum' — light blue long-spurred flowers
0.8 m (2½ ft.)

'Lamartine' — deep blue   *1.1 m (3½ ft.)*

'Naples' — semidouble, the brightest blue   *1.1 m (3½ ft.)*

'Pink Sensation'   *0.8 m (2½ ft.)*
Dutch hybrid with *D. nudicaule* is placed in this group; not a sparkling pink, inclined to be rather weak.

'Wendy' — gentian blue, flecked with purple   *0.8 m (3½ ft.)*

D. elatum common delphinium   *0.9 m (3 ft.)*
This species, which most likely is not in cultivation, is responsible for the modern garden strains developed gradually over the past century. It is a robust perennial 0.6-0.9 m (2-3 ft.) high, with flowers in simple racemes somewhat branched at the base, lower petals are blue with upper petals brownish violet, two-lobed and with a yellowish beard, and long-spurred sepals. Native from the Pyrenees to Mongolia. The history of the delphinium is long and obscure, although some very famous breeders were involved. In Europe, Victor Lemoine, James Kelway, Walkins and Samuel, and more recently Frank Williams and Dr. Legro have performed and are working

wondrous miracles with the race. Dr. Legro is responsible for a number of multicolored hybrids including red and yellow.

In the U.S.A., Luther Burbank, Major Newell Vanderbilt, Charles Barber, and Dr. Leonian are best known for their work on delphiniums. Frank Reinelt is still engaged in improving his stately Pacific hybrids.

See Chapter 2 for further comment on strains and culture.

D. formosum 'Coelestinum'   *0.8 m (2½ ft.)   June, July*
Light blue, long-spurred flowers.

D. grandiflorum butterfly delphinium *0.3-0.9 m(1-3 ft.) June*
This perennial delphinium often blooms the first year from seed. It is distinguished by its finely cut foliage and large single flowers of blue or white. It does not live very long and perhaps is best treated as a biennial.

Cultivars   *June, July*

'Blue Butterfly' — elegant form with brilliant blue flowers
46 cm (18 in.)

'Blue Gem' — deep bright blue   *30 cm (12 in.)*

'Blue Mirror' — gentian blue   *30 cm (12 in.)*

'Connecticut Yankee' — shades of light and dark blue, used mostly as an annual   *0.4 m (1½ ft.)*

'White Butterfly' — white   *30 cm (12 in.)*

D. menziesii   *20 cm (8 in.)   June, July*
Dark blue butterfly tuberous type, native of British Columbia. The tubers dry completely after flowering.

D. nudicaule   *30 cm (12 in.)   June, July*
Short-lived dwarf perennial, striking orange scarlet flowers.

D. tatsienense   *46 cm (18 in.)   July, Aug.*
Dwarf oriental species, various shades of blue and white flowers.

DIANELLA

Extremely interesting Australasian plants, suitable for sheltered areas in the milder parts of British Columbia. Drooping panicles of flowers, grass-like leaves, and highly ornamental fruit. In a mild enough location, plant them in a mixture of peat moss and loam.

D. intermedia   *1.5 m (5 ft.)   June, July*
Free-flowering New Zealand species having white flowers in many branched panicles, followed by bright blue berries.

D. tasmanica   *1.5 m (5 ft.)   June, July*
The better known of the tall vigorous species. Rigid sword-like leaves up to 0.9 m (3 ft.) long; large, pale blue flowers

drooping in well-branched panicles; deep blue berries up to 1.9 cm (¾ in.) long that persist for a very long period.

DIANTHUS pink   (Color photo p. 78)

This genus contains some excellent herbaceous border plants as well as many compact mound-like rockery gems. The selections listed here are some of the more robust and enduring types and the easily grown tufted plants; they all grow well in any good garden soil in a sunny location. Most of them are perfectly hardy, although some hybrids involving *D. plumarius* need protection on the prairies.

D. × allwoodii   30-38 cm (12-15 in.)   June, July
Hybrid strain *(D. plumarius × D. caryophyllus)* with flowers of good texture, color, and fragrance. Rather tender. Good forms of this pink are stunning tufted plants with large single flowers produced in such profusion that no leaves can be seen.

'Little Joe'   15 cm (6 in.)   June-Sept.
A hybrid of *D. plumarius × D. alpinus;* single crimson flowers all summer. Needs some winter protection in colder zones.

D. arenarius   25-30 cm (10-12 in.)   June-Aug.
Mat-forming green shoots bearing white flowers with a green spot and reddish beard.

D. × arvernensis   8 cm (3 in.)   May, June
Hybrid of *(D. monspessulanus × D. sylvaticus)*. Light pink fragrant flowers on gray mats of leaves. Like a miniature Cheddar pink. Synonym: *D. gratianopolitanus*

D. barbatus sweet William   (Color photo p. 71)
See Chapter 3, Biennials, sweet William.

D. caesius see *Dianthus gratianopolitanus*

D. caryophyllus carnation   30-46 cm (12-18 in.)   July, Aug.
This species is mainly responsible for the greenhouse carnation. Some strains of seeds such as Chabaud, Dwarf Vienna, Grenadin, and Marguerite are sold as annual carnations. These will bloom in late summer from seeds sown indoors in March. They will often withstand the first winter in cold areas if given some protection.

D. deltoides maiden pink   15-30 cm (6-12 in.)   June, Aug.
A very vigorous useful pink for the front of flower borders and in large rock gardens. Its widespreading mats of dark green leaves are quite attractive all year, even when not in bloom. The flowers are light rose.

Cultivars   June-Aug.

'Alba' – white flowers   25-30 cm (10-12 in.)

'Brilliant' – crimson flowers, dark green leaves
15 cm (6 in.)

'Erectus' – compact; thick upright stems, rich red flowers
15 cm (6 in.)

'Flashing Light' – bright crimson flowers   15 cm (6 in.)

'Huntsman' – brilliant red flowers   15 cm (6 in.)

'Wisley' – carmine flowers with purple eyes, dark leaves
15 cm (6 in.)

D. gratianopolitanus Cheddar pink 15 cm (6 in.)   June, July

One of the mysteries of horticultural nomenclature is that a name such as *D. gratianopolitanus* should be allowed to supersede the much simpler and better known *D. caesius*. However, the Cheddar pink by any name is a beloved pink with sweetly fragrant pink flowers on mats or tufts of grayish green leaves. Synonyms: *D. caesius* and *D. × arvernensis*

Cultivars   June, July

'Flore Pleno' – semidouble, pink fragrant flowers
15 cm (6 in.)

'Rose Queen' – double, bright rose flowers   15 cm (6 in.)

D. knappi   38 cm (15 in.)   July
Yellow flowers in clusters at the apex of the stems. Not showy, but the only true yellow species.

D. plumarius grass pink, cottage pink
This hardy pink is the main species responsible for many hardy garden pinks. This species along with *D. caryophyllus*, *D. superbus*, and a few other species is responsible for the hardy border pinks, which are extremely popular in English flower borders but in most of Canada are not hardy and at best are short-lived.

The species is similar to Cheddar pink, *D. gratianopolitanus*, but has two flowers on each stem instead of one and they are deeply cut. The winged part of the basal claw of each petal is not as broad in this pink. Typical flowers are white with a dark crimson center.

No distinction can be made between the border pinks more closely associated with this species and those in various groups such as border carnations, highland pinks, and others. A list of the more promising border pinks follows. Some are bred in the U.S.A. and Canada, and some in Europe.

Cultivars

'Candystripe' – large, fully double white flowers, striped and stippled, bright red; gray foliage
30 cm (12 in.) July-Sept.

'Caprice' – fully double pink flowers

'Cheyenne' -- very hardy, fully double bright pink blooms
30 cm (12 in.) June-Sept.

'Cyclops' – red   30 cm (12 in.) June, July

'Dinah' – semidouble Persian rose with maroon center
30 cm (12 in.) June, July

'Dubonnet' – large, double, fragrant, deep wine-colored carnation   38 cm (15 in.) June, July

'Emile Pare' – double, pink, long season of bloom
30 cm (12 in.) June-Sept.

'Emperor' – double red   30 cm (12 in.) June, July

'Evangeline' – soft rose   30 cm (12 in.) June, July

'Excelsior' – a pink form of 'Mrs. Sinkins', the well-known white cultivar   25 cm (10 in.) June-Sept.

'Her Majesty' – one of the best double whites, fragrant flowers   20 cm (8 in.) June, July

'Highland Queen' – vivid scarlet
30 cm (12 in.) June, July

'Inchmery' – double, silvery pink
30 cm (12 in.) June, July

'John Ball' – a very hardy white with red center
30 cm (12 in.) June, July

'Mrs. Sinkins' – very fragrant, white; well known in England where it is used for edging
30 cm (12 in.) June-Sept.

'Old Spice' – salmon orange   30 cm (12 in.) June-Sept.

'Pink Princess' – from USDA, fringed, fragrant, 4 cm (1½ in.) pink, bloom salmon orange
30 cm (12 in.) June-Sept.

'Shadow Valley' – very hardy cultivar, red flowers
20 cm (9 in.) June-Sept.

'Sweet Memory' – red and white
30 cm (12 in.) June-Sept.

D. superbus   38 cm (15 in.) June-Sept.
Lacy pink or white fringed flowers, green spotted at the base. Shortlived.

Cultivars   38 cm (15 in.) June, July

'Blue Loveliness' – blue fringed flowers

'White Loveliness' – lacy white flowers on long stems

DICENTRA bleedingheart, Dutchman's-breeches
(Color photo p. 79)

This genus is best known for the graceful and attractive bleedinghearts, *D. spectabilis*. Many of the other species are as beautiful and may be used in the garden in various ways, for shade, for ground covers, and in full sun for borders. By judicious use of the species and cultivars, the flowers will bloom all summer long. Most species are best left in their location for many years; but some need dividing every 2 or 3 years.

D. chrysantha golden eardrops   1.2 m (4 ft.) July, Aug.
Probably useful only in the milder regions of British Columbia. Large, graceful perennial producing erect racemes of golden yellow flowers in full sunlight. Apparently very difficult to transplant, so it is best to sow seeds.

D. cucullaria Dutchman's-breeches   13 cm (5 in.) May
A native of Canadian woodlands, this exquisite little plant may be grown in shady areas in home gardens. Produces a longer flowering period if supplied with ample water, otherwise it dies down early after flowering. White or pinkish white heart-shaped flowers with yellow tips.

D. eximia plume bleedingheart   46 cm (18 in.) May-Aug.
Drooping bright rosy red flowers on graceful stems, spurs are incurved, and racemes branched. Dissected fern-like, grayish blue leaves. Self sowing and spreading in some areas.

'Alba' – white flowers   30-46 cm (12-18 in.) May, June

D. formosa
western bleedingheart   30-46 cm (12-18 in.) May-Sept.
Graceful leaves and flowers similar to *D. eximia*, but spreading by underground rhizomes. Pink or dull red flowers, somewhat paler than *D. eximia*, but with the same naked scape and branched racemes. This species grows in the shade, and produces a longer flowering period if supplied with ample water.

'Adrian Bloom'   30 cm (12 in.) May, June
Crimson flowers, much brighter than 'Bountiful'.

'Bountiful'   30 cm (12 in.) May, June
Dwarf bleedingheart, neat and compact; similar to *D. formosa* but with deep pink flowers.

'Summer Beauty'   38 cm (15 in.) May-Sept.
Everblooming light pink flowers, soft gray-green leaves.

'Sweetheart'   *38 cm (15 in.)*   *May-Sept.*
Glistening white flowers, light green leaves. Extremely effective with primroses in May.

D. formosa ssp. *oregana*   *20 cm (8 in.)*   *July, Aug.*
Tufted bleedingheart, stout rhizomes, coarsely toothed, silvery leaves, and cream purple-tipped flowers with short blunt spurs. Best in a cool well-drained location and a peaty soil. Synonym: *D. oregana*

D. oregana see *Dicentra formosa* ssp. *oregana*

D. spectabilis bleedingheart   *0.9 m (3 ft.)*   *May, June*
One of the treasures of the flower border, very often the only perennial in many cottage gardens and small home gardens. Much larger leafy-stemmed perennial than the other species. Large lyre-shaped rosy pink flowers on graceful arching branches. Like the other two species, it is likely to become shabby in summer so that plants like baby's-breath should be planted nearby and allowed to overlap and hide the drying leaves. This bleedingheart is best in a sunny position but it will grow in semishade.

    'Alba'   *0.6 m (2 ft.)*   *May, June*
White, somewhat weaker than the species.

DICTAMNUS   (Color photo p. 79)

D. albus gasplant   *0.6 m (2 ft.)*   *June, July*
Attractive plant, dark green leaves, and long arching branches of white flowers. Called gasplant because of its ability to exude a volatile oil from glands produced in the stem, particularly just below the inflorescence. On a hot still day the gas may be ignited by a lighted flame placed at the right spot. The flame will burn brilliantly for a short time and then peter out, leaving the inflorescence uninjured. The same oil that causes the plant to ignite produces an aromatic fragrance.

The gasplant is one of the most permanent of the perennials; each year it produces a specimen plant with glossy dark green pinnate leaves and a good display of flowers. Best in good rich soil, but it will withstand ordinary drought conditions. It is easy to grow from seeds sown outside in the fall; these seeds germinate in April.

    'Caucasicus' see *Dictamnus albus* 'Gigantius'

    'Gigantius'   *0.6 m (2 ft.)*   *June, July*
Neat larger flowers than the species. Synonym: 'Caucasicus'

    'Purpureus'   *0.9 m (3 ft.)*   *June, July*
Larger plant with purplish flowers. Synonym: 'Ruber'

    'Ruber' see *Dictamnus albus* 'Purpureus'

DIERAMA wandflower

These arching wands of pendant pink, purple, or white bells are a marvel of grace. This genus of perennial plants grows only in very mild areas. They are the most graceful of all The white or pink to purple flowers with tubes about 1.3 cm (½ in.) long arising from brownish or brownish-striped drained soils, where their arching stems can be seen to perennials. They do best in deep, rich, moist but well-advantage, such as on the margins of a pool.

D. pendulum   *0.8 m (2½ ft.)*   *June, July*
bracts are at the ends of pendant branches. These iridaceous perennials are native to South Africa, and can survive only in the mildest parts of British Columbia.

D. pulcherrima   *1.2 m (4 ft.)*   *June, July*
Similar habit to *D. pendulum*, but larger and stronger and with larger purple, almost blood-red tubes. The bracts are much longer than in *D. pendulum* and almost white except for a brown base. As in the other species, the plant grows from a corm, which should be fairly deeply planted.

    'Nanum'   *60 cm (24 in.)*   *June, July*
Lovely bells of soft smoky pink.

DIGITALIS foxglove   (Color photo p. 70)

The showier members of this genus are the hybrids of the biennial foxglove, *D. purpurea*, see Chapter 3. All the other species so often mentioned in books on perennials are really insignificant plants that are not worthy of a place, even in a connoisseur's garden. A lot of work should be done on improving the perennial species in this genus.

DODECATHEON shootingstar

Hardy, low-growing perennials that need moist but well-drained soil and part shade. The unusual flowers with their stamens pointing forward and their petals turning backwards somewhat resemble a miniature cyclamen. The plants become dormant in summer, so they need close association with other leafy perennials. They blend well with the spring-flowering native *Phlox divaricata* and primroses.

D. hendersonii mosquito bills   *30 cm (12 in.)*   *May*
Deep green glabrous leaves, lilac purple flowers with yellow-margined anthers. Native to British Columbia.

D. meadia shootingstar   *30-38 cm (12-15 in.)*   *May, June*
Showy rosy lilac flowers like a burst of 20 or 30 shooting stars arising from leafless flower stalks about 30-38 cm (12-15 in.) high. The foliage consists of attractive rosettes of ovate-oblong leaves. Reddish yellow anthers with purple filaments. The showiest and earliest species to grow.

D. pauciflorum see *Dodecatheon pulchellum*

'Madame Mason'　　　　　　　*0.6 m (2 ft.)　May*

Larger and more robust than the species, sparkling yellow flowers 5 cm (2 in.) across borne singly on 61-cm (24-in.) stems. These plants are more compact and much more suitable for flower borders than the species.

D. cordatum　　　　　　　*15 cm (6 in.)　May, June*

A very fine little compact plant for the rock garden and the front of the border; large 8-cm (3-in.) golden daisies on solitary flower heads.

D. pardalianches

great leopard's-bane　　　　*0.6-0.9 m (2-3 ft.)　May, June*

A coarse perennial with tuberous rootstocks, heart-shaped leaves, and clusters of 3 to 5 yellow flowers on strong stout stems.

'Bunch of Gold'　　　　　　*0.8 m (2½ ft.)　May, June*

A much improved and more refined plant than the species. Canary yellow flowers.

'Spring Beauty'　　　　　　*46 cm (18 in.)　May, June*

An outstanding cultivar with fully double deep yellow flowers.

D. plantagineum　　　　　*0.6-0.9 m (2-3 ft.)　May, June*

A rather coarse species with a fibrous rootstock, large toothed heart-shaped leaves, and huge 8-cm (3-in.) golden yellow flowers borne singly on long thick stems.

'Excelsum'

Harper Crewe variety　　　*1.2-1.5 m (4-5 ft.)　May, June*

A much larger, even more robust cultivar with flower heads up to 10 cm (4 in.) across. Best in larger flower borders, where its robust habit is not so overpowering.

D. pulchellum　　　　　　*23 cm (9 in.)　May, June*

Pale lilac flowers with a recurved yellow tube and a deep purple wavy ring in the throat, purple anthers. A similar species, *D. cusickii*, is considered a synonym by some botanists but differs in having sticky toothed leaves and more flowers in the umbel. *D. pulchellum* is not easy to grow, but is best in acid sandy soil. Synonym: *D. pauciflorum*

DORONICUM leopard's-bane

Perennial with simple yellow daisies, valued for its early flowering habit. Excellent for early spring bouquets that last well in water and provide a contrast to flowers of the spring bulbs. The plants are rather coarse. During hot summers their leaves wilt and appear to be about to die until the cool of evening rejuvenates and refreshes them, but this is compensated for by their early summer beauty. Similar to the delphinium, the doronicum will bloom again if cut down after it flowers.

D. caucasicum　　　　　　*46 cm (18 in.)　May, June*

A golden yellow daisy that blooms along with *Mertensia*, white candytuft, *Brunnera*, and primrose to make an excellent spring picture. Leaves are kidney-shaped and a little more resistant to hot summer than leopard's-bane. The plant has a creeping stoloniferous habit.

DRACOCEPHALUM dragonhead

Not usually seen in Canadian gardens, but worth a place in the front of the border because of its purplish snapdragon-like flowers and gray green leaves. All species seem to endure cold winters and hot summers in open locations and poor soils. Flowers from July to August or September, colorful when other perennial flowers are scarce.

D. altaiense see *Dracocephalum grandiflorum*

D. argunense　　　　　　*20 cm (8 in.)　June, July*

The showiest of all dragonheads. Synonym: *D. ruyschiana* 'Speciosum'

D. forrestii　　　　　　*30-46 cm (12-18 in.)　July*

Finely divided leaves, and deep purple flowers in July, borne in whorls on dense spike-like racemes.

D. grandiflorum 15-23 cm (6-9 in.) June, July

Blue flowers 5 cm (2 in.) long in oblong spikes about 8 cm (3 in.) long. Synonym: *D. altaiense*

D. hemsleyanum 30 cm (12 in.) July, Aug.

An erect tufted perennial with hairy, branched, leafy stems and purplish blue flowers in a loose spike.

D. ruyschiana 30 cm (12 in.) July, Aug.

A compact bushy plant with narrow leaves and spikes of purple flowers.

'Speciosum' see *Dracocephalum argunense*

D. wilsonii 0.4-0.9 m (1½-3 ft.) July, Aug.

A very pretty border plant with showy white to lavender violet flowers. Synonym: *Nepeta wilsonii*

DRACUNCULUS

D. vulgaris dragonplant 0.9 m (3 ft.) June

A strange but rather handsome plant that should grow well in milder British Columbia. It has mottled stems and digitately lobed leaves arising from a bow-shaped base. The flowers, similar to Calla lilies, are large bright purple blooms that have a strong disagreeable odor for a short time after opening. Like most of the members of the Araceae this plant is best in peaty moist soil. Synonym: *Arum dracunculus*

DRYAS (Color photo p. 79)

Very hardy mat-forming woody subshrubs. The species included here are suitable for the front of borders and in rock gardens.

D. drummondii 5 cm (2 in.) June, July

A dwarf tufted plant, particularly beautiful for the appendages it produces after flowering.

D. integrifolia 10 cm (4 in.) May, June

A species with smooth-edged and rolled margins. The flowers are slightly smaller than those of the mountain avens.

D. octopetala mountain avens 10 cm (4 in.) May, June

An evergreen creeping shrub with white flowers resembling small single roses.

'Minor' 5 cm (2 in.) June

A diminutive form; all parts of this plant are about half as large as the species.

D. × suendermannii 15 cm (6 in.) May, June

A hybrid *(D. drummondii × D. octopetala)* with creamy buds opening to pure white, slightly nodding flowers, and a habit similar to *D. octopetala*.

ECHINACEA purple coneflower (Color photo p. 79)

A genus of plants native to North America, very closely related to the rudbeckias. Although the wild forms appear to be rather dingy and coarse, the cultivated plants are handsome, easily grown perennials with large dark green leaves and brown stems. They will tolerate dry, sun-baked locations, but are more attractive in good garden soil. In the Prairie Provinces they need some protection in winter.

E. angustifolia 61 cm (24 in.) July, Aug.

A form that grows wild from Saskatchewan to Texas with rose purple drooping ray florets and brownish cones. This species could most likely be improved by selection.

E. purpurea purple coneflower 1.2 m (4 ft.) Aug.-Oct.

Often listed as *Rudbeckia purpurea*. May be muddy, unsightly, and objectionable, depending on the seed. It is best to buy named cultivars.

'Bright Star' (Leuchtstern) 1.2 m (4 ft.) Aug.-Oct.

A very brightly colored form with rich rosy red flowers, brighter than 'The King', but harder to propagate.

'Robert Bloom' 1.2 m (4 ft.) Aug.-Oct.

A new English cultivar with broad-petaled flowers or carmine purple and orange centers. A vigorous and erect plant.

'The King'   *0.9 m (3 ft.)   Aug., Sept.*
The most popular form, having rosy red 10-cm (4-in.)
flowers and brownish cones. Splendid for autumnal
arrangements.

'White Lustre'   *0.8 m (2½ ft.)   Aug., Sept.*
Rather dull reflexing petals and yellow brown cones, but
its large 10-cm (4-in.) flowers are acceptable in the border
during early fall.

ECHINOPS globe thistle   (Color photo p. 79)

A large genus of very hardy herbaceous plants with white
woolly leaves and thistle-like globe heads of flowers. Only a
very few of the species are cultivated, and these are very
easy to grow in any soil and in full sunlight.

E. exaltatus
small globe thistle   *0.9-1.2 m (3-4 ft.)   July-Sept.*
The well-known *E. ritro* and *E. sphaerocephalus* are considered by botanists as selections of the species. A vigorous
plant with silvery blue globe-shaped flower heads, like drum-
sticks or marimbas. These heads of flowers last all summer
and can be dried for winter bouquets. Synonyms: *E. ritro, E.
sphaerocephalus*

'Veitch's Blue'   *0.9 m (3 ft.)   July-Sept.*
Brighter blue flowers.

E. humilis   *0.9-1.2 m (3-4 ft.)   July-Sept.*
Similar to *E. exaltatus*, but with wavy-margined basal
leaves, almost spineless, cobwebby above and hairy beneath.
The flower heads are bluish.

'Blue Cloud'   *1.2 m (4 ft.)   July-Sept.*
Clear blue.

'Nivalis'   *1.2 m (4 ft.)   July-Sept.*
White flowers and silver gray foliage.

'Taplow Blue'   *1.5 m (5 ft.)   Aug., Sept.*
A very vigorous tall globe thistle with light blue glistening heads of flowers with silvery overcast. The finest of
all cultivars tested in Ottawa, it deserves a place in the
border.

E. ritro see *Echinops exaltatus*

E. sphaerocephalus see *Echinops exaltatus*

EPIMEDIUM barrenwort, bishop's-hat

Very daintily poised flowers on slender stems appear with
the young leaves in spring. The flowers are in loose sprays of
buff, yellow, mauve, and white. These delicate flowers and
their superb pale green foliage combine to make the plants
useful in the border or as a ground cover. The flowers last
extremely well in water and make excellent miniature

arrangements. They prefer a light sandy soil and part shade,
but will grow reasonably well in full sun.

E. alpinum 'Rubrum' see *Epimedium coccineum*

E. coccineum   *30 cm (12 in.)   May*
This is a garden hybrid of *E. alpinum* × *E. grandiflorum*
with red early foliage and crimson flowers 3 cm (1 in.) in
diameter, flushed yellow or white. It is more robust than *E.
alpinum* and not as large as *E. grandiflorum*, but has 20 or
30 flowers in a cluster, more flowers than in the clusters of
any other species. Synonyms: *E.* × *rubrum, E. alpinum*
'Rubrum'

E. grandiflorum   *20-38 cm (8-15 in.)   May*
A real gem of a plant, larger flowers than the rest of the
species, 3-5 cm (1-2 in.) in diameter, ternate leaves about
30 cm (1 ft.) long, in a short 6–16-flowered raceme. Outer
sepals of the flowers are red, inner sepals violet, and the
spurred petals white. Synonym: *E. macranthum*

'Rose Queen'   *20-38 cm (8-15 in.)   May*
Crimson carmine flowers and white-tipped spurs.

'Violaceum'   *23 cm (9 in.)   May*
Slightly smaller than the species; flowers with light violet
petals and sepals.

E. macranthum see *Epimedium grandiflorum*

E. niveum see *Epimedium* × *youngianum* 'Niveum'

E. perralderianum   *30 cm (12 in.)   May*
The young leaves with rich bronze markings; flowers bright
yellow, 3 cm (1 in.) across, in simple loose 12–25-flowered
racemes.

E. pinnatum   *23 cm (9 in.)*   *May*

A densely hairy plant with biternate leaves and bright yellow flowers with red nectaries (spurs).

E. pinnatum var. *colchicum*   *30-46 cm (12-18 in.)*   *May*

This is the form usually sold as *E. pinnatum*. It has fewer (3–5 instead of 5–11) leaflets in the compound leaves, and the flowers are the same.

E. × *rubrum* see *Epimedium coccineum*

E. × *versicolor*   *30 cm (12 in.)*   *May, June*

A garden hybrid *(E. grandiflorum* × *E. pinnatum* var. *colchicum)*, small flowers with rose sepals and yellow petals with red-tinted spurs, in 10–20-flowered racemes. The biternate leaves are red when young, 3 or 5 foliate, and spiny-toothed.

E. × *warleyense*   *23-51 cm (9-20 in.)*   *May, June*

A distinct hybrid *(E. alpinum* × *E. pinnatum* var. *colchicum)* with very attractive coppery-red flowers and heart-shaped toothed leaves. The sepals are yellow with blunt red streaked spurs.

E. × *youngianum*   *23-30 cm (9-12 in.)*   *May, June*

A hybrid *(E. diphyllum* × *E. grandiflorum)* with white flowers tinged greenish, few in number, bell-shaped, and nodding. Not commonly grown; the following cultivar is usually preferred.

   'Niveum'   *15 cm (6 in.)*   *May, June*

A very charming form of the hybrid above, with glistening white flowers and bronzy foliage on very compact plants. Synonym: *E. niveum*

EREMURUS foxtail lily

The foxtail lilies are some of the most imposing of all border plants and are well worth the effort of discriminating gardeners who will give them the special care they deserve. Their roots are fleshy and radiate from the crown. They should be handled carefully, because they are easily damaged, which may start an infectious decay. The plants are best set out in the fall in well-drained sandy soil. They should be protected in winter by mounding coarse sand about 15 cm (6 in.) high around the crown of the plant. Without protection the flower bud may be killed. A screen is needed for protection against prevailing winds.

E. bungei see *Eremurus stenophyllus*

E. elwesii   *1.8-2.7 m (6-9 ft.)*   *May, June*

A very noble species with fragrant flesh-colored flowers and leaves 0.9 m (3 ft.) long.

   'Albus' – white flowers   *1.8-2.7 m (6-9 in.)*   *May, June*

Eremurus himalaicus, foxtail lily

E. himalaicus

Himalayan desertcandle   *0.9-1.8 m (3-6 ft.)*   *May, June*

Pure white flowers with orange anthers.

E. × *himbrob*   *1.8-2.7 m (6-8 ft.)*   *May, June*

A hybrid *(E. himalaicus* × *E. robustus)* with magnificent spikes of pale pink flowers.

E. olgae   *1.2 m (4 ft.)*   *June, July*

Dense spikes of white star-shaped flowers tinted rose, long stamens, each petal with a central brown vein.

E. robustus   *1.2 m (4 ft.)*   *May, June*

Similar to *E. elwesii*, but with narrower leaves. Peach-colored flowers on many flowered racemes.

E. × *Shelford*   *1.5-2.4 m (5-8 ft.)*   *June, July*

Very free flowering with graceful spikes 1.5-2.4 m (5-8 ft.) high, yellow, salmon, pink, and bronze flowers. Hybrid of *E. bungei* × *E. olgae*.

E. stenophyllus   *0.9-1.5 m (3-5 ft.)*   *July*

Flowers bright yellow with orange anthers. Oblong dense racemes 10-13 cm (4-5 in.) long. Synonym: *E. bungei*

Cultivars *0.3-0.9 m (1-3 ft.) June*

  'Highdown Gold' — golden yellow
  'Isobel' — pink, shaded orange
  'Magnificus' — larger and brighter than the type
  'Moonlight' — pale yellow
  'Rosalind' — bright pink
  'Sir Arthur Hazelrigg' — coppery orange
  'White Beauty' — white

ERIGERON fleabane (Color photo p. 80)

Aster-like daisies from 15 cm to 0.9 m (6 in. to 3 ft.) tall, summer flowering. The flower stems arise from tufts of foliage rather like the English daisy *(Bellis perennis)*, flowers 5-6 cm (2-2½ in.) across, violet, purple, rose, rosy lavender, and apricot. So easy to grow that many species are weeds that are likely to spread in the garden. Selected species, however, are good perennials that have a place in all flower borders.

E. aurantiacus *23 cm (9 in.) July-Sept.*
A pretty dwarf species with semidouble orange daisies 5 cm (2 in.) across.

E. caespitosus *30 cm (12 in.) June, July*
A very hardy dwarf, compact, native species.

E. leiomerus *10 cm (4 in.) June*
Tiny lavender blue daisies on tufted plants.

E. macranthus *46 cm (18 in.) June, July*
Abundant violet flowers during summer.

E. mucronatus *15 cm (6 in.) June-Aug.*
Tremendously popular, produces myriads of pink and white daisies all summer. Excellent for crevice planting in patios and walks in the milder parts of British Columbia.

E. speciosus *46 cm (18 in.) July, Aug.*
Violet flowers on plants 61 cm (24 in.) across. The selected varieties are far superior to seedlings of the species.

Cultivars *July, Aug.*

  'Azure Beauty' — double, lavender blue *0.6 m (2 ft.)*
  'Double Beauty' — double, light blue *46 cm (18 in.)*
  'Mrs. E. H. Beale' — soft mauvy blue *0.3 m (1 ft.)*
  'Quakeress'— the most popular lavender blue *0.6 m (2 ft.)*
  'Summertime' — white *0.6 m (2 ft.)*
  'Wuppertal' — amethyst violet *0.6 m (2 ft.)*

Miscellaneous cultivars *June-Aug.*

  'Amity' — lilac pink, erect stems *0.6 m (2 ft.)*
  'B. Ladhams' — pink, a branching habit *0.4 m (1½ ft.)*

'Charity' — clear light pink, tall and strong *0.6 m (2 ft.)*
'Darkest-of-All' — very deep violet blue *0.6 m (2 ft.)*
'Dignity' — single-rayed flowers of violet mauve, over 5 cm (2 in.) in diameter *0.6 m (2 ft.)*
'Felicity' — dwarf, erect plants bearing large clear pink flowers *0.6 m (2 ft.)*
'Festivity' — upright stems with flowers 6 cm (2½ in.) across, a clear lilac color *0.6 m (2 ft.)*
'Foerster's Siebling' — very deep pink semidouble flowers, July, Aug. *51 cm (20 in.)*
'Gaiety' — profusion of bright deep pink flowers *0.8 m (2½ ft.)*
'Lilofee' — double blue flowers *0.6 m (2 ft.)*
'Mesa-Grande' — semidouble light blue flowers *0.6 m (2 ft.)*
'Prosperity' — double mauve pink *0.8 m (2½ ft.)*
'Serenity' — very deep violet mauve 5.7 cm (2¼ in.) across *0.8 m (2½ ft.)*
'Violetta' — semidouble violet blue *0.8 m (2½ ft.)*
'Vanity' — pink flowers excellent for cutting *0.8 m (2½ ft.)*

ERYNGIUM sea holly (Color photo p. 80)

Very prickly plants, leaves, stems, and teazle-like flowers surrounded by prickly bracts. However, the beauty of this plant far outweighs any of these inconveniences. The silvery blue leaves, flowers, and stems produce a lovely effect from June to September. A good plant for poor, dry, sunny, sandy soils and at its best by the sea.

'Donard Variety' *0.9 m (3 ft.) June-Sept.*
A cultivar with showy heads of cylindrically shaped steel blue flowers and bracts.

E. agavifolium *1.8 m (6 ft.) July-Sept.*
A tremendous grower with striking rosettes of narrow, spine-toothed leaves up to 1.5 m (5 ft.) long. Hardy only in milder parts of British Columbia.

E. alpinum 'Improved Form' *0.6-0.9 m (2-3 ft.) July-Sept.*
Large blue heads of flowers.

E. amethystinum *0.4 m (1½ ft.) June-Sept.*
Low-branching, one of the best perennials for the flower border. Globose flowers, steel-gray stems and bracts.

E. bourgatii *46 cm (18 in.) June-Aug.*
Striking silvery blue cut-leaved species.

E. *bromeliifolium*   *0.9 m (3 ft.)*   *July*
Greenish white round-headed flowers on very imposing spikes. Foliage similar to bromeliad with large awl-shaped teeth on the leaves.

E. *dichotomum*   *0.6 m (2 ft.)*   *July, Aug.*
Globose heads of blue flowers, very striking foliage.

E. *oliverianum*   *0.9 m (3 ft.)*   *July*
Deep glowing amethyst flowers and stems.

E. *pandanifolium*   *0.8 m (2½ ft.)*   *Aug., Sept.*
A noble half-hardy plant with spine-edged leaves 1.2-1.8 m (4-6 ft.) long. Purple brown flowers on panicles.

E. *planum*   *0.9 m (3 ft.)*   *July, Aug.*
The most common; blue flowers on tall branched stems. A plant more suited to the wild garden than to the border.

E. *tripartitum*   *0.8 m (2½ ft.)*   *July*
Steel-blue flowers with long spreading bracts.

   'Violetta'   *0.8 m (2½ ft.)*   *Aug., Sept.*
   Large intense violet blue flowers.

ERYSIMUM

E. *asperum* see Chapter 3, Biennials.

EUPATORIUM hemp agrimony

This genus contains a large number of predominately weedy plants, but a few useful very hardy ornamentals, which are not widely known or grown. Those of ornamental value have a profusion of delicate flower clusters in late August and September. Although they belong to the Compositae or daisy family, the flower heads of members of this genus are composed of fluffy tubular or disk florets only, no ray florets. Very easily grown in almost any well-drained soil.

E. *ageratoides*   *1.2 m (4 ft.)*   *Aug., Sept.*
Fluffy heads of white flowers.

E. *cannabinum* 'Plenum'
double hemp agrimony   *1.2 m (4 ft.)*   *Sept.*
The double form of the common hemp agrimony has highly ornamental, double, fluffy pinkish rose flowers in broad heads.

E. *coelestinum*
mistflower, hardy ageratum   *0.9 m (3 ft.)*   *Sept., Oct.*
Large flower heads in ageratum-like clusters, light to deep violet. It is one of the most outstanding blue perennials in autumn.

E. *purpureum* Joe-Pye weed   *1.8 m (6 ft.)*   *Sept.*
Tall clustered heads of purplish crimson. Good for naturalizing.

EUPHORBIA spurge   (Color photo p. 80)

A group of plants of very diverse habitats, forms, and uses. The Christmas poinsettia (E. *pulcherrima*), crown-of-thorns (E. *splendens*), and cactus-like plants such as E. *meloformis* are included in the genus. The herbaceous perennials have unquestioned hardiness and freedom from pests and diseases; they thrive in poor soils in full sun. With too much care and rich soil they will tend to spread too rapidly.

E. *corollata* flowering spurge   *0.9 m (3 ft.)*   *Aug.*
This native plant of Ontario to Florida and Texas has fluffy white blooms in umbels that look very much like the annual baby's-breath (Gypsophila *elegans*). The flowers are excellent for cutting and the plants are good for borders, especially when the foliage is wine red in August and September.

E. *cyparissias* Cypress spurge   *30 cm (12 in.)*   *May, June*
A dwarf, to 0.3 m (1 ft.) high, with attractive yellow-green bracts in May and June. Because this species spreads by underground shoots and may become invasive, it is often used as a ground cover to hold the soil on dry banks.

E. *epithymoides* cushion spurge   *46 cm (18 in.)*   *Apr.-June*
Rounded heads of flowers supported by showy golden bracts that are at their best from April to early June along with tulips and daffodils. A very refreshing plant in early spring. Attractive foliage all summer. The plants are so completely uniform that they may be used for small formal hedges. Synonym: E. *polychroma*

E. *griffithi*   *0.6 m (2 ft.)*   *June, July*
Recently introduced from England. Also called 'Fireglow'. Bright orange heads in early summer.

E. *polychroma* see Euphorbia *epithymoides*

E. *sikkimensis*   *0.9 m (3 ft.)*   *June-Aug.*
Purple foliage and yellow flowers. In spring the leaves are purple- and pink-toned and worth growing for this facet alone.

E. *wulfenii*   *1.2 m (4 ft.)*   *Apr.-July*
A handsome shrubby spurge with glaucous green foliage and large heads of yellowish green flowers.

FILIPENDULA meadowsweet

Very beautiful spirea-like herbaceous plants with very finely divided leaves and dainty flower heads composed of many individual florets. Most of the species are particularly effective when planted in or near moist woodlands, where they benefit from cool shade; some species will grow well in an open sunny border.

F. camtschatica   *1.8-2.4 m (6-8 ft.)*   *July, Aug.*
Very large, with white fragrant flower plumes and huge palmate leaves. A moisture-loving giant that needs plenty of water. Synonym: *Spiraea gigantea*

  'Rosea'   *1.8-3.1 m (6-10 ft.)*   *July, Aug.*
  Fragrant pink flowers borne in large corymbs.

F. multijuga   *15 cm (6 in.)*   *July, Aug.*
A diminutive species with soft pink flowers.

F. palmata   *0.6-0.9 m (2-3 ft.)*   *July*
Dainty pale pink plumes of flowers that fade with age, 7–9-lobed terminal leaves and 5-lobed lateral leaves, white hair on the undersides. The plant offered most often as this species is *F. purpurea*, the less hardy Japanese species. Synonym: *Spiraea digitata*

F. purpurea   *0.6-1.2 m (2-4 ft.)*   *July, Aug.*
This Japanese species needs moist soil and some protection in winter in Eastern Canada. A very beautiful plant with large lobed leaves and heavy crimson-stemmed plumes of fluffy carmine pink flowers. Synonyms: *F. palmata, Spiraea palmata*

  'Alba'   *0.6-1.2 m (2-4 ft.)*   *July, Aug.*
  White flowers and pale green leaves.

  'Purpurascens'   *0.6-1.2 m (2-4 ft.)*   *June-Aug.*
  Rosy red heads of flowers and rose-tinted leaves.

F. rubra queen-of-the-prairie   *1.2-2.4 (4-8 ft.)*   *June*
Although probably named for the American prairies, this species is quite hardy on the Canadian prairies if given good moist conditions. It is very attractive with its feathery plumage of deep peach flowers borne on tall branching stems. Synonyms: *Spiraea lobata, S. rubra*

  'Venusta'   *1.2-2.4 m (4-8 ft.)*   *June, July*
  Deep pink or carmine flowers. Synonym: *Spiraea venusta*

F. ulmaria
queen-of-the-meadow   *0.6-0.9 m (2-3 ft.)*   *June-Aug.*
This grand beauty from the moist meadows of England is quite hardy, even at Winnipeg, and has become naturalized as far east as New England. The graceful 3–5-lobed leaves are downy white underneath and the creamy white flowers are packed in dense heads. It needs lots of moisture and some shade. Synonym: *Spiraea ulmaria*

Cultivars   *0.6-0.9 m (2-3 ft.)*   *June-Aug.*

  'Aurea'
  A variegated form with dark green and gold foliage.
  'Flore Pleno' — double flowers

F. vulgaris dropwood   *0.6-1.8 (2-3 ft.)*   *July, Aug.*
A beautiful border subject with fern-like foliage forming a large tuft from which the flower stems arise. The creamy white flowers are borne on strong stiff stems. Grows well in the sun in average garden soil. Synonyms: *Spiraea filipendula, S. hexapetala*

  'Flore Pleno'   *38 cm (15 in.)*   *July, Aug.*
  Dark green foliage and creamy white double flowers. Should be grown in every border.

FUNKIA

F. subcordata see *Hosta plantaginea*

GAILLARDIA blanketflower   (Color photo p. 80)

Very gay perennials and annuals valuable for cutting, named after Gaillard de Marentonneau, a French patron of botany.

G. aristata   *0.6-0.9 m (2-3 ft.)*   *June-Aug.*
Usually best in a well-drained sandy soil with plenty of sunshine. In cold wet clammy soils it often does not survive the winter. The species has yellow ray florets, but it is not seen in gardens, its place is taken by various hybrids with yellow and red flowers.

G. × grandiflora
All the hybrids are listed under this name; some are very short-lived and others are more perennial.

Cultivars   *June-Aug.*

  'Baby Cole'   *15 cm (6 in.)*
  Neat, compact plants, red and yellow blooms all summer.

  'Burgundy' — deep red   *0.6 m (2 ft.)*

  'Goblin' — large, rich golden yellow   *0.6 m (2 ft.)*

  'Ipswich Beauty' — large orange and brown red
        0.9 m (3 ft.)

  'Mandarin' — orange, flame red   *0.8 m (2½ ft.)*

  'Portola' hybrids — red, tipped gold   *0.6 m (2 ft.)*

  'Wirral Flame'   *0.8 m (2½ ft)*
  Dark crimson brown, flecked yellow at the tips.

GALEGA goat's-rue

These perennials are so straggly and untidy that they should be planted toward the back of the border. They have small sweet-pea-like flowers borne in axillary spikes. They prefer a sunny well-drained location in poorer soils, and need protection in winter in the Prairie Provinces.

G. officinalis *0.9-1.5 m (3-5 ft.)* *June, July*
Soft green pinnate leaves and spikes of white or mauve flowers.

Cultivars *0.9-1.5 m (3-5 ft.)* *June, July*
 'Alba' — white flowers
 'Carnea' — flesh pink flowers
 'Hartlandii' — lilac and white flowers
 'Her Majesty' — lilac blue flowers
 'Lady Wilson' — mauve flowers in large sprays

G. orientalis *0.9 m (3 ft.)* *July, Aug.*
Blue flowers on hairy stems. Very attractive in bloom, but has a sprawling habit of growth.

GAURA

G. lindheimeri gaura *0.9-1.2 m (3-4 ft.)* *June-Sept.*
Very attractive plant from Texas, best in very light sandy soil. The white flowers with rose calyxes are odd looking and borne in loose graceful spikes.

GENTIANA

Many of these species are hard to grow except under specific conditions and in certain soil. The species referred to in this publication will grow in ordinary soil; they are better suited for a border than most species. They are very hardy.

G. asclepiadea willow gentian *0.4 m (1½ ft.)* *Aug., Sept.*
Long arching stems with willow-like leaves and spikes of dark purple tubular flowers. Thrives in shade or partial shade, and best in very moist soil. Often naturalizes itself in a location it likes.

 'Alba' white milkweed gentian *0.4 m (1½ ft.)* *Aug., Sept.*
 White flowers.

G. lutea yellow gentian *0.9 m (3 ft.)* *July, Aug.*
Tall leafy spikes of yellow star-shaped flowers arising from wide rosettes of crinkled basal leaves. Needs a rich, very moist spot in a sunny location.

G. septemfida *15-23 cm (6-9 in.)* *June-Aug.*
A very easy to grow dwarf gentian with bright blue trumpets. Thrives in any good soil. Best in a group planting near the front of the border and away from overhanging plants.

GERANIUM cranesbill (Color photo p. 81)

A very widely distributed genus containing some excellent garden subjects that flower freely in any light soil in the sun. The popular bedding geraniums are not referred to here; they are properly named pelargoniums.

G. anemonifolium *0.3-0.6 m (1-2 ft.)* *May*
Single stemmed with rounded finely cut leaves. Pale purple flowers in corymbs.

G. armenum *0.6 m (2 ft.)* *June, July*
Open flowers, 4 cm (1½ in.) in diameter, very vivid magenta color on bushy plants. Synonym: *G. psilostemon*

G. cinereum *15 cm (6 in.)* *June-Aug.*
Gray-green foliage and pale pink flowers.

 'Album' *15 cm (6 in.)* *June-Aug.*
 A very choice white-flowered cultivar.

G. cinereum var. *subcaulescens* *15 cm (6 in.)* *June-Aug.*
Brilliant cerise flowers with darker centers.

G. dalmaticum *23 cm (9 in.)* *June-Aug.*
Hummocks of round, shiny green leaves with pink flowers 3 cm (1 in.) in diameter. A real gem for rock gardens and flower borders.

G. endressi *23-30 cm (9-12 in.)* *June-Aug.*
Small fine-lobed buttercup-like leaves and pale pink cup-shaped flowers lightly marked with red.

 'A. T. Johnson' *30 cm (12 in.)* *June-Aug.*
 A lovely shade of silvery pink.

 'Rose Clair' *46 cm (18 in.)* *June-Aug.*
 White with beautiful purple veining.

 'Wargrave Pink' *38 cm (15 in.)* *June-Sept.*
 A deeper shade of pink.

G. farreri see *Geranium napuligerum*

G. grandiflorum see *Geranium meeboldii*

G. ibericum *0.6 m (2 ft.)* *July, Aug.*
Large violet flowers 5 cm (2 in.) in diameter arise from erect leafy stems. Leaves 5–7-lobed.

 'Album' — white flowers *0.8 m (2½ ft.)* *June, July*

 'Johnson's Blue' — bright blue flowers
       46 cm (18 in.) *June-Aug.*

G. ibericum var. *platypetalum* *0.8 m (2½ ft.)* *June, July*
This variety is quite similar to the species but is taller and has larger leaves. It has deep violet flowers with reddish veins. Synonym: *G. platypetalum*

G. macrorrhizum   38 cm (15 in.)   June

A variable species available usually as the 'Ingwersen's Variety', with pinkish-veined flowers, 5 cm (2 in.) across, on little sprays above the leaves.

G. meeboldii lilac cranesbill   46 cm (18 in.)   June-Aug.

Very large flattish flowers of rich blue, veined with red. Very dense, dainty, rounded 5-lobed foliage forming a bushy plant 46 cm (18 in.) wide. Synonym: *G. grandiflorum*

'Alpinum'   25-30 cm (10-12 in.)   June-Aug.

Much larger flowers on a dwarfer plant.

G. napuligerum   10 cm (4 in.)   May, June

A really delightful alpine with compact tufts of gray foliage and soft pink flowers with black anthers. Synonym: *G. farreri*

G. nodosum   30 cm (12 in.)   May, June

Purplish red veined flowers and notched petals.

G. phaeum mourning widow   0.6 m (2 ft.)   May, June

Not very showy, but planted for its curious deep dusky purple, almost black flowers, with a white spot at the base of each petal. Useful for naturalizing in semishady areas.

G. platypetalum see *Geranium ibericum* var. *platypetalum*

G. pratense
meadow cranesbill   0.4-0.9 m (1½-3 ft.)   July-Sept.

One of the largest flowers belonging to this genus, handsome open blue flowers 3 cm (1 in.) across. Its long stalked leaves are divided into seven lobes.

Cultivars   June, July

'Album' – white   0.9 m (3 ft.)

'Album Plenum' – double, white, very attractive, easily grown   0.6-0.9 m (2-3 ft.)

'Coeruleum Plenum' – double, light blue   0.6 m (2 ft.)

'Purpureum Plenum' – double, deep mauve   0.6 m (2 ft.)

'Silver Queen' – pearly gray   0.9 m (3 ft.)

G. psilostemon Armenian cranesbill   0.6 m (2 ft.)   July

Grows to about 0.6 m (2 ft.) high with open red violet flowers blotched with ebony on each petal. These near-magenta flowers are 4 cm (1½ in.) in diameter. An excellent setting for a few 'Sugar Plum' or 'Plum Dandy' petunias. Synonym: *G. armenum*

G. renardii   20-30 cm (8-12 in.)   June

A clump-forming species, very graceful and palmately lobed, dark olive green leaves with prominent veins. Flowers are white with a violet purple center.

G. sanguineum
bloody cranesbill   0.3-0.4 m (1-1½ ft.)   June-Aug.

Noted particularly for its very long flowering period and intense magenta flowers. The twiggy growth forms mats covered with dark green seven-parted leaves.

'Album'   0.3 m (1 ft.)   June-Aug.

Exquisite white flowers.

'Lancastriense'   15 cm (6 in.)   June-Sept.

Lower growing habit. One of the gems of the rock garden, useful for borders in association with other plants. Its flesh-colored purple-veined flowers are long lasting and continue to produce blooms over a long season.

'Prostratum'   20 cm (8 in.)   June-Aug.

Habit similar to 'Lancastriense', but with magenta flowers more like the species.

G. sylvaticum   0.6 m (2 ft.)   May, June

A native British woodland plant, with bluish purple to rose flowers on bushy plants rather like *G. endressi*.

'Mayflower'   0.6 m (2 ft.)   May, June

A much more refined cultivar with light blue flowers.

G. wallichianum   23-30 cm (9-12 in.)   July-Oct.

Good for the front of the border, because of its spreading prostrate habit. The saucer-shaped purple flowers are produced continuously from July to October.

'Buxton's Blue'   15 cm (6 in.)   July-Sept.

Much better than the species, saucer-shaped 5-cm (2-in.) opal blue flowers with a white center. Somewhat tender.

G. wlassovianum   76 cm (30 in.)   July-Sept.

A rhizomatous species with somewhat woody stems and long-stalked kidney-shaped leaves and purple violet flowers. Very hardy, from Manchuria.

GEUM avens

A family of very easily grown perennials that has been improved considerably during the past decade or so. They include some of the most vivid colors in border plants. Their chalice-shaped flowers are in bloom nearly all summer. They should not be planted near Oriental poppies or penstemon, because their brilliant colors will clash. They grow in any garden soil, in sun or light shade.

G. × *borisii*   30 cm (12 in.)   June, July

A very distinct long-lived mounded plant producing an abundance of orange-scarlet flowers on 23-25-cm (9-10-in.) stems. A hybrid of *G. bulgaricum* × *G. reptans*.

G. bulgaricum Bulgarian avens   0.4 m (1½ ft.)   June-Aug.

Bright yellow nodding flowers and hairy toothed leaves. The fruits have a long feathery plume.

G. chiloense scarlet avens   *0.6 m (2 ft.)   June-Aug.*
A tender species not usually seen in gardens today, but one of the parents of many hardier cultivars. The most famous cultivar is 'Mrs. Bradshaw', and the next is 'Lady Stratheden', both short-lived perennials that can reproduce themselves from seed.

Cultivars

'Dolly North'   *0.6 m (2 ft.) June-Aug.*
Free flowering with yellow orange semidouble flowers.

'Fire Opal'   *0.4 m (1½ ft.) June, July*
Single blossoms of orange overlaid with a warm reddish glow.

'Georgenberg' — new, dwarf, deep yellow flowers
30 cm (12 in.) May-July

'Golden West' — rich golden yellow flowers
0.4 m (1½ ft.) May-Aug.

'Lady Stratheden' — rich golden yellow double flowers
0.6 m (2 ft.) May-Sept.

'Mrs. Bradshaw' — double red flowers
0.6 m (2 ft.) May-Sept.

'Orangeman' — large orange flowers
0.6 m (2 ft.) June-Aug.

'Prince of Orange' — orange flowers
0.6 m (2 ft.) June-Aug.

'Princess Juliana' — glowing copper orange flowers
0.6 m (2 ft.) June, July

'Red Wings' — crimson scarlet flowers
0.6 m (2 ft.) July, Aug.

'Rubin' — semidouble scarlet flowers
0.6 m (2 ft.) July, Aug.

G. × *heldreichii*   *0.4 m (1½ ft.) June-Oct.*
A garden hybrid probably with *G. montanum* as one of its parents. Large orange-red flowers on long branching stems.

'Superbum' — large orange yellow flowers
0.3 m (1 ft.) June, July

G. rossii   *20 cm (8 in.) May-July*
Deeply cleft leaves and yellow flowers about 4 cm (1½ in.) in diameter.

GILLENIA

G. trifoliata Indian physic   *0.6 m (2 ft.) May, June*
White spirea-like panicles and red calyxes. An unusual plant for a semishaded border.

GLAUCIDIUM

G. palmatum   *38 cm (15 in.) Apr., May*
A very beautiful woodland perennial with two large palmate leaves at the upper part of the stem from which arises a large pale mauve saucer-shaped flower.

GLAUCIUM

G. flavum horn poppy   *30 cm (12 in.) June, July*
An English seaside plant with very effective rosettes of silvery leaves and chrome yellow flowers.

GUNNERA

G. manicata   *61 cm (24 in.) July*
Huge rhubarb-like leaves of pale green, often used at the side of pools.

GYPSOPHILA chalkplant   (Color photo p. 81)

Very well known and extremely useful for borders. Three tight airy graceful leathery masses of flowers. They have become almost the traditional perennial for hiding the early decaying leaves of *Papaucy, Dicentra, Anchusa,* and other plants that have untidy after-flowering habits. Their cloud-like branches form masses, which may be pushed in front of these plants to serve as complete replacements or as screening.

These are easily grown plants except for some of the *G. paniculata* hybrids, which are usually grafted onto the roots of seedlings of the type. Some added lime often improves their growth.

G. bodgeri see *Gypsophila paniculata* 'Compacta Plena'

G. manginii   *0.9 m (3 ft.) July, Aug.*
A branching plant with panicles of white flowers suffused pink. The smooth blue-green leaves make an excellent setting for the large flowers.

G. monstrosa   *0.6-0.9 m (2-3 ft.) June, July*
A hybrid of *G. repens* × *G. steveni*, much taller than *G. repens* with more branches and long stems up to 0.9 m (3 ft.) terminating in a cloud of white blossoms.

G. paniculata baby's-breath   *0.9 m (3 ft.) June, July*
The single, white 'Baby's Breath' is a very reliable hardy perennial with narrow pointed leaves and myriads of tiny white flowers in panicles up to 0.9 m (3 ft.) high and almost as wide.

'Bristol Fairy'   *0.8 m (2½ ft.) June-Aug.*
Double, white flowers.

'Compacta Plena'   *30 cm (12 in.)   May, June*
Dwarf, compact, but more vigorous than 'Bristol Fairy', with double, white flowers. Synonym: *G. bodgeri*

'Flamingo'   *0.6 m (2 ft.)   June-Aug.*
Double, rosy pink, suggested as a form of *G. oldhamiana*.

'Perfecta'   *0.8 m (2½ ft.)   June-Aug.*
A Dutch novelty with larger more fully double flowers than 'Bristol Fairy', but similarly somewhat short lived.

'Pink Star'   *23 m (9 in.)   June-Aug.*
Deeper and more compact than 'Rosenschleier'.

'Rosenschleier' rosy veil   *23 cm (9 in.)   June-Aug.*
Dainty, double, pink flowers on semiprostrate plants.

HAPLOPAPPUS

H. coronopifolius see *Haplopappus glutinosus*

H. glutinosus   *23 cm (9 in.)   May-July*
Dwarf, prostrate plants with deep yellow daisies. Synonym: *H. coronopifolius*

H. lyalli   *23 cm (9 in.)   May-July*
Dwarf plant with yellow flowers.

HELENIUM sneezeweed   (Color photo p. 81)

North American plants quite invaluable for late summer and autumn. They are characterized by their very prominent central disks and larger daisy-like flowers. Careful selection of cultivars is needed to avoid the tall straggly old-fashioned types.

Very easy to grow, most of them are quite useful for the back of the border. Tall ones may be kept bushier and dwarfer by pinching back all tips until mid-June.

H. autumnale   *0.6-1.8 m (2-6 ft.)   July-Sept.*
Most of the cultivars arose from this species, which grows wild from Quebec to Florida. Bright yellow flowers on usually tall rangy 1.5-1.8-m (5-6-ft.) plants unless growing in very poor soil, when they may grow only 0.6 m (2 ft.) high. Needs some protection on the prairies.

Cultivars   *July-Sept., except where noted*
'Allgold' — large flowers, yellow with bronze reverse
                                    1.2 m (4 ft.)

'Bruno' — a glowing mahogany red   *0.6 m (2 ft.)*

'Butterpat' — warm, rich, golden yellow flowers
                                    0.9 m (3 ft.)

'Chipperfield Orange' — yellow flowers, streaked and splashed crimson   *1.2 m (4 ft.)*

'Coppelia' — erect, fairly dwarf, brownish red flowers
                                    0.9 m (3 ft.)

'Copper Spray' — coppery red, similar to 'Bruno'
                                    1.1 m (3½ ft.)

'Crimson Beauty' — neat habits, bronze crimson flowers
                                    0.6 m (2 ft.)

'Fountain' — neat plant, bronze flowers in Aug. and Sept.
                                    1.2 m (4 ft.)

'Golden Fox' — very large red orange brown flowers
                                    0.9 m (3 ft.)

'Golden Youth' — large, rich butter yellow flowers
                                    0.9 m (3 ft.)

'Goldlackzwerg' see 'Mahogany'   *0.9 m (3 ft.)*

'Gypsy' — free flowering, bronze and gold flowers
                                    0.9 m (3 ft.)

'July Sun' — golden orange flowers   *0.9 m (3 ft.)*

'Karneol' — rich bronze flowers   *0.8 m (2½ ft.)*

'Madame Canivet' — yellow with a dark brown disk
                                    0.9 m (3 ft.)

'Mahogany' ('Goldlackzwerg') — gold and brownish red flowers   *0.9 m (3 ft.)*

'Moerheim Beauty' — warm glowing bronze red flowers
                                    0.9 m (3 ft.)

'Red Indian' — reddish bronze flowers

'Riverton Beauty' — old, popular, yellow flowers
1.5 m (5 ft.)

'Riverton Gem' — rich crimson flowers, streaked yellow
1.4 m (4½ ft.)

'Spatrot' — the best rich bronzy red   1.2 m (4 ft.)

'The Bishop' — large red, buttercup yellow flowers, dark center   0.9 m (3 ft.)

'Waltraud' — large golden brown flowers   1.1 m (3½ ft.)

'Wyndley' — erect, coppery orange flowers   0.6 m (2 ft.)

H. hoopesii   0.6 m (2 ft.)   May, June
A native from the Rocky Mountains to Oregon and California. It is an early flowering, more or less compact, dwarf plant with bright yellow flowers and golden centers. Useful for growing in containers for the patio.

HELIANTHUS sunflower   (Color photo p. 81)

A very large and coarse group of annuals and perennials; mostly for the rear of the flower border. They will grow almost anywhere and in any soil but are best planted in good heavy garden loam.

H. atrorubens
purple disk sunflower   1.8-2.1 m (6-7 ft.)   Sept., Oct.
Noted particularly for a very fine variety 'The Monarch' with very tall branching stems carrying large 13-15-cm (5-6-in.) golden, semidouble flowers. These will last for weeks when cut and placed in water. It is tuberous and can be stored like a dahlia, if there is a possibility that it might not over-winter, because it does not withstand unusually damp winters. Synonym: *H. sparsifolius*

H. decapetalus 'Multiflorus'
thinleaf sunflower   0.6-1.5 m (2-5 ft.)   July-Sept.
This is the species under which most cultivars are best placed.

Cultivars   July-Sept.

'Badirector Linne' — blood red   1.2 m (4 ft.)

'Capenock Star' — clear yellow flowers on tall bushy plants   1.5 m (5 ft.)

'Capenock Star Supreme' — good semidouble, yellow   1.2 m (4 ft.)

'Flore Pleno' — double, yellow   1.2 m (4 ft.)

'Loddon Gold' — rich double, yellow   1.5 m (5 ft.)

'Morning Sun' ('Morgens Anne') — double, lemon yellow   1.2 m (4 ft.)

'Soleil d'Or'   1.2 m (4 ft.)
Sulfur yellow with fluted and quilled petals like a cactus dahlia.

'Triomphe de Grand'   1.2 m (4 ft.)
Large golden yellow flowers with ball-shaped centers.

H. laetiflorus prairie sunflower   1.8 m (6 ft.)   Aug., Sept.
Very widespreading habit, likely to become a weed if not kept in check. Synonym: *H. scaberrimus*

'Mrs. Mellish'   1.8 m (6 ft.)   Aug., Sept.
Semidouble rich yellow flowers. Just as likely to spread as the species. Do not plant this cultivar unless it is needed to fill a difficult area where not much else will grow.

H. orgyalis see *Helianthus salicifolius*

H. salicifolius   1.8-2.4 m (6-8 ft.)   Sept., Oct.
A late bloomer, small golden yellow flowers produced in large numbers on long sturdy stems. Synonym: *H. orgyalis*

H. scaberrimus see *Helianthus laetiflorus*

H. sparsifolius see *Helianthus atrorubens*

HELIOPSIS orange sunflower

Sunflower-like perennials similar to *Helianthus*, but smaller, more compact, and with brighter daisy-like flowers. Easy to grow in any kind of soil, but needs reasonable moisture.

H. helianthoides   1.2-1.5 m (4-5 ft.)   July-Sept.
A native perennial from Ontario to Florida and west to Mississippi. Inclined to be too rank and weedy, but the cultivar 'Pitcheriana' is only 0.9 m (3 ft.) high and has larger flowers.

H. scabra   1.5 m (5 ft.)   July-Sept.
Native from Maine to New Jersey and Arkansas, the species is a rank grower with numerous flower heads and rough stems and leaves. The cultivars are much better than the species.

Cultivars   July-Sept.

'Ballerina' — warm yellow flowers   0.9 m (3 ft.)

'Gigantea' — large golden, semidouble flowers   1.2 m (4 ft.)

'Incomparabilis' — rich golden yellow, double flowers   0.9 m (3 ft.)

'Goldgefieder' — full double blooms, yellow rays and green centers   1.1 m (3½ ft.)

'Golden Plume' — double flowers of perfect form   0.9 m (3 ft.)

'Gold Greenheart' 0.9 m (3 ft.)
Very popular with double buttercup yellow flowers with green centers.

'Golden Rays' – some double light
orange flowers 0.8 m (2½ ft.)

'Light of Loddon' – bright yellow flowers 0.9 m (3 ft.)

'Patula' – semidouble flowers 0.9 m (3 ft.)

'Summer Sun' – pure orange yellow flowers 0.9 m (3 ft.)

HELLEBORUS Christmas rose, hellebore, Lenten rose
(Color photo p. 81)

Where shade is provided and the climate is a little milder than the Ottawa area, these can be highly desirable for planting in borders or as ground covers. They need a well-drained soil, which is best prepared by digging out the existing soil and making a pocket or bed of good loam and well-rotted leaves, peat moss, and coarse sand. Interplanting with ferns makes quite a good association.

H. corsicus Corsican hellebore 0.6-0.9 m (2-3 ft.) *Apr., May*
A very handsome species with pea green or apple green flowers borne on large clusters holding 15–20 blooms. The leaves are glaucous green and divided into three spiny-toothed leaflets.

H. foetidus stinking hellebore 0.6 m (2 ft.) *Apr., May*
A strong-smelling species with poisonous roots. The blooms are pale green, tipped with purple, and produced very early. The foliage is very deeply cut and divided almost to the center of each leaf.

H. niger Christmas rose,
black hellebore 0.6 m (2 ft.) *Apr., May*
With deep green leathery leaves, divided into oval segments and toothed at the apex. Very clean white saucer-shaped flowers with bright golden anthers. Grows best with some shade in deep soil that does not dry out.

Cultivars 0.6 m (2 ft.)
'Altifolius' *Apr., May*
Longer flower stems and larger 10-13-cm (4-5-in.) flowers.

'Keesen Variety' – a good white form *Apr., May*

'Potter's Wheel' – very large pure white flowers *Mar., Apr.*

H. orientalis Lenten rose 0.6 m (2 ft.) *Mar., Apr.*
A very variable species. The colored perianth may be cream, green, purple, rose, or almost black. Easier to grow than the Christmas rose. Usually sold as *H. orientalis* hybrids.

'Albion Otto' 0.6 m (2 ft.) *Mar., Apr.*
Pure white flowers with purple spots.

H. viridis green hellebore 0.4 m (1½ ft.) *Mar., Apr.*
A real Irish green flower, which is native to parts of Ireland. The flowers are pale green arising from undivided lance-shaped leaves, dull green above, and with prominent veins beneath.

HEMEROCALLIS day-lily (Color photo p. 66)

The day-lily is about the most adaptable of all plants. Few plants surpass it in vigor and beauty with its tall stout flower scapes and graceful arching foliage. All species appear to be hardy and dependable and will grow in practically any soil.

For *Hemerocallis* cultivars, see Chapter 2, Basic Perennials.

H. flava see *Hemerocallis lilioasphodelus*

H. fulva tawny day-lily 0.9 m (3 ft.) *July, Aug.*
This striking day-lily is as much a part of the Canadian roadside scene as many native wild flowers. By its spreading rhizomes it is able to creep some distance from the original garden to ditches and slopes nearby, where it grows unmolested by ploughs and clippers. Large reddish orange flowers on long stalks 1.2 m (4 ft.) high.

H. fulva var. *kwanso* 0.8 m (2½ ft.) *July-Sept.*
A double form of the species. It has given breeders trouble for centuries because of its sterility.

H. gracilis see *Hemerocallis minor*

H. lilioasphodelus lemon day-lily 0.9 m (3 ft.) *May-July*
Vigorous; dark green leaves; short funnel-shaped clear lemon flowers, 8–12 on a scape; spreading underground. The slightly fragrant flowers are showy early in the season and complement the other cultivars in flower at the same time. Synonym: *H. flava*

H. minor dwarf yellow day-lily 30 cm (12 in.) *May, June*
Grass-like deciduous foliage and small lemon yellow flowers. Synonym: *H. gracilis*

H. multiflora 0.9 m (3 ft.) *Aug., Sept.*
Long dark green leaves; short, starry funnel-shaped lemon yellow flowers on short branched scapes. Flowers later than the other species and has been responsible for the later blooming hybrids.

HEPATICA 10 cm (4 in.) *Apr.*

Beautiful anemone-like flowers native to our woodlands. Excellent for naturalizing in the shade.

H. acutiloba   23 cm (9 in.)   Apr., May
Similar to *H. triloba* but with lobes of leaves and acute involucral bracts. White and blue flowers. Synonym: *Anemone acutiloba*

Cultivars   June, July
  'Alba Plena' — double, white flowers   0.6 m (2 ft.)

  'Candidissima' — dwarf, compact;
  white flowers   38 cm (15 in.)

  'Flore Plena' — double, purple flowers   0.6 m (2 ft.)

  'Purpurea'   0.6-0.9 m (2-3 ft.)
  An old-fashioned border plant with fragrant purple flowers.

HERACLEUM

H. mantegazzianum giant
parsnip, cartwheelflower   2.4-3.1 m (8-10 ft.)   July-Sept.
A gigantic perennial for the wild garden with coppery red stems, deeply cut leaves 0.9 m (3 ft.) long, forming a tuft 3.7 m (12 ft.) across. The immense flower heads are said to contain up to 10,000 flowers. This giant perennial occasionally crops up in Canadian gardens, coming perhaps with seeds from Europe.

HESPERIS

H. matronalis sweet rocket   0.6-0.9 m (2-3 ft.)   June, July
An old-fashioned favorite "cottage garden" plant that produces a profusion of fragrant four-petaled single stock-like flowers on long spikes. The seedlings vary in colors of white, mauve, or purple. It is best to plant one of the good garden cultivars listed.

Cultivars *0.6 m (2 ft.) except where indicated*   May-July
  'Bloom's' — rich coral red flowers on graceful sprays

  'Bressingham Blaze' — very free flowering, large coral red flowers

  'Carmen' — carmine-pink flowers, strong stemmed

  'Damask' — sprays of small carmine rose flowers, 46 cm (18 in.)

  'Firebird' — crimson scarlet erect flower stems

  'Freedom' — rose pink flowers, 46 cm (18 in.)

  'Gaiety' — glowing carmine red

  'Garnet' — deep rosy pink

  'Hartsman'— crimson scarlet

  'Jubilee' — rose

  'Mary Rose' — bright pink

  'Oakington Jewel' — coral pink suffused copper

  'Pearl Drops' — pearl white flowers on slender stems

  'Pluie de Feu' — popular old crimson scarlet

  'Red Spangles' — rich scarlet crimson

  'Rhapsody' — large glowing pink

  'Scarlet Sentinel' — a very showy vigorous plant with long-lasting scarlet flowers

  'Scintillation' — bright pink bells, tipped coral and carmine, a large number of graceful spires

  'Snowflakes' — the very best white with large flowers

  'Sparkler' — carmine and scarlet

  'Splendour' — salmon scarlet

  'Sunset' — pink bells with coral red lips

HEUCHERA alumroot, coral bells   (Color photo p. 81)

These are excellent, neat, low-growing perennials with evergreen leaves in milder areas and leaves that turn to bronze in colder areas. Their graceful spikes of tiny bell-shaped or saucer-like drooping flowers last a long time, and the pattern and texture of the leaves of most species is a highly desirable feature. They grow best in good rich moist soil and do not tolerate dry conditions. As a result of the interest in this genus, a large number of hybrids have been developed with bigger flowers, new colors, and more vigorous plants. Because the stems become woody, the plants should be divided at least every 3 years.

H. angulosa loddon blue   10 cm (4 in.)   Mar., Apr.
One of the earliest species to flower.

H. brizoides × *Tiarella cordifolia* see *Heuchera* × *Heucherella tiarelloides*

H. × *Heucherella tiarelloides*   30 cm (12 in.)   May, June
Under the intergeneric hybrids are grouped all the resulting cultivars so obtained. The only selection available commercially is 'Bridget Bloom', which looks more like a heuchera but is smaller and more compact. At Ottawa this hybrid gives a remarkable display of light pink flowers all summer and has withstood many winters. Synonym: *Heuchera brizoides* × *Tiarella cordifolia*

H. sanguinea coral bells
This flamboyant red species from Mexico and Arizona was at one time the only member of this genus used in gardens. Now there are so many hybrids, that it is almost forgotten. The following cultivars have *H. sanguinea* parentage and *H. brizoides* influence. The *H. sanguinea* hybrids usually have larger flowers than the *H. brizoides* hybrids, whose sprays of bloom are more graceful.

H. transilvanica　　　*20-30 cm (8-12 in.)　Apr.*
A species from Romania with three-lobed glabrous leaves, shining beneath when old, and whitish or light blue flowers, larger than *H. triloba*. Best on alkaline soils.

H. triloba hepatica　　　*10 cm (4 in.)　Apr.*
Many forms of this plant may be found in the woodlands with various colors and sizes of blooms. Two cultivars are 'Alba' and a double pink form 'Rubra Flore Pleno'.

HIBISCUS　　(Color photo p. 82)

H. moscheutos　　　*1.2 m (4 ft.) Sept.*
The hardiest of the herbaceous hibiscus with white or cream flowers and purple red centers. From this and the other species have been bred a large number of hybrids, which if protected in winter will survive in Eastern Canada.

They are very large plants, up to 2.4 m (8 ft.) high with a spread of several feet, so they need a great deal of room. They need very wet soil or frequent watering. Sunny or partly shaded locations are best. The plants are very late starting in spring and are likely to be given up for lost. When all other plants are growing vigorously, they are just beginning to sprout. Their late blooms are often spoiled by early frosts in the Ottawa area, but in a warm fall or with some protection they blossom luxuriously. Synonym: *H. palustris*

Cultivars	*Sept., Oct.*
'Annie Hemming' — flamboyant red	*1.5 m (5 ft.)*
'Bessie Ross' — white with red eye	*1.5 m (5 ft.)*
'Poinsettia' — red flowers	*1.2-1.5 m (4-5 ft.)*
'Satan' — velvet rose flowers	*1.5 m (5 ft.)*
'Southern Belle' — early flowering strain (Aug., Sept.), white and pale rose	*0.9 m (3 ft.)*
'Super Rose' — very large rose	*1.2 m (4 ft.)*
'The Clown' — salmon pink, flushed white	*0.9 m (3 ft.)*

H. palustris see *Hibiscus moscheutos*

HIERACIUM hawkweed

The best of these are good perennial plants with dainty heads of small yellow or orange daisy-like flowers with ground-hugging silvery foliage. But some species can become very weedy; in fact, some of our worst farm weeds are hawkweeds.

H. bombycinum　　　*23 cm (9 in.)　June-Sept.*
Very woolly leaves and yellow flowers.

H. villosum shaggy hawkweed　　*30 cm (12 in.)　June-Aug.*
Silvery leaves and shaggy bright yellow flowers.

HOLLYHOCKS　　(Color photo p. 70)

See Chapter 3, p. 29.

HOSTA plantain-lily, funkia　　(Color photo p. 82)

Extremely attractive plants with particularly beautiful foliage. Because of their tolerance for shade they are best planted under trees and in shaded areas. Very useful in special positions where foliage texture and form are needed, such as beside a woodland seat, or flanking a pool. Their large showy leaves are marbled, crimped, variegated, and bluish gray or dark green. Best in good rich moist soil and shade, because the full sun is likely to burn the foliage. Much confusion surrounds the nomenclature of these plants. This publication follows Nyls Hylander *The Genus Hosta in Swedish Gardens*, Vol. 16 (No. 11), Acta Horti Bergiani, Uppsala, 1954, which seems to cover the genus very well and is being followed by many horticulturists.

H. albo-marginata　　　*0.6 m (2 ft.)　Aug.*
A small plant similar to *H. lancifolia* but with more spreading growth. Leaves pure green on both sides with a very narrow pure white border. Flowers funnel-like, with white-margined segments, deep violet stripes, and very dark tips; flower tube purple. Synonym: *H. lancifolia* var. *albo-marginata*

H. caerulea see *Hosta ventricosa*

H. crispula   0.6 m (2 ft.)   May, June
Very dense, slow growing; fine green leaves with a broad
white border from which narrow stripes run toward the
middle. Broadly funnel-shaped lavender flowers. Synonym:
H. fortunei 'Marginata Alba'

H. decorata   0.6 m (2 ft.)   Aug.
Compact, not over 0.6 m (2 ft.) tall; oval blunt-tipped leaves
abruptly narrowed to the stalk; small dark lilac blossoms;
spreading by stolons.

    'Marginata Alba' (known as Thomas Hogg)
    White-margined leaves.   0.6 m (2 ft.)   Aug.

H. elata   0.8 m (2½ ft.)   May, June
Dense, large groups having many vegetative shoots. Light
bluish violet funnel-shaped flowers. Synonym: *H. fortunei*
'Gigantea'

H. fortunei plantain-lily   0.6 m (2 ft.)   May, June
The reddish violet inflorescence lasts much longer than the
leaves. The funnel-shaped flowers are wide open at the
mouth. Leaves with distinctly winged petioles.

    'Gigantea' see *Hosta elata*

    'Marginata Alba' see *Hosta crispula*

H. glauca see *Hosta sieboldiana*

H. lancifolia lance-leaved plantain-lily   0.6 m (2 ft.)   Aug.
A small species with tufted growth and narrow leaves forming a clump of about 2 ft. (0.6 m). Pale lilac not very showy
flowers.

    'Fortis' see *Hosta undulata* 'Irromena'

    'Tardiflora' see *Hosta tardiflora*

H. lancifolia var. *albo-marginata* see *Hosta albo-marginata*

H. media picta see *Hosta undulata*

H. plantaginea fragrant plantain-lily   0.6 m (2 ft.)   Aug.
Very large heart-shaped bright green leaves and fragrant
white trumpet-like flowers on stems 0.6 m (2 ft.) high. The
fragrance of these flowers is very noticeable. Synonym:
Funkia subcordata

    'Grandiflora'   0.6 m (2 ft.)   Aug.
    Has much longer flowers than the type, up to 10 cm (4 in.)
    long.

H. sieboldiana   46 cm (18 in.)   June
A species with very ornamental corrugated glaucous or blue
green leaves 30 cm (12 in.) long. Its flowers are washy lilac-
white mostly hidden by the foliage, but its leaves are
valued for their distinctive texture and color. Synonym:
H. glauca

H. tardiflora   30 cm (12 in.)   Aug.
Densely tufted, pointed green leaves, and small dense flower
spikes. Synonym: *H. lancifolia* 'Tardiflora'

H. undulata wavy-leaved plantain-lily   0.8 m (2½ ft.)   Aug.
Very wavy undulating leaves, variegated white on green, 15
cm (6 in.) long and 9 cm (3½ in.) wide. Very effective in
arrangements. Pale lilac flowers about 5 cm (2 in.) long on
scapes 76 cm (30 in.) high. Synonyms: *H. media picta,*
H. variegata

    'Irromena'   0.9 m (3 ft.)   Aug.
    Pure green almost flat leaves. Taller than the type.
    Synonym: *H. lancifolia* 'Fortis'

    'Univittata'   0.8 m (2½ ft.)   Aug.
    A white central stripe on each large wavy leaf.

H. variegata see *Hosta undulata*

H. ventricosa blue plantain-lily   0.9 m (3 ft.)   July, Aug.
Elongated heart-shaped leaves about 23 cm (9 in.) long and
15 cm (6 in.) wide. The blades are deep bluish green, and
the flowers lavender blue, striped with a paler shade, on
large scapes up to 0.9 m 3 (ft.) long. Synonym: *H. caerulea*

IBERIS candytuft

Dwarf evergreen subshrubs with dark green foliage. These
are extremely useful for the front of the border, where they
may display their gleaming white blossoms with the tulips.

I. corifolia see *Iberis saxatilis* 'Corifolia'

I. gibraltarica   30 cm (12 in.)   May-July
A somewhat tender species, with lilac to lilac purple flowers
that last well into summer. Often self-sows, a method by
which it perpetuates itself in colder areas.

| I. *saxatilis* | 15 cm (6 in.) | May-July |

Very small, compact subshrubs 8-15 cm (3-6 in.) high, very neat plants with glistening white flowers that turn purple when they fade.

| 'Corifolia' | 15 cm (6 in.) | May-July |

Glabrous leaves and white nonfading flowers. Synonym: *I. corifolia*

I. *sempervirens* perennial
candytuft   30 cm (12 in.)   May-July
A very common border edging plant or rock garden plant with evergreen leaves and chalk white flowers on 20 cm (8 in.) stems.

| 'Christmas Snow' | 30 cm (12 in.) | May-July |

Tends to rebloom again in fall.

| 'Little Gem' | 20 cm (8 in.) | May-July |

Compact, densely flowered plant in clumps 20 cm (8 in.) across.

| 'Purity' | 30 cm (12 in.) | May-July |

A superior cultivar with larger flowers and compact plants.

| 'Snowflake' | 30-38 cm (12-15 in.) | May-July |

A very desirable form with large clusters of flowers and deep green leaves forming mats 0.6-0.9 m (2 or 3 ft.) wide.

INCARVILLEA trumpetflower   (Color photo p. 82)

These exotic-looking perennials have flowers resembling large gloxinias and striking shiny pinnate leaves. They like good rich garden soil and full sun with some protection from prevailing winds. In Eastern Canada, they need a covering of straw during winter.

| I. *compacta* dwarf trumpetflower | 0.3 m (1 ft.) | May, June |

Shy blooming, compact habit, bright rose pink trumpetflowers about 6 cm (2½ in.) long.

| I. *delavayi* Chinese trumpetflower | 0.4 m (1½ ft.) | May, June |

The most popular and vigorous species, aster-like leaves and large bright rosy red flowers, 5-6 blooms on each stem.

| 'Bees Pink' | 0.4 m (1½ ft.) | May, June |

Has blooms of a paler shade.

| I. *grandiflora* | 0.3 m (1 ft.) | May, June |

Dwarf species, with extremely large rich rosy red flowers with yellow throats, often 8-10 cm (3-4 in.) across.

| 'Brevipes' | 0.3 m (1 ft.) | May, June |

Much brighter deep rosy purple flowers.

INULA

This is a genus of easy-to-grow yellow daisy-flowered perennials, some coarse and straggly, others compact choice garden plants. Best in full sun in a good moist soil and some mulching could be a great advantage.

| I. *ensifolia* | 0.3 m (1 ft.) | Aug. |

A neat little bush with good foliage bejewelled with golden daisies borne singly on stiff stems.

| I. *grandiflora* | 0.6 m (2 ft.) | June |

The earliest flowering cultivar, similar to but much smaller than *I. helenium*, with orange yellow solitary flowers.

I. *grandulosa* see *Inula orientalis*

| I. *helenium* elecampane | 0.9-1.2 m (3-4 ft.) | June-Aug. |

A very vigorous plant with large stems and coarse leaves. The flowers resemble small bright yellow sunflowers. Will grow in extremely wet soils or can be planted in the border with other foliage plants.

| I. *hookeri* | 0.3-0.6 m (1-2 ft.) | Aug.-Oct. |

Yellow-rayed flowers on bushy plants.

| I. *magnifica* | 1.5-1.8 m (5-6 ft.) | June-Aug. |

Very robust, to 1.8 m (6 ft.) tall, purple striations on the rough hairy stems. Leaves are oval up to 0.3 m (1 ft.) long and 18 cm (7 in.) wide. Its bright golden yellow flowers are borne in terminal heads.

| I. *Oculus-Christii* | 0.3 m (1 ft.) | July |

A smaller plant than the other species bearing shaggy-petaled golden yellow flowers with the central cushion-like disc covered with a whitish wax when young. This gives the flower the eye-like appearance from which it gets its specific name.

140

I. orientalis Caucasian inula   *0.6-0.9 m (2-3 ft.)   June-Aug.*
This well-known species has large wavy-edged orange flowers 10-13 cm (4-5 in.) across and large striking leaves. Synonym: *I. grandulosa*

   'Laciniata'   *0.6-0.9 m (2-3 ft.)   June-Aug.*
   Laciniated ray florets.

I. orientalis var. *superba*   *0.6-0.9 m (2-3 ft.)   June-Aug.*
Much larger flowers than the type.

I. royleana   *0.6 m (2 ft.)   Aug.-Oct.*
A very striking Himalayan species bearing huge flowers with ray petals so long that they droop. The buds are black with green collars surrounding them. Needs a rich deep soil for permanence.

IRIS   (Color photos pp. 67, 82)

A genus of rhizomatous and bulbous perennials known to most gardeners through the large and striking flowers of the German or tall bearded kinds. The wealth of good garden materials among the other species is almost entirely forgotten by those who strive for larger and more distinctive blooms in the one widely known group. During recent years through the efforts of a few enlightened breeders, there has undergone a change in perspective and many of the almost forgotten types are being improved to meet modern demands. Beautiful as the tall bearded iris is, its use in a garden is limited and so it is wise often to choose another variety. Because the iris family comprises so many large groups and each group requires different cultural treatment, it is divided here into various groups according to their habits of growth and methods of cultivation.

Tall Bearded Iris

These beautiful plants are perhaps the most popular perennials in North America; their closest rival is the day-lily. In a few short years growers have brought this iris to an extremely high level. No other perennial has such a wide diversity of soft flower colors and forms as the tall bearded iris. They are characterized by the bearded appendages on the surface of the drooping petals, called falls. Some dwarf and intermediate cultivars are also bearded, so this group is distinguished by the term "tall." The flower consists of standard petals, which stand erect from the flower, fall petals, and beards. On some cultivars all three types of petals may have separate colors on the same flower.

The tall bearded iris was at one time under the species *I. germanica*, or *I. barbata*, but the plant has been hybridized so often with other species such as *I. pallida, I. variegata, I. cypeara, or I. trogana,* and tetraploids have been developed by crossing widely separated groups, that the plant now bears very little resemblance to this species.

For further notes on the culture and cultivars, see Chapter 2, Basic Perennials.

Dwarf Bearded Iris

This is a large group that is getting even larger as crosses are made with tall and intermediate irises to develop larger and better proportioned flowers in a wider array of colors. Most of these plants are early flowering, usually in April or May, and grow from 8 to 38 cm (3 to 15 in.) high. The miniature dwarf bearded irises are under 25 cm (10 in.) and the standard dwarf bearded irises are 25-38 cm (10-15 in.). Their culture is similar to the other bearded irises, but they may be planted in spring or fall. They are most satisfactory in narrow beds, on gentle slopes, and in the front of a border.

I. arenaria   10 cm (4 in.)   Apr., May
Very diminutive plant, bright yellow flowers, and a prominent orange beard on a stem under 10 cm (4 in.) high. Synonym: *I. flavissima* 'Arenaria'

I. chamaeiris   8-15 cm (3-6 in.)   Apr., May
Very often mistaken for and often sold as *I. pumila*, this species has a true stem on the flower, whereas *I. pumila* has a long lobe and no stem. The flowers of hybrids of this species are blue, purple, yellow, or whitish, and all have bright orange beards.

I. flavissima 'Arenaria' see *Iris arenaria*

I. mellita   10 cm (4 in.)   Apr., May
Related to *I. pumila*, but with sharply heeled narrow leaves, and red-veined smoky brown flowers. Also a yellow-flowered form.

I. pumila   8-13 cm (3-5 in.)   Apr., May
Densely spreading habit; mauve, yellow, cream, or reddish purple flowers.

Some of the best cultivars of the miniature dwarf bearded (MDB) and standard dwarf bearded (SDB) iris follow.

'Ablaze' — yellow and reddish bicolor,* bordered yellow
MDB 15 cm (6 in.)

'Angel Eyes' — white self, blue spot   MDB 13 cm (5 in.)

'April Morn' — flax blue self   MDB 10 cm (4 in.)

'Atomic Blue' — clear sky blue self   MDB 10 cm (4 in.)

'Baria' — barium and citron-yellow bitone†
SDB 25 cm (10 in.)

'Blue Mascot' — blue self   MDB 10 cm (4 in.)

'Brassie' — chrome yellow blend   SDB 36 cm (14 in.)

'Bright White' — white self   MDB 25 cm (10 in.)

'Butterball' — sulfur yellow self   MDB 23 cm (9 in.)

'Cherry Spot' — white and cherry red bicolor
MDB 17 cm (6½ in.)

'Cup and Saucer' — mahogany red purple self
MDB 13 cm (5 in.)

'Dale Dennis' — white with orchid markings and crests, plicata   SDB 25 cm (10 in.)

'Dark Fairy' — plum purple bitone   SDB 33 cm (13 in.)

'Dear Love' — pale blue self   SDB 30 cm (12 in.)

'Easter Holiday' — yellow chartreuse bitone
SDB 25 cm (10 in.)

'Flaxen' — light blue self   MDB 10 cm (4 in.)

'Golden Fair' — golden yellow self with a white blaze
SDB 30 cm (12 in.)

'Green Spot' — white self, green spot on falls
SDB 25 cm (10 in.)

'Honey Bear' — yellow and chestnut brown overlay
MDB 13 cm (5 in.)

'Knotty Pine' — tan and brown blend   SDB 33 cm (13 in.)

'Lemon Flare' — lemon cream, ivory self
SDB 25 cm (10 in.)

'Pam' — light yellow   SDB 25 cm (10 in.)

'Promise' — violet red self   MDB 20 cm (8 in.)

'Red Gem' — near oxblood red self   MDB 19 cm (7½ in.)

'Sparkling Eyes' — white and blue violet bicolor
MDB 18 cm (7 in.)

'Veri-Gay' — bright deep yellow and red bicolor
MDB 13 cm (5 in.)

Intermediate Bearded Iris

These are appreciated very much more now, and they will be even better known when the Median section of the American Iris Society finishes developing this group and the standard dwarf bearded and border types of iris. These are thin-stemmed plants with smaller flowers than the tall bearded group.

In the intermediate class of median irises, many of the cultivars are tough and persistent and will bloom abundantly in the same spot for years. Their history is as obscure as that of the other groups. Some of them are believed to have originated with *I. albicans*, a white iris of very ancient lineage; *I. kochii*, a red purple iris; and the German and Florentine irises. From the progeny of these species that have been crossed and intercrossed have come the present-day types, which show a very wide range of patterns and colors, all with perfectly formed blooms that stand up to all kinds of weather and grow with ease in any soil.

---

*bicolor — two colors
†bitone — two tones of the same color

Some of the best cultivars follow.

'Alien' – tan and brownish red to lavender blue bicolor
46 cm (18 in.)

'Allah' – blue lavender bitone   38 cm (15 in.)

'Cloud Fluff' – pure white self   46 cm (18 in.)

'Easter Bunny' – white self   46 cm (18 in.)

'First Lilac' – lilac self   46 cm (18 in.)

'Kiss Me Kate' – sea-foam green with a margin of wisteria
violet   51 cm (20 in.)

'Lilli Hoog' – blue self   46 cm (18 in.)

'Lillipinkput' – golden apricot self   30 cm (12 in.)

'Pink Debut' – light pink self   51 cm (20 in.)

'Pink Fancy' – light shell pink self   46 cm (18 in.)

'Sugar' – pale ivory and ivory yellow bitone   51 cm (20 in.)

'Yellow Dresden' – lemon yellow self   66 cm (26 in.)

'Zwanenburg' – brown and yellow brown variegata
51 cm (20 in.)

Japanese Iris   (Color photo p. 82)

The Japanese iris needs acid soil and plenty of moisture. *Iris kaempferi* will grow in moist soil or in a well-prepared border, but *I. laevigata* is a bog-garden plant that needs its roots under water most of the time. The hybrids have been selected because of their tolerance for drier conditions, but they will thrive in a place beside a lily pond, where their majestic floral effect is multiplied by their reflection in the water. Cultivars of the Mar-Higo or Higo strains have been successfully grown at Ottawa in a nursery.

The Japanese irises are the flat tops or, as Eleanor West-meyer, a well-known irisarian, says, "the flying saucers" of the iris world. Single blossoms may measure 25-36 cm (10-14 in.) in diameter, and the petal tracings of intricate design and texture resemble the most exotic brocades, satins, and velvets. They are extremely beautiful flowers and well worth cultivating, even by the most discriminating gardener.

Cultivation of these magnificent flowers is quite different from methods used for the common iris. These exotic jewels love moisture, ample food, and organic matter, and they are at their worst in limey soils. The common iris, on the other hand, must not be overfed and will revel in copious supplies of old mortar, rubble, and lime.

In Japan, the Japanese irises are often grown in discarded rice paddies, which can be flooded during the growing season and drained for the winter. Ideal conditions for their successful growth are a good rich, very moist soil in summer and well-drained dry conditions in winter.

Because these are not rhizomatous plants but have fibrous roots, they should be planted in the usual way with their roots firmly embedded in the soil. If planting can not be done in August, it is better to wait until early spring.

During summer the plants should receive copious, almost flood-like, amounts of water, and lots of mulching with peat moss or pine needles.

Cultivars

'Blue Coat' – medium blue double flowers, tailored form

'Diamond Night' – marbled dark bluish purple double flowers

'Imperial Palace' – velvety beet-red double flowers, well branched

'Imperial Robe' – very large, imperial purple double flowers

'Ivory Glow' – large, warm-tinted double flowers with cream styles

'Ocean Mist' – soft sky blue, double flowers with white center

'Pink Frost' – light pink, heavily ruffled, double flowers

'Red Titan' – bright red violet ruffled double flowers

'Rose Tower' – big ruffled rose red flowers with white center

'Royal Pageant' — large, pink flowers

'Snowy Hills' — frilled double flowers with good substance

Siberian Iris (Color photo p. 82) *0.6-1.2 m (2-4 ft.)* *May, June*

The Siberian iris has long been considered the Cinderella of the iris world, but in recent years its popularity has so greatly increased that it probably will become a rival of the more ornamental and more glamorous tall bearded type. These stalwart perennials have always been admired by more experienced gardeners, but they are now being grown in increasing numbers by homeowners, who are beginning to appreciate their lasting qualities and usefulness.

It is unfair to compare these irises with the bearded types because they perform quite a different function in the garden. Although the flowers are smaller, they are exquisitely refined and yet not as fragile as the bearded kinds. High winds merely mass the flowers together for a while and never scar even the widest open floret.

Whereas the tall bearded iris flower can stand alone, elegant and refined in every detail, the flowers of the Siberian irises must have company. Their special purpose in the garden is to present collectively a bright blue, purple, sky blue, or white sea of bloom as a reward for their inclusion.

Clumps of these irises, 0.6-1.2 m (2-4 ft.) high, are attractive from spring to fall, because, unlike the bearded ones, their foliage does not deteriorate right after flowering, but continues to grow unchecked. The lush green leaves are slender and grass-like and suitable for poolside planting.

Cultivation is extremely simple, because Siberian irises are very resistant to insects, disease, and even rough treatment from children, pets, and mechanical equipment.

Meadow plants by nature, these irises do best in moist locations, although they will grow reasonably well in sandy soils.

Besides poolside plantings, they are well suited for massing in drifts by themselves. Miss Isabella Preston, formerly a plant breeder at the Central Experimental Farm who was responsible for many new introductions of these perennials, was so fascinated by the water-like mass effect of these plants that she named all her varieties after Canadian rivers and lakes.

They may also be planted in perennial borders, beside a tree or gate or at the margins of rock gardens, or they can be used as a hedge-like divider between gardens.

Although these plants do best in full sun, they will tolerate some shade and most soils if they are well drained. Allow them plenty of space, because they grow fast and can be left for 8-10 years before needing to be divided.

The best time to plant Siberian irises is September. They become established quickly and large clumps will produce a few blooms the first year. No matter when you order the young plants, most nurserymen will ship them only in September. Single young plants are very small and slow in starting to grow, so, if you want an immediate effect, select fairly large clumps.

Prepare the soil well and enrich it with humus such as compost or peat moss and a complete fertilizer (6-9-6). Dig a large enough hole for each clump so that the roots are not crowded, water them well, and firm the soil around them. Set each group of plants at least a foot apart to allow for future growth.

A year or two after they have been planted, the clumps become so strong that weeds rarely get a foothold. However, you must keep weeding them until this time arrives. As they get older, it is wise to fertilize them in early spring and again in July.

There are many good standard cultivars, and some exciting new ones are being introduced. Among the best irises under test at the Plant Research Institute's trial gardens are 'White Swirl', the best new white; 'Blue Herald', a deep royal blue; and 'Tunkhannock', a purple lavender.

Older cultivars of merit are 'Snow Crest', white; 'Tycoon', bluish white; 'Royal Ensign', aster purple with a blue patch; 'Eric the Red', reddish magenta; 'Gatineau', light blue; 'Mountain Lake', bright clear blue; 'Cool Spring', two-toned blue; and 'White Dove', a very hardy white.

Two breeders in England have introduced yellow and spotted varieties to this class of iris, so soon we may expect a wider choice of colors. The recently introduced lavender and white cultivar 'Mandy Morse' is one of the best tested in recent years and is well worth using in your border.

Spuria Iris

This is a comparatively new group of hybrid beardless iris originating from hybrids involving the species *I. ochroleuca, I. monnieri, I. crocea, I. graminea, I. sintensi, I. songarica,* and many others that closely resemble *I. spuria,* the name given to a number of forms of beardless irises with six-ribbed fruit, a round stem, and nearly linear leaves.

The modern hybrid species are very much like the Spanish iris *(I. xiphium),* but the constricted neck in the iris that separates the haft from the fall petals is shorter and the oval blade larger. This standard is wider and more striking. Some new hybrids are nearly 1.8 m (6 ft.) high, although the average is 1.2 m (4 ft.). There is a tremendous variation in the type of flower and the flower color from yellow to brown, lavender, blue, and purple. Many kinds are hardy at Morden, Manitoba.

Other Beardless Irises

I. aurea   1.2 m (4 ft.)   June
Golden yellow flowers; strong growing, needs plenty of moisture. Two sessile clusters of flowers on each stout stem.

I. bulleyana   0.4 m (1½ ft.)   June, July
Pale lavender blue, reticulated yellow flowers; needs very moist soil.

I. cristata crested iris   15 cm (6 in.)   May
Soft fringed blue and spotted orange, a crest of petaloid tissue on the falls instead of a beard.

I. delavayi   0.9-1.2 m (3-4 ft.)   June
Deep violet purple, *I. sibirica* type, needs very moist almost swampy soil.

I. dichotoma vesper iris   0.6 m (2 ft.)   Aug.
Small purple, white, or greenish white flowers with a spotted purple haft. Its common name refers to its very fleeting bloom that opens in the afternoon and fades before nightfall.

I. douglasiana   23 cm (9 in.)   May, June
A showy dwarf Californian species that forms thick mats of grassy foliage. The flowers are 4-5 cm (1½-2 in.) deep, very variable in color, but mostly blue with golden penciling.

'Margot Holmes'   0.4 m (1½ ft.)   June
An outstanding hybrid *(I. douglasiana × I. chysographea)* with rich plum purple flowers emphasized by deep golden markings on the falls. This hybrid needs moist soil.

I. fimbriata see *Iris japonica*

I. flavescens   0.8 m (2½ ft.)   May
A flag iris, with pale yellow flowers.

I. forrestii Yunnan iris   30-38 cm (12-15 in.)   July
Good clear yellow flowers resembling a dwarf *I. sibirica,* but not so rugged or hardy and needs a great deal more moisture.

I. graminea   30 cm (12 in.)   May, June
Fragrant reddish blue and violet flowers in 25-cm (10-in.) stems, partly hidden by the leaves.

I. innominata   30 cm (12 in.)   May, June
A charming dwarf; variable apricot, orange, and mauve flowers. Dr. Riddal of Portland, Oregon, has a fine selection of unnamed seedlings that should be available soon.

I. japonica   0.4 m (1½ ft.)   May, June
A very attractive, distinct iris producing many flowered inflorescences of lavender blue. Synonym: *I. fimbriata*

I. lacustris   8 cm (3 in.)   May
A tiny flag iris, with pale blue flowers.

I. minuta   10-20 cm (4-8 in.)   May
A very small species, having narrow grass-like leaves, small solitary yellow flowers, and falls edged with brown.

I. ochroleuca yellowband iris   1.5 m (5 ft.)   June, July
A very large plant, wide strap-shaped leaves, several tiers of large yellow and white flowers on long stems.

'Queen Victoria'   0.9 m (3 ft.)   July
Creamy white flowers with an orange blotch.

I. orientalis   1.1 m (3½ ft.)   June
Similar to *I. sibirica,* but not so hardy and more demanding. Needs rich, very moist soil and some winter protection in cold areas. Violet blue flowers with conspicuous crimson spathes. Synonym: *I. sanguinea*

'Nana'   30 cm (12 in.)   June
Typical Siberian iris type with violet blue flowers.

I. pseudacorus yellow flag iris   0.9 m (3 ft.)   May, June
Although native to Europe and Western Asia, this species is often found naturalized in North America. An excellent subject for growing alongside streams and pools with its roots in the muddy banks or in a good moist border. Its almost scentless bright yellow flowers are clustered at the top of stout 0.9-m (3-ft.) stems.

I. sanguinea see *Iris orientalis*

I. setosa   10-30 cm (4-12 in.)   May, June
This species and a number of hybrids are grown in Alberta, where they are quite hardy and drought resistant.

I. tectorum roof iris of Japan   0.3 m (1 ft.)   June
Beautiful wavy blue flowers, crested white. Gained fame because of its being planted on the thatched roofs of some Japanese houses. Needs a sunny sheltered well-drained location. Not adaptable for growing on the roofs of Canadian houses.

I. versicolor   0.6 m (2 ft.)   May, June
The native blue-flowered flag iris. A very good border subject. Many beautiful variations in flower color occur in nature.

I. wilsonii   0.6 m (2 ft.)   June
An *I. sibirica* type, light yellow blooms with veined brown throat.

KIRENGSSHOMA

K. palmata yellow waxbells   *0.9-1.2 m (3-4 ft.)*   *Sept., Oct.*
A very choice and distinct Japanese perennial with very beautiful maple-like palmate leaves and pendulous, tubular waxy canary-yellow flowers, produced in the fall. Needs a very moist, partly shady location and must never be allowed to become dry in the summer.

KNIPHOFIA torch-lily, red-hot-poker

Where this plant survives the winter it is of great value for bold effects, especially beside pools, lakes, and streams. Its drooping, elongated scarlet, orange, or yellow flowers are borne close together at the top of the stem, and resemble a flaming or glowing torch, which accounts for its common name. The torch-lilies may be a lot hardier than is assumed. In Britain, they were thought to be tender until it was found that the methods of handling them during dormancy accounted for excessive winter losses. The plants hold their foliage during winter, but they spread out over the ground exposing their crowns to the elements. During showers these crowns fill with moisture, which freezes in severe weather and expands the membranes, eventually killing the plant. In England, gardeners tie the leaves together with a rubber band or plant tie and cover the vital crown.

If the plant is protected in this manner, it may be hardy in southern Ontario, milder parts of British Columbia, and perhaps even in Nova Scotia and parts of Newfoundland. These plants are so lovely that, with newer and hardier strains forthcoming, they are worth a try in various parts of the garden and with different kinds of covering. Some of these plants have withstood a few winters in the Ottawa area when they were grown near the foundation and at the side of the house where the snow collected. Some gardeners, too, have successfully stored the plants in the basement during the winter. Synonym: *Tritoma*

K. aloides see *Kniphofia uvaria*

K. caulescens   *0.9-1.2 m (3-4 ft.)*   *Aug.-Oct.*
Very glaucous leaves arising from a woody trunk, and reddish salmon to yellowish white flowers. Remarkably similar to a yucca in foliage habits.

K. foliosa   *0.6-0.9 m (2-3 ft.)*   *Aug.*
Similar to *K. caulescens*, but growth arises from the base on distinct stems instead of from a trunk. Red-tinged yellow flowers on long racemes.

K. galpinii   *46 cm (18 in.)*   *Aug.-Oct.*
A very small torch-lily for the front of the border. Needs a sheltered location, even in Britsh Columbia. Quite distinctive vivid flame-colored flowers.

K. tubergeni   *0.8 m (2½ ft.)*   *Aug., Sept.*
Light yellow flowers on very neat plants.

K. uvaria red-hot-poker
This is the specific epithet under which most of the hybrids are placed. The common red-hot-poker has dense spikes of orange, scarlet, and yellow flowers, very conspicuous in July and August. Synonym: *K. aloides*

Cultivars

'Alcazar' — fiery red flowers   *0.8 m (2½ ft.)*   *Aug.*

'Bees' Lemon' — very bright lemon yellow flowers
    0.8 m (2½ ft.)   *Aug.*

'Bees' Sunset' — orange flame flowers   *0.9 m (3 ft.)*   *Aug.*

'Buttercup' — yellow flowers, free flowering and vigorous
    1.2 m (4 ft.)   *Aug.*

'Coral Sea' — coral pink-flowers   *0.9 m (3 ft.)*   *Aug.*

'Earliest of All' — orange scarlet flowers
    0.9 m (3 ft.)   *Aug.*

'Maid of Orleans'   *1.5 m (5 ft.)*   *July-Sept.*
Dense spikes of pale primrose flowers, opening to ivory white, withstands more severe winters and blooms over a longer period than the other cultivars.

'Mount Etna' — intense scarlet flowers
    1.5 m (5 ft.)   *Aug., Sept.*

'Royal Standard' — bright red and yellow flowers
    1.1 m (3½ ft.)   *July*

'Springtime' — coral red and ivory flowers
    0.9 m (3 ft.)   *July*

'Summer Sunshine' — clean flame red flowers
    0.9 m (3 ft.)   *July*

'White Fairy' — cream flowers   *0.8 m (2½ ft.)*   *July*

LAMIUM dead-nettle

Mostly perennial weeds, but the following are attractive in the front of the border and will grow in full sun or light shade in any garden soil. If sheared mildly in midsummer, the plant will be neater and more compact.

L. galeobdolon 'Florentium'   *46 cm (18 in.)*   *July*
Yellow flowers in dense whorls.

L. maculatum spotted dead-nettle   *30 cm (12 in.)*   *July*
A very hardy semitrailing plant with purple flowers, square stems, and leaves with bold white stripes and blotches covering the main veins. A useful ground cover. Some of the varieties have white and pink flowers.

'Aureum'   *30 cm (12 in.)*   *May, June*
The golden-spotted foliage is much more attractive than the species. A good edging plant in moist soils.

L. orvala giant dead-nettle   *0.6 m (2 ft.)*   *May-July*
Very bright rosy purple flowers in early summer, and leaves often more than 6 cm (2½ in.) wide.

L. veronicaefolium   *46 cm (18 in.)*   *May-Aug.*
A little known native of Spain, with large bright rosy pink blooms on 15-cm (6-in.) stems and rusty green leaves that are coarser than *L. maculatum*. Flowers most of the summer.

LATHYRUS

Perennial version of the annual sweet pea, but with smaller and many more blooms on each stem. They are useful for trailing over fences at the back of the flower border.

L. grandiflora everlasting pea *1.5-1.8 m (5-6 ft.)*   *June-Aug.*
A perennial climber with rose-colored flowers similar to and almost as large as the annual sweet pea. The flowers only last a day, but the plant is so floriferous that it is seldom out of bloom. It does not appear to produce seed, so it must be divided, a slow process, which probably explains why it is so seldom seen in gardens.

L. latifolius perennial pea   *0.6-2.4 m (2-8 ft.)*   *July-Sept.*
An excellent climber for a corner site or the back of a border. Neat elliptical leaves and long-stemmed sprays of large multi-flowered rosy blooms. During the summer, some supports are needed for the plant to climb on. A few improved cultivars are available commercially:

    'Pink Beauty' — deep pink flowers
    'Roseus' — rich rose flowers
    'Snow Queen' — white flowers
    'White Pearl' — pearly white flowers

L. rotundifolia   *0.9 m (3 ft.)*   *May, June*
A species with bright brick red flowers and rounded leaves; it requires a warm location.

LAVANDULA

A subshrub that is grown mostly in flower borders, rock gardens, on top of walls, and as edging in places where perennials are mostly used.

L. spica common lavender   *0.3-1.2 m (1-4 ft.)*   *June-Aug.*
Under this name *L. officinalis*, *L. vera*, and all the cultivars are included. In mild climates this becomes a shrub 0.9-1.2 m (3-4 ft.) high, with square stems and whitish leaves. The inflorescence is up to 0.4 m (1½ ft.) long with terminal spikes of crowded exquisitely fragrant pale greeny blue flowers.

Cultivars   *all summer*
    'Alba' — white flowers   *0.8 m (2½ ft.)*

    'Gruppenhall' (Old English, Gigantea) — long blue spikes
                               0.9 m (3 ft.)

    'Hidcote Blue' — dwarf, with red purple flowers
                             38 cm (15 in.)

Lavatera cachemirica, Cashmerian mallow

'Jean Davies' — pale pink flowers   *38 cm (15 in.)*

'Munstead' — neat growing, deep blue flowers
                             30 cm (12 in.)

'Twinkle Purple' — neat growing, a very good color
                             0.6 m (2 ft.)

LAVATERA tree-mallow   (Color photo p. 83)

Vigorous strong-growing annuals, biennials, and perennials. Some of the perennials are shrubby.

L. cachemirica Cashmerian mallow
         1.2-1.5 m (4-5 ft.)   *all summer*
Very easy to grow from seed. Large bush plants with a profusion of pink flowers all summer. The heart-shaped leaves have five crenate lobes, nearly glabrous above, and long stalks.

L. olbia tree lavatera   *1.2-1.5 m (4-5 ft.)*   *all-summer*
A shrubby perennial, extremely useful for a dry sunny place or a corner of the garden. The soft grayish leaves are quite prominent on the wide bush. Flowers are single pink mallows, 8 cm (3 in.) in diameter. Plants are short lived.

    'Rosea'   *1.5-1.8 m (5-6 ft.)*   *all summer*
    Pure satiny rose, not shrubby; excellent for the back of the border.

L. thuringiaca   *1.2-1.5 m (4-5 ft.)*   *all summer*
Very similar to *L. cachemirica*, but with rosy-colored flowers.

LIATRIS blazingstar, gayfeather   (Color photo p. 83)

Very unusual, striking tuberous-rooted plants with stiff spikes 0.6-0.9 m (2-3 ft.) high for the border in late summer and autumn. The fluffy flowers open first from the top of the spike and continue progressively downward. Best in sandy well-drained soil. *L. punctata*, *L. scariosa*, and *L. spicata* are hardy at Morden, Manitoba, and in Alberta, so it is quite likely that these and the other species will grow well on the rest of the prairies.

L. callilepis see *Liatris spicata*

L. graminifolia grassleaf gayfeather
0.6-0.9 m (2-3 ft.)   Aug.-Oct.

Extremely narrow foliage and bright rose flowers resembling rose-colored pokers. A very good plant in dry soil.

L. punctata dwarf gayfeather, dotted blazingstar
0.3-0.8 m (1-2½ ft.)   Aug.

A dwarf species, native from the Canadian Prairies to New Mexico.

L. pycnostachya Kansas feather
0.9-1.5 m (3-5 ft.)   Aug.-Oct.

A very tall species, with dense spikes of purple flowers 15-20 cm (6-8 in.) long.

L. scariosa   0.6 m (2 ft.)   Aug.-Oct.
Resembles *L. pycnostachya*, but with rather narrow spoon shaped leaves and light purple hemispherical heads of 15 to 45 flowers.

Cultivars   Aug.-Oct.
'Alba' — white flower heads   0.6 m (2 ft.)

'Nana' — a rare cultivar, reddish purple flowers
20 cm (8 in.)

'September Glory' — all the purple flowers open simultaneously   0.9 m (3 ft.)

'Silver Tips' — silver tipped pink flowers   0.6 m (2 ft.)

'White Spire' — a sport of 'September Glory' with the same habit   0.9 m (3 ft.)

L. spicata spike gayfeather   0.9 m (3 ft.)   Sept.
The most commonly grown species, it will withstand wetter conditions that the other species. The long-lasting flowers are reddish purple. Synonym: *L. callilepis*

Cultivars
'Alba' — white flowers   0.6 m (2 ft.)   Aug.-Sept.

'Kobold' — a deeper color than the species
0.6 m (2 ft.)   Aug.-Oct.

LIGULARIA   (Color photo p. 83)

Very handsome bold-leaved plants related to and often included with the genus *Senecio*. Of tremendous value for waterside planting or in moist borders, where their richly colored flowers and heavily textured striking leaves show to advantage.

L. clivorum see *Ligularia dentata*

L. dentata golden groundsel   1.2 m (4 ft.)   July-Sept.
Very large heart-shaped leaves to 0.3 m (1 ft.) across, branching stems, and 8-cm (3-in.) orange yellow flowers with brown centers. Synonyms: *L. clivorum*, *Senecio clivorum*

Cultivars   July-Sept.
'Desdemona' — very striking purple foliage, orange flowers
1.2 m (4 ft.)

'Gregynog Gold'   0.9 m (3 ft.)
Golden flowers with bronzy centers; a hybrid of *L. dentata* × *L. veitchiana* that originated at Gregynog Hall, Newton, Wales.

'Othello' — leaves veined and tinted purple, flowers orange
1.2 m (4 ft.)

L. × hessei　　　　*1.5-1.8 m (5-6 ft.)　July-Sept.*
An interspecific hybrid *(L. clivorum × L. wilsoniana)* with large spikes of rich yellow flowers.

L. hodgsonii　　　　*0.6 m (2 ft.)　June, July*
Dark foliage and orange flowers.

L. intermedia　　　　*0.4 m (1½ ft.)　Aug.*
Kidney-shaped or broadly cordate glabrous leaves on long stalks, four or five yellow flowers on each flower head.

L. japonica giant ragwort　　*1.5 m (5 ft.)　Aug., Sept.*
A very large expansive species with leaves almost a foot across, divided into 7-11 segments. The thick flower stems carry 8-cm (3-in.) orange flowers. This plant needs plenty of water and is best planted at the edge of a pool. Synonym: *Senecio japonica*

L. stenocephala 'Globosa'　　*1.5 m (5 ft.)　Aug., Sept.*
Pure yellow flowers.

L. tussilaginea　　　　*0.6 m (2 ft.)　Aug.*
A very handsome perennial with a creeping underground rootstock and nearly round much-toothed leaves. Light yellow flowers on very woolly branched stems.

Cultivars　　　　　*0.6 m (2 ft.)　Aug.*
　'Argentea' – white-margined leaves

　'Aureo Maculata' leopardplant – golden, white, pink variegated leaves; not as hardy as the species

L. veitchiana　　　*0.9-1.8 m (3-6 ft.)　July-Sept.*
A very bold massive plant with heart-shaped leaves almost 0.6 m (2 ft.) across. Deep yellow flowers in large clusters.

L. wilsoniana　　　　*1.5-1.8 m (5-6 ft.)　July*
Another very vigorous large-leaved plant, but earlier flowering than most of the other species. Golden yellow flowers borne densely on columnar spikes.

LIMONIUM sea-lavender

A genus of annual and perennial plants long known and even better known as *Statice*. The perennials are long-lived plants provided they are located in the sun and given adequate drainage. Their paper-textured flowers last a long time and are valuable for drying. Most of the plants in this genus withstand unusual saline soils such as those bordering the sea and salty conditions.

L. bellidifolium　　　*30 cm (12 in.)　Aug.-Oct.*
A miniature species with tiny violet purple flowers.

L. cosyrense　　　　*15 cm (6 in.)　July-Sept.*
Dense tufts of leaves and clusters of tiny mauve flowers.

L. dumosum see *Limonium tataricum* 'Angustifolium'

L. elatum　　　　　*0.6 m (2 ft.)　July*
The spiny-tipped basal growths of this species are composed of long stick-like bright green leaves. Branched flower heads bear small blue flowers in profusion.

L. eximium　　　　　*0.6 m (2 ft.)　Aug.*
An uncommon species with winged leaf stalks. The flower panicles are downy but not winged. Rosy lilac flowers crowded into separate lateral spikes.

　'Album' – white flowers　*0.4-0.6 m (1½-2 ft.)　July-Sept.*

L. felicularis　　　　*0.6 m (2 ft.)　July, Aug.*
Similar to *L. latifolium*, but with narrower leaves and lilac pink flowers.

L. gmelini　　　　　*0.6 m (2 ft.)　Aug.*
Rosettes of oval oblong leaves and branched heads of tiny pinkish tubular flowers.

L. latifolium
common sea-lavender　*0.6-0.9 m (2-3 ft.)　Aug., Sept.*
The most popular species for gardens with its cobwebby masses of deep lavender blue flowers and rosettes of long leaves. At its best, it is a very showy border plant with a large number of light airy flowers forming a canopy more than a yard wide. Easy to grow and very hardy.

Cultivars　　　　　*0.8 m (2½ ft.)　July-Sept.*
　'Blue Cloud' – large mauve florets

　'Chilwell Beauty' – very large deep violet blue flowers

　'Elegance' – deep blue flowers

　'Grandiflora' – deep lavender blue flowers

　'Grittleton Variety' – large sprays of light lavender mauve flowers

　'Violetta' – deep violet flowers

L. minimum　　　　*10 cm (4 in.)　July, Aug.*
A rock garden miniature bearing clusters of flowers with white calyxes and reddish lilac corollas. This effect is enhanced by glistening gray green foliage. A very dainty alpine.

L. tataricum　　　　*0.8 m (2½ ft.)　Sept.*
Sometimes forming a mass of leaves and flowers 1.5 m (5 ft.) wide. A woody rootstock. The triangular panicles of flowers are 0.3-0.4 m (1-1½ ft.) high and 15 cm (6 in.) wide, thinly branched, but produce an effective showing of ruby red flowers.

Cultivars
　'Angustifolium'　　　*0.6 m (2 ft.)　Sept.*
A billowy mass of silvery gray lavender flowers. Synonym: *L. dumosum*

　'Nanum' – fluffy pinkish flowers *23 cm (9 in.)　Aug., Sept.*

LINARIA

A few species are tall enough to be useful as border plants and will tolerate dry hot summer weather. They do not live long, but replenish themselves by seeding, a somewhat weedy trait and one that should be watched closely.

L. dalmatica Dalmatian toadflax   *0.6 m (2 ft.)   June-Sept.*
A rather graceful and elegant snapdragon-like plant with large yellow flowers and blue green foliage.

L. purpurea purple toadflax   *0.6 m (2 ft.)   July-Sept.*
Erect branches covering slender spires of purplish blue flowers, with white throats. Flowers are produced over a long period.

'Canon J. Went'   *0.9 m (3 ft.)   July-Sept.*
Tall spires of pink flowers.

L. triornithophora   *0.8 m (2½ ft.)   June-Sept.*
A very attractive species with glaucous foliage and comparatively large lilac purple flowers with yellow throats.

LINDELOFIA

L. longiflora   *0.4 m (1½ ft.)   May-Aug.*
A Himalayan borage with rough long lanceolate leaves and small bright blue forget-me-not-like flowers. Grows well in ordinary garden soil. Should be hardy in the Niagara region and coastal British Columbia.

LINUM flax   (Color photo p. 83)

The linums are extremely valuable for border planting with their abundant showy flowers, mostly in shades of blue. They need full sun, because in the shade and on dull days their flowers remain closed. Well known agriculturally for the production of flax.

L. alpinum   *15 cm (6 in.)   June-Aug.*
A real alpine gem, with large, pale China blue flowers.

L. austriacum   *0.3-0.6 m (1-2 ft.)   June, July*
A neat erect plant from the Austrian Alps, with soft blue flowers.

'Album' – white flowers   *0.3-0.6 m (1-2 ft.)   June, July*

L. flavum yellow flax   *0.4 m (1½ ft.)   June, July*
Bright, open, golden yellow flowers borne very profusely on branching stems. Needs some protection in colder areas, and is short lived but easily raised from seed.

Cultivars   *June, July*
'Cloth of Gold' – similar to 'Compactum' but smaller
    23 cm (9 in.)

'Compactum' – more compact and free flowering than the species   *0.3 m (1 ft.)*

L. narbonense   *0.4 m (1½ ft.)   May-July*
The slender arching branched stems are covered with large red blue panicles of flowers. Erect, small, narrow glaucous leaves.

Cultivars   *0.4 m (1½ ft.)   May-July*
'Heavenly Blue' – soft pale blue flowers

'June Perfield' – much larger flowers and more abundant than the species

'Peto's Variety' – the finest of all flaxes; evergreen in milder climates

'Six Hill's Variety' – sky blue flowers

L. perenne   *0.6 m (2 ft.)   May-July*
This lovely sky blue flax is less attractive than *L. narbonense*, but bears large quantities of flowers.

'Album' – white flax   *0.6 m (2 ft.)   May-July*

L. perenne var. *lewisii*   *0.8 m (2½ ft.)   May-July*
This North American variety is taller and has longer pedicels than the species.

L. salsoloides   *20 cm (8 in.)   June, July*
This rather choice little trailing alpine plant has a woody base, narrow evergreen leaves, and produces large pearly white flowers in abundance.

'Alpinum'   *20 cm (8 in.)   June, July*
More compact than the species.

LIRIOPE

A genus of interesting little rhizomatous plants formerly better known as *Ophiopogon*. These plants form compact clumps of dark evergreen onion-like foliage and produce slender, small blue flowers in autumn. In milder parts of Canada and the United States they are used successfully as a ground cover. They need well-drained soil, and tolerate some shade.

L. muscari grape-hyacinth liriope
    38 cm (15 in.)   Aug., Sept.
Spikes of deep purple flowers reminiscent of a small-flowered grape-hyacinth.

L. spicata   *46 cm (18 in.)   Aug., Sept.*
A creeping perennial with a slender perennial rootstock. A little taller and less ornamental than the other species and with pale lilac flowers.

'Majestic'   *23 cm (9 in.)   Aug., Sept.*
Violet purple spikes of bloom.

LITHOSPERMUM

L. canescens puccoon   25 cm (10 in.)   June
A very hardy perennial, with bright orange flowers.

L. intermedium heavenly blue   23 cm (9 in.)   June-Aug.
A bright blue edging and rock plant.

LOBELIA   (Color photo p. 83)

The perennial lobelias are very majestic plants that need deep moist soil. Never allow the roots to become dry during the growing season. *L. cardinalis* and its hybrids need an almost constant stream of moisture at their roots during the summer.

L. cardinalis cardinalflower   0.9 m (3 ft.)   Aug., Sept.
A handsome native perennial for waterside planting or for bedding in cool moist parts of Canada. Showy brilliant scarlet blooms on leafy spikes.

L. fulgens   0.9 m (3 ft.)   Aug.-Oct.
A very beautiful perennial with long, narrow, rich wine-red leaves and scarlet flowers. Not as hardy as *L. cardinalis*, but, in areas where it survives the winters, it is easier to grow. The stems are thicker and downier and its flowers have broad hanging petals.

L. siphilitica
blue cardinalflower   0.6-0.9 m (2-3 ft.)   Aug.-Oct.
When well grown in a cool moist soil, it makes a very beautiful plant with long spikes of light blue flowers and narrow oblong leaves. Where it is hardy, it is much easier to grow and associates well with species of *Eupatorium* and *L. fulgens*.

L. × speciosa   0.9-1.2 m (3-4 ft.)   Aug.
A hybrid of *L. cardinalis × L. siphilitica*, under which many so-called *L. cardinalis* hybrids belong. These include some of the most striking cultivars originated by Dr. W. Bowden of the Plant Research Institute.

Lobelia cultivars

According to Dr. Bowden, these cultivars are so complex they cannot be aligned with any particular species, although many show the influence of *L. fulgens*. 'Illumination' is a hybrid of *L. fulgens × L. cardinalis*. 'Queen Victoria' has coarse stems, reddish leaf blades, and large fulvent orient red corollas with the three lower lobes up to 3.5 cm (1.4 in.) across. A tetraploid form with much larger flowers and greener leaves was raised from the cultivar by Bowden.

Cultivars   Aug., Sept.
 'Gloire de Ste. Anne'   0.9-1.2 m (3-4 ft.)
   Narrow leaf blades, densely fine pubescent leaves, and bright red flowers.

 'Hustoman' — bright red flowers and dark red foliage
                            0.9 m (3 ft.)

 'Purple Emperor' — violet purple flowers   0.9 m (3 ft.)

 'Queen Victoria'   0.9 m (3 ft.)
Large corollas of purple red, leaf blades and stems reddish.

 'The Bishop'   0.9 m (3 ft.)
Large, showy velvety vermilion scarlet flowers with reddish leaf blades.

Dr. Bowden has suggested that 'Queen Victoria' arose from crosses made between *L. cardinalis* and several introductions from Mexico. He also believes *L. × speciosa* might have been involved.

LUNARIA

L. annua honesty   0.9-1.2 m (3-4 ft.)   July, Aug.
A biennial with scentless mauvy lilac flowers and very attractive silvery circular seedpods, which resemble thin sheets of silvery parchment. These pods are often used in dried arrangements, after the seeds have been shed and the outer valves removed. Synonym: *L. biennis*

L. biennis see *Lunaria annua*

L. rediviva perennial honesty 0.6-0.9 m (2-3 ft.)   July, Aug.

LUPINUS   (Color photo p. 83)

L. polyphyllus lupine, lupin   0.9-1.5 m (3-5 ft.)   June, July
Lupines are the strangest perennials. Where they thrive, their beautiful long columnar spikes of pea-shaped flowers are unsurpassed by any other perennials. But they are capable of destroying a man's faith in his own gardening ability, because the tenderly nurtured plants may lack color, luster, and vigor, and yet an old abandoned garden given no attention whatsoever will produce a dazzling display. One reason is that nearly all the plants available now are descended from the famous Russell strain, which is not as hardy or vigorous as the older kinds. Another reason is that lupines will not grow in soil that has the slightest content of alkalinity. Also, and probably of the greatest significance, a lupine, similar to a few other plants, needs close company to shade its roots. Although it flourishes in the sun, its roots are easily dried out if not protected. Therefore it is best to apply a mulch in early July. Lupines that flower in old gardens are usually growing through weeds. They have grown from seeds that dropped to the ground and have developed deeply penetrating roots.

They should be planted in the middle of the border in groups of three to five, with a cultivar comprising one group. They need light sandy soil and excellent drainage.

The Russell strains have not the same stamina as the older types, which are very hard to get nowadays, but their large blossoms, shapely spikes, and extensive range of colors provide a brilliance not previously known in perennial borders.

In his search for newer colors and larger blooms, Mr. George Russell obtained lupines from sources throughout the world. Although no exact record was kept of his method of breeding, it is fairly certain that some annual species were used, which accounts for the short duration of the strain.

To keep the plant living as long as possible, never allow the spikes to form seeds, and remove the dead flower heads as soon as they fade. Grow the plants in sandy loam with good drainage, and work in lots of peat to provide acid conditions. Plant the lupines firmly in the sun or light shade and be prepared to renew the plants every third or fourth year from seeds or cuttings taken with a piece of the old rootstock in April. The following are 25 of the best cultivars, 0.8-0.9 m (2½-3 ft.) high, that flower in June and July.

Cultivars

'Betty Astel' — deep pink flowers

'Blue Jacket' — deep blue flowers with white standards

'Blushing Bride' — ivory white flowers with a flush of rose

'Celandine' — clear yellow flowers

'Charmaine' — flaming orange flowers

'Commando' — deep violet and creamy yellow flowers

'Elsie Waters' — pink and cream flowers

'Fireglow' — vivid orange flowers with gold on the standards

'Flaming June' — self-colored orange flowers

'Fred Yale' — deep terra-cotta and yellow flowers

'George Russell' — coral pink flowers with cream standards edged pink

'Guardsmen' — fine deep pink flowers

'Happy Days' — salmon pink and yellow flowers

'Josephine' — slate blue and yellow flowers

'Lady Fayne' — soft rose flowers, shaded coral

'Lilac Time' — rosy lilac and light mauve flowers

'Monkgate' — deep blue and white flowers

'Mrs. Micklethwaite' — salmon pink flowers with pale gold standards

'Mrs. Noel Terry' — fleshy pink flowers, shaded cream

'Patricia of York' — bright yellow flowers

'Radiant' — crimson flowers

'Rapture' — bright rose pink flowers, flushed ivory

'Susan of York' — yellow and terra-cotta flowers

'Sweetheart' — orange and pink flowers

'Thundercloud' — violet blue flowers with standards flushed rosy mauve

LYCHNIS campion   (Color photo p. 84)

A genus containing many vividly colored perennials with wide open dianthus-like flowers. All grow well in porous soil, rich in humus. Quite a few species are invasive, some are weedy.

L. alpina Arctic campion   *25 cm (10 in.)   June, July*
An Arctic alpine plant with dense rosettes of foliage and pink or white flowers. Synonym: *Viscaria alpina*

L. × arkwrightii   *30 cm (12 in.)   June, July*
An interspecific hybrid *(L. chalcedonica × L. × haageana)* much like *L. × haageana*, but with very large intense scarlet flowers. Somewhat short lived in Eastern Canada.

L. chalcedonica Maltese cross   *0.6-0.9 m (2-3 ft.)   July*
A very ancient border plant with large roundish heads of vivid scarlet flowers and rough oval leaves. Consider carefully before you plant this perennial in your border, because its brilliance is likely to clash with other flower colors. It is an admirable plant for growing in wet locations.

Cultivars   *0.6-0.9 m (2-3 ft.)   July*
   'Alba' — white flowers, not very popular

   'Alba Plena' — double, white flowers

   'Rubra Plena' — double, red flowers

   'Salmonea' — unusual pastel salmon flowers. Synonym: 'Salmon Queen'

L. coronaria rose campion   *0.4-0.6 m (1½-2 ft.)   July, Aug.*
This showy little perennial is short lived, but often reproduces itself abundantly by self-seeding. It has thick, silvery, soft velvety hairs on its stems and foliage.

The brilliant rose flowers contrast with the silvery leaves. Because it is short lived and perpetuates itself by seeding, it often appears in the border at the Plant Research Institute without a label, which leads to inquiries about its identity. Synonym: *Agrostemma coronaria*

Cultivars   *46 cm (18 in.)   June-Aug.*
   'Abbotswood Rose' — very vivid pink flowers all summer; a little more compatible with other colors

   'Alba' — white flowers

   'Atrosanguinea' — bright crimson magenta flowers

   'Flore Pleno' — fully double flowers

L. flos-jovis flower-of-Jove   *30 cm (12 in.)   June-Aug.*
Purple and scarlet flowers, silvery leaves.

'Hort's Variety'   *30 cm (12 in.)*   *June-Aug.*
Bright pink, much better than the species.

L. × haageana   *30 cm (12 in.)*   *June, July*
A large-flowered brilliantly colored orange scarlet hybrid
(*L. fulgens × L. coronata* 'Sieboldii') campion that comes
remarkably true from seed. It is not a long-lived perennial,
but is perfectly hardy at Morden, Manitoba. In milder areas
it may need protection from wet winters.

L. viscaria German catchfly   *46 cm (18 in.)*   *May-July*
Very sticky stems, long grassy leaves, and abundant showy
rosy red flowers.

Cultivars   *26 cm (18 in.)*   *June, July*
    'Splendens' – magenta rose flowers

    'Splendens Flore-Pleno' – double magenta rose flowers,
0.3 cm (1 in.) in diameter

LYSIMACHIA loosestrife   (Color photo p. 84)

This genus includes some good herbaceous perennials and
some splendid ground-covering plants. These plants need
moist soil. They are easily divided. Some are invasive.

L. brachystachys   *0.9 m (3 ft.)*   *July, Aug.*
Erect leafy spikes that form a dense bush with short tapering spikes of small white flowers.

L. ciliata   *0.9-1.2 m (3-4 ft.)*   *July*
Vigorous creeping roots and small showy yellow flowers on
1.1-m (3½-ft.) branching stems. Invasive if allowed un-
restricted growth. Synonym: *Steironema ciliata*

L. clethroides gooseneck loosestrife
      0.6-0.9 m (2-3 ft.)   *July-Sept.*
Long, nodding, dense gooseneck spikes of white blooms and
oval leaves, which are highly colored in the fall. Best in
semishaded location in moist rich soil, where it will spread
rapidly.

L. ephemerum   *0.9 m (3 ft.)*   *July, Aug.*
Outstanding, with long slender terminal racemes of small
white flowers, sometimes tinged purple, abundant very
glaucous foliage.

L. fortunei   *0.6-0.8 m (2-2½ ft.)*   *July-Sept.*
An erect, sparsely branched perennial related and similar to
L. clethroides, with slender stems and broad narrowly lanceolate alternate leaves. White flowers similar to *L. clethroides*
but smaller.

L. nummularia moneywort, creeping Jenny
      15 cm (6 in.)   *June-Aug.*
One of the best ground covers for damp clay slopes and the
tops of walls. Because of its very vigorous spreading habit,
it is likely to become weedy unless used in restricted areas.
It does well in sun and shade, and should be considered when
choosing a plant for a location where nothing else will grow.

Its bright yellow flowers are quite effective and will brighten
a shady area. This species and the cultivar 'Aurea' are
often used as trailing plants for window boxes and baskets.

'Aurea' golden moneywort   *0.3 m (1 ft.)*   *June-Aug.*
Golden yellow leaves, no less invasive than the species.

L. punctata dotted-leaved loosestrife
      0.8 m (2½ ft.)   *June, July*
A vigorous growing plant in ordinary soil. Stiff leafy spikes
of bright yellow star-like flowers. Named for its dotted
leaves. Rather invasive in rich moist soils.

L. vulgaris yellow loosestrife   *0.9 m (3 ft.)*   *July-Sept.*
Terminal panicles of deep yellow and orange flowers. Needs
more moisture and shade than the dotted-leaved loosestrife.

LYTHRUM purple loosestrife   (Color photo p. 84)

A roadside weed that has been carefully bred to produce
bright gay flowering perennials in early fall. They are easy
to grow, even in poor soils, in sun or shade. It is one of the
few perennials that will thrive in wet shaded areas.

L. salicaria purple loosestrife *0.6-1.5 m (2-5 ft.)*   *Aug., Sept.*
A tall roadside perennial native of Great Britain and naturalized in Australia and North America. The species with
rosy purple flowers is useful· for naturalizing in wild and
woodland gardens. Many good cultivars have been derived
from the species.

Cultivars   *July-Sept.*
    'Brightness' – bright clear rose pink flowers, dark foliage
      0.9-1.2 m (3-4 ft.)

    'Lady Sackville' – bright rose deepening to rose pink
flowers   *0.9-1.2 m (3-4 ft.)*

    'Morden's Gleam' – bright rose flowers   *0.9-1.2 m (3-4 ft.)*

    'Robert' – deep pink flowers, compact habit, dwarf
      0.6 m (2 ft.)

    'The Beacon' – intense rose red flowers, dense spikes
      1.1 m (3½ ft.)

L. virgatum wand loosestrife   *0.6-0.9 m (2-3 ft.)*
Purple flowers in threes in the leaf axils of twiggy racemes.
Grows more stiffly than *L. salicaria*, with narrower leaves
and slender tapering species.

Cultivars   *July-Sept.*
    'Dropmore Purple' – deep purple flowers   *0.8 m (2½ ft.)*

    'Morden Pink'   *0.9 m (3 ft.)*
    A bud sport producing trim spikes of rose pink flowers all
summer.

    'Rose Queen' – long branching spikes of bright rose
flowers   *0.4 m (1½ ft.)*

    'The Rocket' – taller but brighter and deeper than 'Rose
Queen'   *0.8 m (2½ ft.)*

MACLEAYA plume-poppy   (Color photo p. 85)

Very handsome, impressive plants with beautiful foliage, but so tall that they must be planted at the back of the border, or isolated against a wall or background of evergreens. Do not plant in a small border, because its underground shoots can be invasive. Its handsome leaves are fig-like, bronzy green above, and silvery beneath. The flowers are tiny, but borne in great profusion in tall feathery spikes 1.5-2.7 m (5-9 ft.) high. They grow in various kinds of soil, in sun or light shade. Synonym: *Bocconia*

M. cordata   1.8 m (6 ft.)   July, Aug.
Large rounded leaves and white or buff white tubular flowers, each with 8-12 stamens. Synonym: *Bocconia cordata*

M. microcarpa   2.7 m (9 ft.)   July, Aug.
Flowers are bronzy or golden brown on taller spikes than *M. cordata* and have 24-28 stamens; the leaves are more conspicuously veined.

'Coral Plume'   1.8 m (6 ft.)   July, Aug.
Very handsome silvery foliage and coppery plumes.

MALVA musk mallow

Most mallows are weedy and rather coarse, but a few are worth cultivating for their hollyhock-like flowers and distinctive foliage.

M. alcea   0.9-1.2 m (3-4 ft.)   July-Oct.
Large soft rosy purple flowers on terminal and axillary flower spikes.

'Fastigiata'   0.9-1.2 m (3-4 ft.)   June-Aug.
Better than the species, pink saucer-like flowers on erect spikes. The plants are short lived and do best in hot dry soils.

M. moschata musk mallow   0.3-0.6 m (1-2 ft.)   May-Oct.
Finely cut buttercup-like musk-scented foliage and wide soft pink flowers. Another short-lived species that needs replanting with self-sown seedlings every 3-4 years.

'Alba'   0.3-0.6 m (1-2 ft.)   May-Oct.
White flowers, more pleasing than the species.

MALVASTRUM

M. coccineum see *Sphaeralcea coccinea*

MECONOPSIS   (Color photo p. 85)

A genus of mostly monocarpic plants or, at the very best, short-lived perennials. Most of them require very deep rich soil, some shade, and moist humid conditions.

M. baileyii see *Meconopsis betonicifolia*

M. betonicifolia
Himalayan blue poppy   0.6-1.2 m (2-4 ft.)   June, July
The famous 'Blue Poppy' is a fabulous plant, and one with which all expert gardeners crave to succeed. The flowers are sky blue, 5 cm (2 in.) across, round, and lightly crisped like taffeta. The base of yellow stamens in the center of the flowers intensifies the blue. The leaves and stems are very hairy. If not for the uncanny success achieved with this plant by the late Cleveland Morgan in his garden near Montreal, it might be considered impossible to grow in Eastern Canada. The plant usually flowers itself to death when it does flower, so perhaps it is best treated as a monocarpic plant. Always keep more seedlings, to counteract its early demise. Do not let it flower the first year, because its beautiful blue flowers are variable. Seed should be saved only from selected plants. Synonym: *M. baileyii*

'Pratense'   0.6 m (2 ft.)   June, July
Much longer lived than the species, but its flowers and spikes are not as large.

M. cambrica Welsh poppy   0.3 m (1 ft.)   June-Sept.
The Welsh poppy is very similar to the alpine poppy (*Papaver alpinum*) with its deeply cut grayish leaves in tufts, from which arise yellow and orange flowers, one on a stem. In a suitable soil and location it seeds itself freely and soon becomes naturalized.

'Flore Plena'   0.3 m (1 ft.)   June-Aug.
Double, rich orange or yellow flowers, but a little harder to grow than the species.

M. quintuplinervia harebell poppy *0.6 m (2 ft.) May, June*
A true perennial with creeping roots and nodding lavender blue bells on 30-cm (12-in.) stems. From Tibet and Western China, it is likely to be hardy only in the milder parts of British Columbia.

MERTENSIA bluebells   (Color photo p. 85)

A genus with a wide and diversified range of blue-flowered plants that should be more widely grown. The common Virginia bluebells is very well known in most Canadian gardens. Best in moist semishade, but quite adaptable.

M. ciliata mountain bluebells   *0.6 m (2 ft.) May, June*
A native North American species from Montana and Oregon to New Mexico with bright blue bell-shaped flowers on long pendulous sprays. The buds, as in other blue boraginaceous plants, are pink.

M. echioides Himalayan bluebells   *23 cm (9 in.) June*
A good ground cover in mild areas, because of its habit of forming creeping tufts of growth. Its sprays of bright blue flowers are 20-23 cm (8-9 in.) tall and make a good show in early summer. Synonym: *M. primuloides*

M. lanceolata prairie bluebells   *25 cm (10 in.) Apr., May*
Good in very dry soils, because it produces sky blue flowers under the most adverse conditions.

M. longiflora   *0.3 m (1 ft.) May, June*
From the Rockies, this species produces clusters of bell-shaped blue blossoms on very neat dwarf plants. The foliage dies down in summer in dry locations.

M. paniculata   *0.9-1.2 m (3-4 ft.) May, June*
Similar to *M. ciliata*, but taller and with smaller flowers. It is native from Quebec to Alaska and south to Washington, where it grows 0.9 m (3 ft.) high in dry areas and 1.5 m (5 ft.) high in good rich moist soils. It is used in some gardens as a ground cover in very poor dry soils, a use to which it is very adaptable. The dark green leaves stay on the plant all summer, and the blue flowers develop on sturdy stems. Regarded as tender at Morden, Manitoba.

M. primuloides see *Mertensia echioides*

M. sibirica Siberian bluebells   *0.4 m (1½ ft.) May, June*
A hardy rugged species, beautiful in habit and color. Its small bell-shaped flowers in pendulous clusters are rose pink to purple blue. Best in moist soil. The broad foliage lasts all summer long.

M. virginica Virginian bluebells   *0.6 m (2 ft.) May*
A very hardy species with arching sprays of funnel-shaped clear blue bells. The fresh leaves are purple in spring; as they unfold they change to light green. This is a perfect plant for interplanting with pale pink tulips. In the border it is better interplanted with hostas, vincas, candytufts, or some other plant to provide the foliage for that lost by this plant in June. Virginia bluebells prefer moist soil and part shade.

Planted among shrubs in the Macoun Memorial Garden at the Plant Research Institute, and interplanted with Siberian squills, these plants make a perfect spring picture.

MIMULUS monkeyflower

Highly colored snapdragon-like flowers and simple hairy leaves. They need lots of moisture and some mulching in summer. Best for planting close to pools.

M. × burnetii   *0.3 m (1 ft.) June, July*
Yellow flowers heavily stained and spotted bronze. A hybrid of *M. cupreus × M. luteus.*

M. cardinalis scarlet monkeyflower   *0.6 m (2 ft.) July*
A vigorous bush with white, crimson, or pink flowers, depending on the variety. This species has vivid scarlet flowers on erect stems.

Cultivars
 'Cerise Queen' – dwarf, with bright cerise flowers
     23 cm (9 in.) June, July

 'Grandiflorus' – larger scarlet flowers than the species
     0.4 m (1½ ft.) June, July

 'Rose Queen' – bronzy scarlet flowers
     0.3-0.9 m (1-3 ft.) June-Aug.

M. cupreus   *0.3 m (1 ft.) June-Sept.*
A very beautiful Chilean species closely related to *M. luteus,* but more tufted and with yellow flowers turning copper colored.

The cultivars must be propagated from cuttings to keep their true identity. They need lots of moisture at their roots, and even in the milder parts of British Columbia they need some protection.

Cultivars   *0.3 m (1 ft.) June-Sept.*
 'Bee's Dazzler' – brilliant pillar-box scarlet flowers
 'Bonfire' – bright scarlet flowers
 'Chelsea Pensioner' – scarlet flowers
 'Fireflame' – flame red flowers
 'Leopard' – yellow flowers spotted with orange brown
 'Plymtre' – cherry pink flowers
 'Queen's Prize' – pink, salmon, rose, and yellow flowers
 'Red Emperor' – bright crimson scarlet flowers
 'Whitecroft Scarlet' – scarlet flowers

M. langsdorfii see *Mimulus luteus* 'Guttatus'

M. lewisii red monkeyflower   *0.4 m (1½ ft.) June-Aug.*
Native from British Columbia to California, closely related to *M. cardinalis,* but with pointed leaves and rosy red flowers. Needs lots of moisture.

 'Alba' – white flowers   *0.3 m (1 ft.) July, Aug.*

M. luteus monkey musk   *0.3 m (1 ft.) May-Aug.*
Found along riversides, in mountain meadows, and on moist plains from Alaska to New Mexico. Needs more moisture than the other species and must not be allowed to dry out at any time. The high coloring, free flowering habit, and large blooms have been useful in bedding schemes in Britain. Some of the cultivars are particularly ornamental.

Cultivars   *0.3 m (1 ft.) May-Aug.*

'Alpinus' – densely matted stolons

'A. T. Johnson' – a fine form, heavily spotted with red

'Duplex' – a hose in hose, one flower inside another

'Guttatus' – purple or brown spots on yellow flowers. Synonym: *M. langsdorfii*

'Rivularis' – only one brown spot on each yellow flower

'Youngeanus' – a brown spot on each lobe

M. moschatus muskflower   *15 cm (6 in.) May*
A spreading, clammy woolly perennial with ovate pale yellow leaves, lightly dotted and splashed brown. This describes the muskplant of Victorian days, when they were grown in pots in cottage windows in England mostly because of their musk odor. With no reason except for having larger flowers on more robust plants, the cultivar 'Harrisonii', with no musk odor, supplanted this plant, and by 1914 the 'Mimulus Musk' was lost to cultivation. It is quite possible that the scented plant was a clone, which was passed from house to house, that reigned supreme until superseded by the newer clone. No *M. moschatus* has been found since then that has the same strong agreeable musk odor.

M. primuloides   *8 cm (3 in.) May*
A tufted downy perennial forming carpets of dainty yellow flowers on wiry stems.

M. ringens   *0.6-0.9 m (2-3 ft.) June*
A well-branched plant with 3-cm (1-in.) violet flowers. Needs lots of moisture.

MOLTKIA

M. × intermedia   *23 cm (9 in.) June-Aug.*
A bright blue plant for edging or rock gardens.

MONARDA bee balm, bergamot   (Color photos pp. 13, 85)

The striking red flowers of the bergamot have long been a standby in old gardens. Now the less flamboyant but more restrained species of beautiful pink and lavender with their more refined habit are widely used in decorating our borders in summer. The species grow mostly along the banks of streams, but the cultivars appear to grow in any good soil in the sun.

The plants have square stems, aromatic nettle-like leaves, and brightly colored flowers borne in close heads or whorls on stems 0.6-0.9 m (2-3 ft.) high. They are best associated with Siberian irises and day-lilies, and can be grouped effectively with anthemis, heliopsis, phlox, and delphiniums.

M. didyma Oswego tea   *0.6-0.9 m (2-3 ft.) June-Sept.*
Native from Quebec and Michigan south to Georgia, where it grows in moist woodlands and along riverbanks. A robust useful plant for a sunny border or for naturalizing in light shade. Bright scarlet flowers in solitary or twin terminal whorls.

M. fistulosa wild bergamot   *0.6-1.5 m (2-5 ft.) June-Sept.*
Very robust, with variable flowers from white to light red. Although not especially attractive, it has contributed vigor and tolerance of sun to most of the following cultivars.

Cultivars   *June-Sept.*

'Adam' – cerise red flowers   *0.8 m (2½ ft.)*

'Blue Stocking' – violet purple flowers   *0.9 m (3 ft.)*

'Cambridge Scarlet' – more brilliant scarlet flowers than the species   *0.8 m (2½ ft.)*

'Croftway Pink' – rich rose pink flowers   *0.8 m (2½ ft.)*

'Granite Pink' – soft pink flowers, very vigorous
    1.1 m (3½ ft.)

'Granite Purple' – cyclamen purple flowers   *0.9 m (3 ft.)*

'Melissa' – soft pink flowers   *0.9 m (3 ft.)*

'Pillar Box' – bright red flowers   *0.9 m (3 ft.)*

'Prairie Glow' – bright salmon red flowers *0.8 m (2½ ft.)*

'Prairie Night' – purple flowers   *0.9 m (3 ft.)*

'Salmon Queen' – salmon pink flowers   *0.9 m (3 ft.)*

'Snow Maiden' – very effective white flowers
    0.8 m (2½ ft.)

'Souris' – reddish purple flowers, originated at Brandon, Manitoba   *0.8 m (2½ ft.)*

'Sunset' – purple flowers with smoky overcast *0.9 m (3 ft.)*

MORINA

M. longifolia
Himalayan whorlflower   *0.8 m (2½ ft.) June, July*
A very handsome Nepalese perennial not usually seen even in British gardens and rarely in British Columbia. Large spiny thistle-like leaves with long spikes of flowers around the stem in whorls. The individual pink florets have long tubular blooms with flared mouths. The plants need full sun and well-drained sandy loam.

MYOSOTIS forget-me-not

There are two different groups of *Myosotis:* one is perennial, containing plants described here that last for a number of years; the other is biennial, including the well-known forget-me-not that must be grown from seed sown every 2 or 3 years.

M. alpestris   *20 cm (8 in.)   May, June*
An arctic plant, native of Europe and North America, with dense tufts of light blue flowers. Most plants sold under this name are the biennial forget-me-not, *M. sylvatica.*

Cultivars   *15 cm (6 in.)   May, June*
  'Pink Beauty' — shell pink flowers
  'Ruth Fisher' — dark foliage and azure flowers

M. palustris see *Myosotis scorpioides*

M. scorpioides true forget-me-not   *0.3 (1 ft.)   May, June*
This is the true forget-me-not of Europe and Asia. It is sometimes called scorpion-grass in England, because of its twisted stems that resemble a scorpion's tail. The creeping stems bear long loose branchless racemes of little blue flowers with yellow eyes. Synonym: *M. palustris*

M. scorpioides var. *sempervirens*   *0.3 m (1 ft.)   May-Sept.*
A prolonged flowering season from May until September.

M. sylvatica wood forget-me-not   *0.3 m (1 ft.)   May, June*
See Chapter 3, Biennials.

NEPETA catmint

A genus of over 100 species of perennial or annual plants, but only a very few are worth growing. They need well-drained light soil in full sun or part shade.

N. cataria catnip   *0.6-0.9 m (2-3 ft.)   June, July*
Some gardeners may want to grow this plant for the enjoyment of their cat. Plants are easy to grow from seed, obtainable from dealers who handle unusual plants. This is a spready, sprawling plant with white flowers in dense many-flowered whorls.

N. × faassenii catmint   *0.3 m (1 ft.)   July*
This very popular perennial is a hybrid of *N. nepetella × N. mussinii*, which has been cultivated for years under the name *N. mussinii*, a much inferior plant with heart-shaped leaves and one that seeds itself frequently. *N. × faassenii* has smaller, toothed silvery gray leaves, and it seldom sets seeds. The silvery foliage of the leaves and the pale lavender spikes of flowers are quite effective when planted near summer phlox, lilacs, and delphiniums. Some people find the strong aromatic qualities of the leaves pleasant. This hybrid is an excellent edging plant and may be sheared closely in the fall to produce a more shapely edging the following year. Cuttings taken in July are very easy to root, or you may divide the plant in early spring. Referred to by commercial gardeners as *N. mussinii*.

N. grandiflora
A very hardy perennial with almost heart-shaped leaves and loose clusters of blue flowers. Most plants grown as *N. ucranica* belong under this name.

N. hemsleyana see *Dracocephalum hemsleyanum*

N. mussinii see *Nepeta × faassenii*

N. tartarica Tartarian catnip   *1.2 m (4 ft.)   June-Aug.*
A tall species, dense, forming a leafy bush with spikes of lilac blue flowers.

Cultivars   *0.6 m (2 ft.)*
  'Blue Beauty'   *June, July*
  Green foliage, and large spikes of lavender blue tubular flowers. Synonym: 'Souvenir d'André Chaudron'

  'Six Hills Giant' — fringed light blue flowers, gray leaves
                                              June-Aug.

N. ucranica see *Nepeta grandiflora*

N. wilsonii see *Dracocephalum wilsonii*

OENOTHERA evening-primrose   (Color photo p. 86)

The evening-primroses are among the showiest plants. Most species are either biennial or monocarpic (that is, they die after flowering), but the perennials mentioned here last many years, have a long flowering period, and, unlike the biennials, which only open their flowers at night, they are effective during the daytime. Their yellow, white, and light rose poppy-like flowers blend well with all other garden plants but are best when planted in groups of three or five plants.

O. caespitosa   *0.3 m (1 ft.)   May-Aug.*
A stemless plant with dandelion-like leaves and white flowers, which turn pink or lavender. Because of its long flowering period, it is suitable for planting in the front of the border.

  'Eximea' — taller, with stems   *46 cm (18 in.)   May-Aug.*

O. fruticosa var. *youngii* see *Oenothera tetragona*

O. missouriensis Ozark sundrops   *23 cm (9 in.)   June-Sept.*
One of the best and hardiest species for gardens. In July it produces very large 8-13-cm (3-5-in.) cups of crepe-textured golden flowers and red-tipped buds. It forms a spreading plant with reddish stems and lance-shaped leaves. Probably the hardiest of all species.

O. speciosa wind-primrose,
showy evening-primrose   *46 cm (18 in.)   June-Aug.*
Narrow leaves, and many large flat basin-shaped fragrant white flowers with a pale greenish center that turn soft rose as they mature. A true perennial for the front of the border and rock gardens.

Oenothera tetragona 'W. Cuthbertson', evening-primrose

O. *speciosa* var. *childsii*   *46 cm (18 in.)*   *June-Aug.*
A geographical variety from the Gulf of Mexico and Texas. Beautiful pink flowers; much less hardy than the species.

O. *tetragona*   *0.3-0.6 m (1-2 ft.)*   *all summer*
This species and its varieties are the most attractive of all lemon-colored border flowers. This species includes many of the forms that are better known under *O. fruticosa*, which is an inferior plant with fewer flowers and smaller leaves. Synonym: *O. fruticosa* var. *youngii*

O. *tetragona* var. *fraseri* see *Oenothera tetragona* ssp. *glauca*

O. *tetragona* ssp. *glauca*   *0.3-0.6 m (1-2 ft.)*   *all summer*
A well-known plant with an abundance of deep golden flowers, reddish stems, and thick leathery leaves; usually called *O. glauca* var. *fraseri.* Synonym: *O. tetragona* var. *fraseri*

Cultivars   *all summer*
  'Highlight' — new, with large yellow flowers *46 cm (18 in.)*

  'Illumination'   *30 cm (12 in.)*
  Red buds, deep yellow flowers on thickset plants. Synonyms: 'Fireworks', 'Fyrwekeri'

  'Riparia' — bright yellow flowers   *30 cm (12 in.)*

  'W. Cuthbertson'   *30 cm (12 in.)*
  Very showy, red buds, golden yellow flowers, and bronzy leaves.

  'Yellow River'   *38 cm (15 in.)*
  Much larger blooms than the species and well-spaced stems.

OMPHALODES navelwort

Attractive borage-like plants with small blue flowers and simple foliage. It is best at the front of borders, preferably in a semishaded location in well-drained soil.

O. *cappadocica*
dogwood, navelseed   *30 cm (12 in.)*   *June-Aug.*
A neat little species with 15-cm (6-in.) sprays of very bright deep blue forget-me-not-like flowers and oval or heart-shaped leaves on long stalks.

  'Anthea Bloom'   *15 cm (6 in.)*   *May, June*
  A new introduction with a neater habit, more abundant intense sky blue flowers. Grayish green leaves, showier than the species.

O. *verna* blue-eyed Mary   *15-23 cm (6-9 in.)*   *Apr., May*
Running underground roots and violet-like leaves on long stalks. Bright blue flowers with a white throat, borne in small sprays. A good ground cover for light shady areas.

  'Alba' — white flowers   *15-23 cm (6-9 in.)*   *Apr., May*

OSTROWSKIA

O. *magnifica* giant bellflower   *1.2-1.8 m (4-6 ft.)*   *July*
A truly magnificent border perennial, but very challenging to grow. When growing well, stems are 1.2-1.8 m (4-6 ft.) high with attractive silvery foliage in whorls all the way up the stem and bearing terminal racemes, large 10-15-cm (4-6-in.) cup-shaped pale lilac suffused white flowers on long pedicels. These stately plants need good drainage and alkaline sandy soil. In areas where the winter is wet, they should be covered with glass to prevent their crowns from rotting.

PACHYSANDRA (Color photo p. 86)

These plants are mostly used as ground covers. They are well adapted to deep shade, where few plants will thrive. Their chief beauty lies in their attractive glossy foliage. They must be planted in good soil and not under trees where roots are likely to be thick.

P. *procumbens*   *20 cm (8 in.)*
A spreading plant with semievergreen grayish green leaves. Much slower growing than *P. terminalis* but its foliage is more suitable for rock gardens. Probably less hardy than the other species. The white or pinky white flowers are quite inconspicuous.

P. *terminalis* Japanese spurge   *30 cm (12 in.)*
A very hardy and vigorous species with glossy deep green persistent foliage and greenish white inconspicuous flowers. Very widely used as a ground cover in shade.

  'Variegata'   *23-30 cm (9-12 in.)*
  Very handsome leaves, more beautiful than those of the species, but the plant grows much slower.

Papaver nudicaule, Iceland poppies

PAEONIA peony   (Color photo p. 69)

This genus contains many species and cultivars, some of which are perhaps the largest of all hardy perennials and perhaps the hardiest of all popular garden flowers. They are attractive at all stages of growth. The foliage is dark glossy green, often turning to bronze in autumn. The flowers are very large, showy, and useful both for garden ornamentation and for cut flowers.

The favorite cultivars are from two groups: those that arose from *P. lactiflora*, formerly known as *P. albiflora;* and the so-called hybrid peonies, which have risen in recent years from simple and complex interspecific crosses.

See Chapter 2, Basic Perennials.

P. anomala   *46 cm (18 in.)   May*
Very early flowering; bright crimson single flowers and deeply divided leaves.

P. emodi   *0.3-0.8 m (1-2½ ft.)   Apr.*
Fine, large, single white flowers 10-13 cm (4-5 in.) across with white stamens.

P. mlokosewitschi lemon peony   *0.6 m (2 ft.)   Apr., May*
Very beautiful, early flowering; fresh glaucous green foliage and bowl-shaped glistening light yellow flowers with coral stamens; followed later by very ornamental open seedpods with a double row of black seeds.

P. officinalis common red peony   *0.6-0.8 m (2-2½ ft.)   May*
Very long-living; solitary single scarlet red flowers and deeply cut foliage.

Cultivars   *May*
   'Albo Plena' – double, white flowers   *0.6-0.8 m (2-2½ ft.)*

   'Rosea Plena' – double, rose flowers   *0.6-0.8 m (2-2½ ft.)*

   'Rubra Plena' – double, dark red flowers   *0.8 m (2½ ft.)*

P. russii Russo's peony   *0.4 m (1½ ft.)   May*
Pale green leaves and rose red flowers with rounded petals.

P. × smouthii   *0.6 m (2 ft.)   Apr., May*
A very early flowering hybrid *(P. lactiflora × P. tenuifolia)* with large, single, crimson, rose-scented flowers and deeply cut foliage.

P. tenuifolia fernleaf peony   *0.3-0.6 m (1-2 ft.)   May*
Very distinct, finely cut fern-like foliage and deep crimson cup-shaped flowers about 8 cm (3 in.) across.

   'Plena'   *0.3-0.6 m (1-2 ft.)   May*
   A double form of the species with deep crimson flowers.

P. veitchii   *0.6 m (2 ft.)   June*
Deeply cut foliage and single magenta flowers.

   'Woodwardii'   *0.3 m (1 ft.)   June*
   Much dwarfer; rose pink flowers.

P. wittmanniana   *0.6-0.9 m (2-3 ft.)   May*
A very attractive species, with creamy white single flowers, showy red stigmas and filaments, and handsome glaucous leaves broadly and deeply cut.

PAPAVER   (Color photo p. 68)

All the true perennial poppies are so much overshadowed by the hybrids of *P. orientale* that any discussion of their merits would be superfluous.

P. nudicaule   *23-38 cm (9-15 in.)   May-Oct.*
A very fine but short-lived perennial poppy, best treated as a biennial. See Chapter 3, Biennials.

P. orientale Oriental poppy   *0.6-0.8 m (2-2½ ft.)   May, June*
The species has bright scarlet flowers with a prominent purple black blotch at the base of each petal. For cultivation of this and all cultivars see Chapter 2, Basic Perennials.

PELTIPHYLLUM

P. peltatum umbrellaplant   *0.8 m (2½ ft.)   May, June*

The umbrellaplant is so-called because of its spectacular circular umbrella-like leaves at the tips of 0.9-m (3-ft.) stems. The pale pink blooms of this plant appear in early spring before the foliage has formed. They are much like flowers of *Bergenia crassifolia*, a closely related genera. The plant grows best in damp ground preferably beside a pool.

PENSTEMON beardtongue   (Color photo p. 86)

This genus includes an unusually large number of species. All the species are ornamental, but only a few of the hardy species are reliable. The new hybrid races that have more vigor and greater hardiness should be grown in preference to the species. The hybrids are less sensitive to their environment, and often live much longer.

P. acuminatus   *0.6 m (2 ft.)   June*
An erect, stiff perennial with oblong or ovate leaves and lilac flowers to 3 cm (1 in.) long on crowded cymes.

P. alpinus   *46 cm (18 in.)   June, July*
From the Rocky Mountains, a small species with clear blue flowers and a white throat on a short few-flowered thyrse.

   Flathead late hybrids   *0.6 m (2 ft.)   June, July*
   A collection of hybrids involving *P. alpinus, P. cobaea, P. glaber*, and *P. strictus* mostly with red flowers.

P. barbatus   *0.9 m (3 ft.)   July, Aug.*
A hardy species with small bright carmine tubular flowers with yellow throats and rather small leaves on 0.9-m (3-ft.) stems.

Cultivars   *July, Aug.*
'Coccineus' — scarlet flowers   *0.8 m (2½ ft.)*

   'Rosy Elf'   *0.6 m (2 ft.)*
   Fine cultivar; radiant deep rose flowers; excellent neat perennial worthy of a place in any border.

   'Torreyi'   *0.9-1.2 m (3-4 ft.)*
   Scarlet flowers, but lacking the usual beard on the lower petals as in most penstemons.

P. cobaea   *0.6 m (2 ft.)   July*
A somewhat variable species with very large bell-shaped flowers up to 5 cm (2 in.) long and wide. Flowers may be dull reddish purple to whitish. Hardiness usually depends on the source of seed, for example, seed from Missouri is hardier than seed from farther south.

   'Ozark' — splendid bright purple flowers
                    0.4 m (1½ ft.)   Aug.

P. crandallii   *20 cm (8 in.)   June*
A good form for rock gardens with upturned sky blue or lilac flowers on a low-spreading plant about 15-20 cm (6-8 in.) high. Needs perfect drainage.

   'Glabescens'   *15 cm (6 in.)   June*
   Larger leaves, stems more upright than the species; pale blue flowers crowd the stems in early summer.

P. diffusus see *Penstemon serrulatus*

P. digitalis   *1.2 m (4 ft.)   June, July*
A robust species with lilac white or bluish white flowers abruptly inflated just above the calyx. Best used for naturalizing along the perimeter of woodlands. Synonyms: *P. laevigatus* var. *digitalis, Chelone digitalis*

P. glaber blue penstemon   *0.6 m (2 ft.)   June*
Erect glaucous unbranched stems and sprays of bright blue to purple flowers.

   'Roseus' — pink to deep rose flowers   *0.6 m (2 ft.)   June*

P. × gloxinoides
A group of very showy named hybrids grown by florists on the west coast of the United States and in England. As for the so-called hardy chrysanthemums, these are propagated annually from cuttings, because they seldom survive even a few degrees of frost. An interspecific hybrid involving *P. hartwegii × P. cobaea*.

Cultivars   *0.6 m (2 ft.)   all summer*
   'Alice Hindly' — pale blue flowers, shaded rose

   'Cherry Red' — good cerise red flowers

   'Firebird' — brilliant crimson flowers the same size as a foxglove; more adaptable for bedding than some cultivars and much hardier

   'Garnet' — wine red trumpet, hardier and longer lived than most of the others except 'Firebird'

   'Majestic' — violet purple flowers

   'Newberry' — light bluish purple flowers

   'Ruby' — deep crimson flowers

   'White Bedder' — white flowers

P. grandiflorus
shell-leaf penstemon   *0.9 m (3 ft.)   all summer*
A stocky plant with thick glaucous spoon-shaped leaves and large 5-cm (2-in.) lavender purple two-lipped flowers.

   Fate hybrid strain   *0.9 m (3 ft.)   all summer*
   A hybrid seedling strain of *P. grandiflorus* and *P. murrayanus* originated by F. Fate, Columbia, Missouri.

   Seeba hybrid strain   *0.9 m (3 ft.)   all summer*
   Another seedling strain with the same parentage as the Fate hybrid strain, but introduced by Lena Seeba, Cook, Nebraska. Both these hybrids grow best under hot dry prairie conditions, where they often reach 1.2 m (4 ft.) high in colors of white to pink, red, and purple.

P. laevigatus var. *digitalis* see *Penstemon digitalis*

P. heterophyllus   *1.2 m (4 ft.)   July*
A short-lived perennial, not very hardy in most of Canada, with dark green bushy growth and spikes of brilliant blue flowers.

P. murrayanus   *1.2 m (4 ft.)*   *Aug.*
Tall, with fiery red flowers and glaucous blue foliage.

P. ovatus   *0.6-0.9 m (2-3 ft.)*   *July*
Very pretty, with erect stems bearing blue flowers that later change to purple. Its longevity appears to be in doubt, although our records show it to have grown vigorously for 10 years before it diminished in size and finally succumbed.

P. palmeri   *0.6-0.9 m (2-3 ft.)*   *July*
A good perennial penstemon with large pink-tinted creamy white flowers on lone twiggy peduncles on sturdy stems 0.6-0.9 m (2 or 3 ft.) high.

P. serrulatus   *46 cm (18 in.)*   *Aug., Sept.*
A late-flowering rather graceful little plant with lilac purple flowers in dense cymes in a leafy panicle. Synonym: *P. diffusus*

P. unilateralis   *0.6 m (2 ft.)*   *May, June*
Bright blue flowers provide a good display in the border during late May and June.

P. utahensis Utah penstemon   *0.6 m (2 ft.)*   *July*
Very vigorous, with leafy red-tinted stems. It produces large well-shaped flowers of blue (suffused mauve, when fully open).

Unaffiliated cultivars   *all summer*
   'Prairie Dusk' — dark purple-blue flowers and green basal foliage   *0.6 m (2 ft.)*

   'Prairie Fire' — scarlet flowers on strong stems
       36-46 cm (15-18 in.)

Six Hills hybrid   *23 cm (9 in.)*   *June-Aug.*
Dwarf, bushy, with pale green leaves and rosy lilac flowers.

PHLOMIS Jerusalem sage

A genus of plants seldom cultivated, but very hardy, trouble-free, and easy to grow in full sun or part shade. Interesting because of the silvery green, heavily textured foliage and curious giant nettle-like inflorescences bearing tubular yellow, purplish, or white flowers.

P. cashmiriana   *0.6 m (2 ft.)*   *July*
A very striking plant with densely woolly stems and leaves and bearing whorls of lilac purple flowers.

P. fruticosa Jerusalem sage   *0.6-1.2 m (2-4 ft.)*   *June*
A neat bush covered with soft green woolly leaves and whorls of buff yellow flowers. This is strictly a shrub, but it is best suited to a flower border. Not as hardy as the herbaceous types.

P. russelliana   *0.6-0.9 m (2-3 ft.)*   *July*
Clusters of yellow flowers on spikes well above the foliage. Synonym: *P. viscosa*

P. samia   *0.9 m (3 ft.)*   *May, June*
A very strong-growing, long-lived plant with very woolly sage-like leaves and small whorls of light yellow flowers. Best in hot dry locations, grows well on the Prairies. Native to North Africa and Greece.

P. tuberosa   *0.9 m (3 ft.)*   *June, July*
Heart-shaped leaves with wavy margins, larger than most of the other species. The purplish rose and white flowers are borne in whorls on long stems. This species, as its name suggests, is tuberous.

P. viscosa see *Phlomis russelliana*

PHLOX   (Color photos pp. 25, 86)

This genus contains many beautiful species including perennial phlox, one of the best groups of basic perennials for a flower border.

P. adsurgens
periwinkle phlox   *15-30 cm (6-12 in.)*   *Apr., May*
Slender shiny leaves and creeping stems forming neat mats with dome-like clusters of rosy white to salmon flowers. Best in acid soil and cool, shaded conditions.

P. amoena hairy phlox   *30 cm (12 in.)*   *May*
Semiprostrate, with vivid rosy pink flowers that do best in poor dry soil in full sun.

Cultivars   *15 cm (6 in.)*   *May, June*
   'Rosea' — deeper rosy pink flowers
   'Variegata' — pink flowers and pretty variegated foliage

P. carolina see *Phlox glaberrima*

P. divaricata wild blue phlox   *30 cm (12 in.)*   *May, June*
A sparkling, light blue, fragrant native plant that fits in well with spring flowers in partial shade. Will make a good ground cover in moist leafy soils.

P. divaricata var. *canadensis*   *30 cm (12 in.)*   *May, June*
The eastern form of *P. divaricata*, with light blue flowers and notched corolla lobes.

P. divaricata var. *laphamii*   *25 cm (10 in.)*   *May, June*
The western form of *P. divaricata*, with rounded corolla lobes and periwinkle blue flowers; showier than the eastern form.

P. glaberrima   *0.6-0.9 m (2-3 ft.)*   *July, Aug.*
An eastern North American native phlox with rose or purple flowers in dense panicles. Corolla petals are undulate at the margins. Very hardy. Often mistaken for *P. carolina*, which differs by having very hairy stems.

   'Buckeye' — showy rosy purple flowers
       46 cm (18 in.)   *July, Aug.*

P. hoodii Hood's phlox   *5-8 cm (2-3 in.)   May, June*
Native of Alberta. Best in dry soils. White flowers on trailing plant.

P. maculata meadow phlox   *0.9-1.2 m (3-4 ft.) Aug.-Oct.*
Found in moist, sunny meadows and fields of eastern North America, with violet, purple, or white flowers on terminal oblong panicles.

P. nivalis trailing phlox   *25 cm (10 in.)   June*
Very similar to *P. subulata*, differing only by having entire corolla lobes, very short styles, and a much looser habit. It flowers a little later than the moss pink, and its hardiness range is more limited: in areas north of Zone 6 (see Map of Plant Hardiness Zones in Canada, C. E. Ouellet and L. C. Sherk, Can. Dep. Agric., 1967) it needs winter protection.

Cultivars   *June*
  'Camla'   *25 cm (10 in.)*
  Very neat; 23-cm (9-in.) high clusters of salmon pink flowers.

  'Camla Alba' — flushed white flowers   *30 cm (12 in.)*

  'Dixie Brilliant' — tender; large light red flowers
                                 30 cm (12 in.)

  'Gladwynne' — large creamy white flowers   *30 cm (12 in.)*

  'Sylvestris' — large 3.2-cm (1¼-in.) rosy red flowers
                                 30 cm (12 in.)

P. ovata mountain phlox   *38 cm (15 in.) June, July*
Medium-sized smooth leaves, little branched ascending stems, dense clusters of rosy purple to reddish purple flowers, occasionally reblooming in late September.

P. ovata var. *pulchra*   *30 cm (12 in.) June, July*
A geographical variety from Alabama that has more beautiful soft pink flowers than the species, but needs some shade and deep moist soil.

P. paniculata summer phlox   *0.9-1.5 m (3-5 ft.) July, Aug.*
This very popular herbaceous perennial plant is native from New York to Georgia and Arkansas. Many good garden cultivars have been derived from this species mostly by European breeders.

See Chapter 2, Basic Perennials.

P. pilosa   *0.4 m (1½ ft.)   May-July*
Erect, with hairy stems and leaves; pale purple flowers in loose corymbs.

P. stolonifera creeping phlox   *23 cm (9 in.) Apr., May*
An eastern North American species that forms large amounts of foliage from which arise small cymes of purple or violet flowers. Because of its stoloniferous and mat-forming habit, it is an excellent ground cover and particularly suited to shady places.

Cultivars   *23 cm (9 in.) Apr., May*
  'Blue Ridge' — erect stems, masses of large soft blue flowers
  'Lavender Lady' — large lavender blue flowers

P. subulata moss pink, ground pink *10-15 cm (4-6 in.) May*
The moss pink grows wild in many parts of eastern North America from New York, Ontario, and Michigan to North Carolina. Tufted prickly leaves and many downy ascending stems bearing masses of pink, white, or lavender purple flowers.

Cultivars   *15 cm (6 in.) May*
  'Alexander's Beauty' — pink flowers
  'Alexander's Surprise' — pink flowers
  'Autumn Rose' — soft rose flowers with dark eye, blooming again in September
  'Betty' — clear pink flowers, 10 cm (4 in.) high
  'Blue Hills' — violet blue flowers
  'Bonita' — delicate lavender flowers, violet center
  'Brightness' — a neat pink tuft
  'Brilliant' — magenta flowers
  'Coreale' — rounded mats, rosy red flowers
  'Eventide' — pale blue flowers
  'Exquisite' — pink flowers, deeper center
  'Fairy' — palest mauve flowers, compact
  'G. F. Wilson' — delicate mauve flowers
  'Samson' — good warm pink flowers
  'Scarlet Flame' — brilliant scarlet ruby near true red flowers
  'Sensation' — deep pink flowers, sturdy growth
  'Starglow' — rosy crimson flowers

'Star of Heaven' — pale mauve with deeper center
'Temiscaming' — vivid rosy red flowers, 10 cm (4 in.) high
'The Bride' — pure white flowers, very floriferous
'Vivid' — rose pink flowers

PHYGELIUS

P. capensis Cape figwort   *0.9-1.5 m (3-5 ft.)*   *Aug.*
A shrub, often cut down to ground level in winter, so best treated as a perennial. Needs some protection in British Columbia. A beautiful bush with thick, tangled branches from which arise candelabra of drooping tubular scarlet flowers. Needs a sunny location and preferably light sandy soil.

PHYSALIS Cape gooseberry, Chinese lantern

Most gardeners avoid growing this plant because its roots run vigorously. Not especially ornamental until late fall, when the orange red inflated calyxes, which enclose the red fruit, have a decorative lantern-like quality. If you grow these plants for the beauty of their calyxes, choose a corner of the garden where they are likely to be less invasive.

P. alkekengi bladder cherry   *0.3 m (1 ft.)*
Scarlet edible fruits surrounded by the characteristic papery calyxes. Synonym: *P. franchetii*

Cultivars
  'Gigantea' — very large scarlet lanterns   *0.9 m (3 ft.)*

  'Monstrosa'   *0.4 m (1½ ft.)*
Cut and fringed deep-orange calyxes, canoe-shaped sepals.

  'Nana' — similar to the species in all parts, but smaller
       23 cm (9 in.)

  'Orbiculari' — the calyxes are spread out like saucers
       0.4 m (1½ ft.)

P. franchetii see *Physalis alkekengi*

PHYSOSTEGIA   (Color photo p. 87)

P. virginiana
false dragonhead, obedientplant   *0.9 m (3 ft.)*   *Sept.*
A very useful sturdy border perennial for late fall, with leafy stems and spikes of tubular flesh-colored or purple flowers. The individual flowers possess the very curious trait of remaining in the position on the stem in which they are manually placed, hence the British common name of obedientplant.

This beautiful perennial grows equally well in moist or dry soil. It can become rampant and needs dividing frequently. Its stiff habit and restricted color range prevent its wide-spread use, but an occasional grouping in the flower border may be an asset.

Cultivars
  'Alba' — white flowers   *0.9 m (3 ft.)*   *Sept.*

  'Bouquet Rose'   *0.8 m (2½ ft.)*   *July-Sept.*
Bright rose flowers, lighter than those of 'Vivid'.

  'Grandiflora' — tall, with bright pink flowers
       1.5 m (5 ft.)   *Sept.*

  'Nana' — dwarfer than the species, less rampant
       0.8 m (2½ ft.)   *Sept.*

  'Rosy Spire'   *1.1 m (3½ ft.)*   *Sept.*
Rosy pink spikes, taller than 'Vivid'. Synonym: 'Grandiflora' × 'Vivid'

  'Speciosa'   *0.8 m (2½ ft.)*   *Sept.*
Sharply toothed leaves, dense paniculate rosy pink racemes.

  'Speciosa Rosea' — lighter rose flowers
       0.8 m (2½ ft.)   *Sept.*

  'Speciosa Rubra' — deep rose flowers *0.8 m (2½ ft.)*   *Sept.*

  'Summer Glow'   *1.1 m (3½ ft.)*   *Aug.*
Rosy crimson flowers, earlier flowering.

  'Summer Snow' — pure white flowers *0.8 m (2½ ft.)*   *Sept.*

  'Summer Spire'   *0.8 m (2½ ft.)*   *Sept.*
Slender tapering spikes of rosy pink flowers.

  'Vivid' — bright rosy pink, free flowering
       0.9 m (3 ft.)   *Sept.*

PLATYCODON   (Color photo p. 87)

P. grandiflorum balloonflower,
Chinese bellflower   *0.3-0.6 m (1-2 ft.)*   *July, Aug.*
The balloonflower gets its common name from the shape of the buds before they open. It is related to the campanulas, but is more upright in growth and has stronger stems and stouter leaves. The blue flowers open out into a broad bell, which is implied by its generic name. Best in sandy loam in full sun or part shade. Care should be taken at planting time to set the roots 3 cm (1 in.) below soil level.

Cultivars   *July, Aug.*
  'Album' — white flowers   *0.4 m (1½ ft.)*

  'Bristol Bell' — lavender flowers   *0.6 m (2 ft.)*

  'Bristol Bluebird' — deep blue flowers   *0.6 m (2 ft.)*

  'Bristol Blush' — flesh pink flowers   *0.6 m (2 ft.)*

  'Bristol Bride' — white flowers   *0.6 m (2 ft.)*

  'Clonmere Rose' — semidouble, pale pink flowers
       0.6 m (2 ft.)

var. *japonicum* – double, blue flowers 0.6 m (2 ft.)

var. *mariesii* – dwarf, with large blue or white flowers 0.4 m (1½ ft.)

'Mother of Pearl' – single, pale pink flowers 0.6 m (2 ft.)

'New Alpine' – true blue variant, breeds true from seeds 0.6 m (2 ft.)

PODOPHYLLUM

P. emodi Himalayan mayflower 0.3 m (1 ft.) Apr., May
A Himalayan version of the American mayapple plant (*P. peltatum*) but with a larger number of handsome, larger lobed, umbrella-like leaves, over 30 cm (1 ft.) wide, bronzy red in spring, and marked with brown in summer. Only a few shell pink flowers, followed later by a scarlet oval seed-pod, which is said to be edible.

POLEMONIUM Jacob's ladder

Because their leaves are distinctively pinnate and fresh light green and their flowers are blue, the polemoniums are useful plants for the border. Some of the dwarf May-flowering species are extremely effective when used as a foil for yellow and pink tulips. They grow well in sun and partial shade and have no special soil requirements.

P. boreale dwarf Jacob's ladder 0.3 m (1 ft.) May-Aug.
A very hardy, low plant, with short spikes of blue and purple blue flowers on medium-sized clusters, growing from small rosettes of green fern-like foliage. Synonyms: *P. lanatum*, *P. richardsonii*

'Superbum' 0.3 m (1 ft.) June-Aug.
The best form with larger heads, but short-lived unless divided frequently.

P. caeruleum Jacob's ladder 0.4-0.9 m (1½-3 ft.) June-Aug.
The pinnate leaves consisting of up to 27 leaflets are supposed to represent Jacob's ladder. The plant produces a large number of erect stems bearing panicles of drooping open bell-shaped blue flowers.

Cultivars
'Blue Pearl' 0.3 m (1 ft.) May, June
Fine cultivar; cobalt blue flowers with yellow eyes and mound-like hummocks of growth.

'Lacteum' – white flowers 0.6 m (2 ft.) June, July

'Sapphire' – tall, erect, abundant bright blue flowers 0.8 m (2½ ft.) June, July

P. carneum 38 cm (15 in.) May, June
A western North American species, which is not easy to grow except in cool moist climates. Light flesh-pink flowers and smooth divided leaves.

P. confertum see *Polemonium viscosum*

P. lanatum see *Polemonium boreale*

P. pauciflorum 0.6 m (2 ft.) July, Aug.
A tender species with yellowish orange flowers. Best in rich moist soil and some shade.

P. pulcherrimum 25 cm (10 in.) May, June
A very beautiful, graceful little plant from western North America. Light blue flowers and golden centers on 15-cm (6-in.) stems. Extremely beautiful fine-textured leaves.

P. reptans creeping Jacob's ladder 0.4 m (1½ ft.) May-July
It does not creep as its specific name implies, but its habit appears creeping by its sprawling weak stems. Soft blue flowers with white eyes.

P. richardsonii see *Polemonium boreale*

P. viscosum 25 cm (10 in.) June-Aug.
A very graceful little species for Colorado, with distinctive small narrow leaflets on pinnate leaves. The rich blue flowers form dense rounded terminal clusters that stand high above the tufted foliage. It grows well in light shade. Synonym: *P. confertum*

POLYGONATUM Solomon's seal (Color photo p. 87)

These plants have graceful arching stems, pendulous green and white flowers, and lily-like foliage, which are ideal for shady areas. They thrive in moist woodsy soils and, although they spread by underground roots, never become invasive. The common name refers to the roots, which when cut transversely show scars that resemble the seal of Solomon, an Arabic six-pointed star. In the Prairie Provinces they are best grown near the foundation of the house because they are slightly tender.

P. commutatum 1.2 m (4 ft.) June
All parts of the plant including the white flowers are much larger than the common Solomon's seal. Some European nurserymen sell them as a giant lily-of-the-valley. Synonym: *P. giganteum*

P. giganteum see *Polygonatum commutatum*

P. multiflorum Solomon's seal 0.8 m (2½ ft.) June
Tall, with arching stems bearing opposite oval oblong leaves and clusters of hanging tubular greenish white bells.

Cultivars June
'Flore Pleno' – double flowers 1.4 m (4½ ft.)
'Striatum' – striped variegated leaves 0.8 m (2½ ft.)

POLYGONUM knotweed, fleeceflower (Color photo p. 87)

A variable genus of mostly weedy and invasive plants and also some woody plants and climbers that vary in size from 0.3 to 5.5 m (1 to 18 ft.). A few perennial species have a place in the border and as ground covers. All species succeed on any good garden soil, some are useful in very poor dry soils.

P. affine Himalayan fleeceflower   *0.3 m (1 ft.) Aug.-Oct.*
Close tufts of spear-shaped leathery leaves and abundant spikes of closely packed small pink flowers. In late summer the foliage is bronze. An excellent ground cover.

Cultivars   *July-Sept.*
  'Darjeeling Red'   *25 cm (10 in.)*
  Much deeper pink, almost crimson, poker-like flowers.

  'Lowndes Variety' — finer foliage, an excellent ground cover   *23 cm (9 in.)*

P. alpinum alpine knotweed   *0.6 m (2 ft.) June*
A beautiful plant with loose sprays of white flowers, which can be used for cutting. Best in a moist location.

P. amplexicaule
mountain fleeceflower   *1.2 m (4 ft.) July-Oct.*
A very robust, fine flowering species with dark green leaves and a profusion of flower spikes from much-branched 1.2-m (4-ft.) stems. Not invasive.

Cultivars   *July-Oct.*
  'Album' — white flowers tinged pink   *0.9 m (3 ft.)*

  'Atrosanguineum' — very deep red flowers. Synonym: 'Speciosum'   *1.1 m (3½ ft.)*

  'Firetail' — tall; carmine scarlet flowers   *1.2 m (4 ft.)*

P. bistorta 'Superbum'   *0.9 m (3 ft.) June, July*
Tends to become weedy. This is an improved form with spikes of dainty soft pink flowers and oblong oval leaves.

P. campanulatum
Himalayan knotweed   *0.6-0.9 m (2-3 ft.) July-Oct.*
Flowers in clusters rather than spikes, as in most of the other species. These are soft pearly pink and showy. In rich moist soil the plant tends to become invasive.

  'Roseum' — deeper pink flowers than the species
                    0.6-0.9 m (2-3 ft.) July-Oct.

P. campanulatum var. *lichiangensis*
                 0.6-1.2 m (2-4 ft.) July-Sept.
A Chinese geographical variety with lanceolate leaves, green above and gray beneath, and creamy white flower sprays.

P. cuspidatum
Japanese knotweed   *1.8-2.4 m (6-8 ft.) Aug., Sept.*
A very impressive but very invasive perennial with leathery-textured and wavy-margined leaves. The stiff erect stems are bamboo-like and reddish. The inflorescences form large sprays of creamy white blossoms. Very common in gardens and as an escape in Ontario, Quebec, and New England. Synonym: *P. sieboldii*

  'Compactum'   *0.6-0.9 m (2-3 ft.) Aug., Sept.*
  Much smaller than the species, with darker green crimped leaves. The branching stems are grooved and spotted with purple. Flowers are creamy white with an aborted ovary and the stamens are longer than the perianth. Recently this form has been named *P. reynoutria*, a name which is sometimes given as a synonym of *P. cuspidatum*. The form that has for so long carried the name *P. reynoutria* is the female version of *P. c.* 'Compactum', which is *P. c.* 'Compactum Femina' (*P. reynoutria* of gardens, not *P. reynoutria* Makin.).

  'Compactum Femina'   *0.3-0.6 m (1-2 ft.) Aug., Sept.*
  A form with aborted stamens but well-developed ovaries, and an enlarged and reddened perianth, which make it much more ornamental than the species. This form is often used as a ground cover for rocky areas, where it can be allowed to spread with abandon. Care must be taken not to plant it in borders, because it is extremely invasive. This plant is commonly known as *P. reynoutria*, but it is not the true *P. reynoutria* described by Makinoi.

P. sachalinense   *3.1-3.7 m (10-12 ft.) Aug., Sept.*
An even more vigorous plant than *P. cuspidatum* and with longer, broader leaves, heart-shaped at the base; erect angular reddish stems and greenish white flowers, carried in the leaf axils. Needs severe restriction because it spreads rapidly.

P. sieboldii see *Polygonum cuspidatum*

P. vaccinifolium   *15-30 cm (6-12 in.) Aug., Sept.*
A prostrate shrubby plant, useful in mild climates for rock gardens and the front of borders. The showy, bright rose red flowers are produced abundantly in erect spikes.

POTENTILLA cinquefoil (Color photo p. 87)

Many of the perennial cinquefoils are long-living showy border plants, some are invasive, and others are objectionable weeds. They are noted particularly for their bright flowers and finely cut finger-like leaves mostly with silvery undersides. Most species do best in poor, dry soils, where they flower abundantly. They will not withstand the cold winter of the Prairies.

P. argyrophylla
Kashmirian cinquefoil   *0.6-0.9 m (2-3 ft.) May-Sept.*
A yellow-flowered species with stems and heart-shaped palmate leaves covered with fine silky hairs. A parent of some excellent hybrids.

'Atrosanguinea' see *Potentilla atrosanguinea*

P. atrosanguinea 0.6 m (2 ft.) June-Sept.
Brightly silvered foliage and dark rich crimson flowers. Synonym: *P. argyrophylla* 'Atrosanguinea'

P. concolor 0.3 m (1 ft.) July-Sept.
A Chinese species that has large deep flowers with orange blotches at their base.

P. emarginata 23 cm (9 in.) May, June
Yellow flowers and soft green hairy strawberry-like leaves. Synonym: *P. fragiformis*

P. fragiformis see *Potentilla emarginata*

P. × hopwoodiana 0.4 m (1½ ft.) May-July
A group of hybrids with red and pink flowers edged with white. A hybrid of *P. nepalensis × P. recta*

P. nepalensis 0.6 m (2 ft.) all summer
A very attractive species that flowers all summer with bright cherry red flowers on branching stems. Flowering is often so profuse that the plant is short lived.

Cultivars 0.4 m (1½ ft.)
  'Master Floris' June-Oct.
  Lives a lot longer than the other cultivars; biscuit-colored flowers, suffused with brick red.

  'Miss Willmott' — see *P. willmottiae*

  'Roxana' — orange scarlet flowers; similar habit to 'Miss Willmott' July, Aug.

P. recta 0.4 m (1½ ft.) June, July
Neat bushy habit; terminal sprays of bright yellow flowers; good for borders.

Cultivars 0.4 m (1½ ft.) June, July
  'Sulphurea' — light yellow flowers

  'Warrensii' — much larger bright yellow flowers than the species

P. willmottiae 0.3 m (1 ft.) all summer
A cultivar with bright rosy crimson flowers on branching stems that spray out from the base. Synonym: *P. nepalensis* 'Miss Willmott'

Cultivars
Many unclassified hybrids have been bred from *P. argyrophylla* and other species that are quite different in character and cannot be allied to any species. Some of these are included in the list that follows. Best in a sunny well-drained location.

'Arc en Ciel' June, July
4 cm (1½ in.) wide, semidouble crimson 0.6 m (2 ft.)
flowers, blotched with gold.

'California' — semidouble golden yellow flowers 0.6 m (2 ft.)

'Etna' 0.8 (2½ ft.)
Deep crimson flowers, taller, and having more silvery leaves.

'Flambeau' — semidouble crimson flowers 0.6 m (2 ft.)

'Flamenco' — large, single, intense red flowers, very vigorous 0.4 m (1½ ft.)

'Gibson's Scarlet' — very popular; blood-red flowers 0.6 m (2 ft.)

'Glory of Nancy' 0.4 m (1½ ft.)
Grayish foliage, outstanding semidouble orange crimson flowers.

'Jupiter' — semidouble deep crimson flowers 0.4 m (1½ ft.)

'M. Rouillard' — double blood-red and yellow flowers 0.6 m (2 ft.)

'Star of the North' — crimson flowers 0.4 m (1½ ft.)

'White Beauty' — erect growth, fine white flowers 38 cm (15 in.)

'Wm. Rollison' 0.4 m (1½ ft.)
Semidouble glowing orange red flowers, with yellow centers.

POTERIUM

P. canadensis see *Sanguisorba canadensis*

P. obtusum see *Sanguisorba obtusa*

PRIMULA primrose (Color photo p. 88)

Primroses are among the most charming and graceful flowers, especially in a woodland glade with light checkered shade. Most of the best species are only hardy in British Columbia, but they may be nurtured into perfection in southern Ontario, especially in the Niagara district. A few very showy types are hardy in Ottawa and on the Prairies.

In general, primroses need a shady moist location and soil that is rich in organic matter. Some of the species do best under boggy conditions.

P. acaulis see *P. vulgaris*

P. alpicola Tibetan primrose 0.4 m (1½ ft.) May, June
Long green serrated leaves and drooping umbels of white, yellow, purple, or violet fragrant flowers; all covered with meal.

Cultivars 0.4 m (1½ ft.) May, June
  'Alba' — white flowers
  'Violacea' — violet purple flowers

P. anisodora 0.6 m (2 ft.) May-Sept.
An aromatic plant with funnel-shaped purple flowers with green centers arranged in candelabra design.

P. aurantiaca   *0.3 m (1 ft.)*   *July*

A sweetly scented plant with toothed leaves and brownish purple flowers with green eyes, arranged in whorls on very strong stems.

P. auricula auricula   *15-23 cm (6-9 in.)*   *Apr., May*

A very handsome species with stout mealy stems and powdery oval leaves. The flowers are bell-shaped, fragrant, and borne in umbels. The flowers of this species are yellow, but there are many mutants and hybrids under either this specific name or *P. × pubescens.*

Auriculas grow easily in most soils, but they must not be allowed to dry out. To help keep the soil moist, add lots of organic matter. In Eastern Canada and the Prairies some winter protection is needed, particularly to prevent the large swollen roots from heaving.

Cultivars   *23 cm (9 in.)*   *Apr., May*

'Blue Velvet' – deep velvety blue flowers with a white eye

'Celtic King' – large clear yellow flowers with a white eye

'Dusty Miller' – yellow flowers, quite farinose, and richly scented

'Jean Walker' – mauve flowers with a cream eye, large, vigorous, and fragrant

'Red Dusty Miller' – red flowers

'W. A. Cook' – deep red flowers, mealy leaves

P. beesiana   *0.6 m (2 ft.)*   *May, June*

A very strong growing species with candelabra type of flower heads bearing light purple velvety flowers. Needs good, rich moist soil, preferably in a waterside location.

P. bulleyana   *0.8 m (2½ ft.)*   *May, June*

A very fine vigorous species with stout stems bearing whorls of deep orange flowers.

P. burmanica   *0.6 m (2 ft.)*   *June*

Reddish purple flowers with a yellow eye; forming tufts of wrinkled primrose-like foliage.

P. chionantha   *0.3-0.6 m (1-2 ft.)*   *May*

An unusual species from Yunnan with fragrant white tubular flowers arranged in whorls on strong stems.

P. cockburniana   *0.3 m (1 ft.)*   *June*

This species has withstood a few winters at Ottawa in a moist woodland location where snow was heavy. Deep orange flowers in whorls on rather slender stems.

P. cortusoides   *0.4 m (1½ ft.)*   *May*

A very hardy Siberian species with rose-colored flowers on many flowered umbels. The crenate softly hairy leaves are rather like those of *P. sieboldii* but much smaller. A good primrose to grow where most others do not succeed.

P. denticulata   *0.3 m (1 ft.)*   *Apr., May*

A really hardy little primrose that grows in all kinds of soil and withstands drier conditions than most of the others. The showy light violet or mauve blooms, borne in globe-shaped clusters, provide a pleasant spring picture when planted in bold groups in the front of the border.

P. denticulata var. *cachemiriana*   *0.3 m (1 ft.)*   *May*

A later flowering variety from Kashmir with purple to lilac rose flowers with yellow throats. Because it flowers later than the species, its leaves are more advanced and provide a better setting for the blooms.

Cultivars   *0.3 m (1 ft.)*   *Apr., May*

'Alba' – white flowers

'Bengal Rose' – fuchsia purple flowers

'Pritchard's Ruby' – rich ruby red flowers

'Purple Beauty' – dark violet purple flowers

'Red Emperor' – bright rose flowers

'Stormonth's Red' – vivid orchid purple flowers

P. elatior oxlip

Very closely resembling the cowslip *P. veris,* but with larger, wider open yellow flowers on 5-12 flowered umbels. A parent of the polyantha hybrids, it grows where polyanthas do not flourish. Best in a moist sunny location.

P. florindae Himalayan cowslip   *0.9 m (3 ft.)*   *July, Aug.*

One of the most vigorous and tallest of the primulas. The sulfur yellow fragrant flowers are produced in large cowslip-like umbels on long stems up to 0.9 m (3 ft.) high. Best in wet boggy soils or by the waterside, but successful with more restricted growth under drier conditions.

P. helodoxa   *0.6 m (2 ft.)*   *June, July*

A vigorous species with large 3-cm (1-in.) golden yellow flowers in 12-20 flowered whorls. Needs moisture.

P. japonica   *0.6-0.9 m (2-3 ft.)*   *May, June*

The most popular and widely grown member of the so-called candelabra primulas with flowers in tiers of umbels. This species is somewhat easier to grow than the other primulas, and has a wide range of colors. In an amenable location it seeds itself and produces colonies. Its habit and growth requirements are more suited to a moist semishaded location than to an open border.

Cultivars   *May, June*

'Miller's Crimson' – dark crimson purple flowers

                               0.6-0.9 m (2-3 ft.)

'Postford White' – white flowers   *0.6 m (2 ft.)*

'Rose Dubarry' – rose pink flowers   *0.6-0.9 m (2-3 ft.)*

P. × juliae Caucasian primrose   *10 cm (4 in.)*   *Apr., May*

The most satisfactory primrose for cultivation in Eastern Canada and generally considered as a good plant in rock gardens and for low carpeting. It is more tolerant of drier conditions than most primulas, and it will grow in shaded areas, where it is useful as a ground cover in dry shady borders. The species forms dense mats of foliage from which masses of lilac purple flowers arise in early spring. This species has been crossed with *P.* × *polyantha* to give a number of brilliantly colored hybrids.

Cultivars

'Alba' — white flowers
'Betty' — rich crimson flowers
'Dorothy' — pale yellow flowers
'E. R. James' — vivid cherry red flowers
'Gloria' — crimson purple flowers
'Gold Jewel' — brilliant yellow flowers
'Jewel' — crimson purple flowers
'Kinlough Beauty' — rose pink flowers
'Mrs. McGillivray' — old rose flowers
'Old Port' — wine purple flowers
'Our Pat' — double, dark purple flowers
'Pam' — rich garnet red flowers
'Snow Cushion' — white flowers
'Snow White' — white flowers
'Wanda' — claret purple flowers

P. × *polyantha* polyanthus primrose

This name covers all the complex hybrids known as polyanthus primroses, which involved crossing and recrossing species such as *P. elatior*, *P. veris*, and *P. vulgaris* to produce a race with a fantastic range of color, size, and beauty. The plants are moderately hardy in Eastern Canada, though they do not live long. They need a sheltered location, adequate drainage, good rich soil, and mulching during the summer.

Some strains are hardier than others; in Ottawa more plants of the Barnhaven strain than any other have survived. Also, Clarke and Ellen Carden hybrids are good American strains. The newest hybrids from Sakata of Japan, known as the Pacific Giant strain, have a tremendous color range and extra large blooms, up to 8 cm (3 in.) in diameter, but their hardiness in our area has not yet been assessed. They are available in at least 11 true breeding colors including deep reds, yellows, blues, and white.

P. pulverulenta
silverdust primrose   *0.8-0.9 m (2½-3 ft.)   June, July*
Very robust; long narrow silvered leaves and strong flower scapes carrying tiers of large cowslip-like purple blossoms. An excellent plant for moist woodlands, a shady border among ferns, or a bog garden.

Of the many hybrid strains, perhaps the most common is the predominantly pink-flowered Bartley strain. The Lissadell hybrids result from a cross of *P. pulverulenta* × *P. cockburniana*, and contain mostly purple scarlet flowers.

Cultivars   *0.8-0.9 m (2½-3 ft.)   June, July*
'Arleen Aroon' — bright red flowers
'Lady Thursby' — bright rose pink flowers, yellow eyes
'Mrs. R. V. Berkeley' — white flowers with orange eyes
'Red Hugh' — vivid red flowers

P. sieboldii Siebold primrose   *23 cm (9 in.)   May, June*
A very satisfactory Japanese species that is more tolerant of sun, heat, and drought than most of the other species. Large 4-cm (1½-in.) fringed pink, rose, white, or purple flowers in late spring. The heart-shaped crinkled leaves form tufts, from which the slender flower spikes arise.

P. veris cowslip   *20 cm (8 in.)   May*
Loose nodding umbels of bell-shaped bright yellow flowers with 20-cm (8-in.) scapes.

P. vulgaris common primrose   *15-23 cm (6-9 in.)   Apr., May*
Small tufts of wrinkled foliage from which arise solitary yellow flowers with a deeper yellow eye. Many deep blue forms are available, which are often preferred to the species. They are excellent companions for the yellow-flowered kinds. Synonym: *P. acaulis*

PULMONARIA lungwort   (Color photo p. 88)

A very hardy easy-to-grow perennial that does best in some shade and moderate soil conditions. The drooping clusters of bell-shaped flowers appear in spring at the same time as the forsythias, and their large spotted leaves add interest to any shady area. The plant seems to thrive on neglect and can always be relied upon to produce an abundance of blue or pink flowers in very early spring.

P. angustifolia cowslip, lungwort   *0.3 m (1 ft.)   Apr., May*
Very rough, unspotted narrow leaves and bunches of drooping funnel-shaped flowers that open pink and then turn bright blue.

Cultivars   *0.3 m (1 ft.)   Apr., May*
'Johnston's Blue' — gentian blue flowers
'Mawson's Variety' — rich blue flowers
'Munstead Blue' — intense blue flowers
'Salmon Glory' — clean coral salmon flowers, very beautiful

P. montana   *0.3 m (1 ft.)   Apr., May*
Bright green softly hairy unspotted leaves and bright brick red flowers, never blue or purple. Synonym: *P. rubra*

P. officinalis
Jerusalem cowslip, lungwort   *0.3 m (1 ft.)   Apr., May*
The more commonly naturalized species. Coarser and less profuse than the other species. Bristly white-spotted cordate leaves and rosy flowers that turn blue.

P. rubra see *Pulmonaria montana*

P. saccharata Bethlehem sage   *0.4 m (1½ ft.)   Apr., May*
Bristly elliptic leaves, profusely white spotted. The rosy flowers turn blue.

Cultivars   *Apr., May*
'Bowles Red' — intense red flowers   *23 cm (9 in.)*
'Mrs. Moon' — very striking dark red flowers   *0.3 m (1 ft.)*
'Pink Dawn' — very refined coral pink flowers   *0.3 m (1 ft.)*

PULSATILLA see ANEMONE

RANUNCULUS buttercup

A very large genus containing a few good border perennials that grow well under ordinary garden conditions. Some are invasive and must be watched carefully.

R. aconitifolius
white bachelor's-button   *0.4 m (1½ ft.)*   *May, June*
A mountain meadow plant with glossy palmate 3-5-lobed deeply toothed leaves and white flowers.

Cultivars   *May, June*
  'Flore Pleno' fair-maid-of-France   *0.4 m (1½ ft.)*
  Small, double white rosette-like flowers.

  'Platanifolius'   *0.6 m (2 ft.)*
  Stronger growing than the species; single white flowers.

R. acris 'Flore-pleno'
yellow bachelor's-button   *0.6 m (2 ft.)*   *June, Aug.*
This double form is worth growing in the front of a border. The yellow button-like rosette flowers are produced abundantly in June and again in August. The species is a weed.

R. gramineus   *0.3 m (1 ft.)*   *May, June*
A grassy-leaved plant with slender erect stems, and flowers similar to but much larger than *P. acris*. Much more refined than *R. acris*, because of its blue grass-like foliage and showy spreading habit.

R. repens 'Pleniflorus'
The double form of *R. repens* is much less weedy than the species and is useful for a ground cover on sunny slopes and in a border where it can be kept in check.

RHEUM rhubarb

A few species of rhubarb are vigorous herbaceous plants with fine-textured bold foliage. They are very useful as isolated specimens on lawns and to produce bold effects near water.

R. australis
Himalaya rhubarb   *1.8-3.1 m (6-10 ft.)*   *June, July*
A very effective plant where heavy texture and diversity of foliage color is desired. Broadly ovate wavy-margined leaves.

R. palmatum sorrel rhubarb   *1.5 m (5 ft.)*   *June, July*
Very large deeply cut rounded heart-shaped leaves and large panicles of creamy flowers.

  'Tanguticum'   *1.2 m (4 ft.)*   *June, July*
  The leaves of this cultivar are longer and more deeply lobed than those of the species.

RODGERSIA

Fine foliage plants for moist garden borders. They have long-stalked large simple or compound leaves and large terminal clusters of flowers.

R. pinnata feathered bronzeleaf   *0.9-1-2 m (3-4 ft.)*   *July*
A very robust species with pinnate leaves and leaflet 13-23 cm (5-9 in.) long. The reddish flowers are in a much-branched panicle.

Cultivars   *0.9 m (3 ft.)*   *June, July*
  'Alba' — white flowers
  'Elegans' — rose pink flowers and reddish stems
  'Rubra' — deeper red flowers than the species

R. podophylla   *0.9-1.2 m (3-4 ft.)*   *June, July*
A species with very heavily netted leaves divided into three sharply toothed lobes. The yellowish white flowers are on a large leafless panicle.

R. tabularis   *0.9-1.2 m (3-4 ft.)*   *July, Aug.*
A distinct species with rounded leaves on thick bristly stems. The white flowers are well above the leaves, similar to an astilbe.

ROMNEYA

Magnificent herbaceous perennials for the milder parts of British Columbia. When well established, they grow to 2.4 m (8 ft.) high, have spreading rootstocks, branching stems with fine blue gray leathery leaves, and huge solitary poppy-like flowers 10-15 cm (4-6 in.) in diameter. These plants have a prominent central boss of golden stamens, which seem to accentuate the purity of the petals.

R. coulteri Matilija poppy   *1.2-1.8 m (4-6 ft.)*   *July-Sept.*
A huge poppy with flexuous stems, glaucous blue leathery foliage, and large fragrant white flowers with golden stamens. The fruits are covered with rigid golden bristles.

R. trichocalyx tree poppy   *2.4 m (8 ft.)*   *July, Aug.*
More erect and less branched than *R. coulteri*. Though there are more flowers, they do not have the same pleasant fragrance. This species appears to be easier to establish than the other.

ROSCOEA

A small genus of Asiatic plants with very fleshy roots, lanceolate leaves, and blue, yellow, or purple flowers as in most of the Zingiberaceae. The flowers are very unusual, funnel-shaped, widening at the throat into an uneven-shaped flower; the lower lip is large (over 3 cm (1 in.) long) and hanging with a frilly edge, and the upper part arches to form a kind of hood. The plant does best in rich well-drained soil and some shade.

R. cautleoides   *0.3 m (1 ft.)   all summer*
The more commonly grown species. Glossy green, slender, pointed, stemless leaves; close spikes of 6 or 7 clear yellow flowers.

R. humeana   *15-23 cm (6-9 in.)   all summer*
Smaller than the previously mentioned species and with spectacular deep violet-purple blooms.

RUDBECKIA coneflower   (Color photo p. 88)

A genus of reliable, somewhat coarse, herbaceous perennials that are useful for cutting and for occasional spotting in the border for midsummer and early autumn bloom. The more modern cultivars are extremely pleasing and should be combined with perennial asters and lythrums to present a striking picture. They are very easily grown, but benefit from a good rich soil.

R. fulgida var. *speciosa*   *0.8 m (2½ ft.)   July-Sept.*
The species has golden yellow florets and a black disk on compact plants. The leaves are rather rough and narrower than most of the other species. Improved varieties are superior. Synonym: *R. newmannii*

Cultivars   *0.8 m (2½ ft.)   July-Sept.*
   'Goldquelle' — double deep yellow flowers

   'Goldsturm' — much larger flowers and slightly twisted rays

R. laciniata coneflower   *2.1 m (7 ft.)   July-Sept.*
A very tall, rampant, and weedy North American perennial with large unevenly divided leaves, large single yellow flowers, and a few drooping ray petals.

   'Golden Glow'   *2.1-2.4 m (7-8 ft.)   July-Sept.*
   The old familiar rudbeckia that adorns many otherwise barren locations, and perhaps should be entirely confined to such places, because it soon becomes weedy in a fairly rich border. It has fully double flowers, 8 cm (3 in.) across, which are good for cutting but are usually regarded with about the same esteem as a dandelion.

   'Herbstonne' autumn sun   *1.2 m (4 ft.)   Aug., Sept.*
   Much more refined than the species, but with similar yellow flowers.

R. newmannii see *Rudbeckia fulgida* var. *speciosa*

R. nitida   *1.5-1.8 m (5-6 ft.)   Aug.-Oct.*
Yellow reflexed ray florets and yellow disks; rounded lance-shaped leaves.

R. subtomentosa sweet coneflower   *1.2 m (4 ft.)   July-Sept.*
Soft grayish foliage and yellow flowers with a dark purple center.

R. triloba brown-eyed Susan   *0.6-0.9 m (2-3 ft.)   Aug.-Oct.*
A biennial or short-lived perennial; quite useful in the border if it is seeded every few years. The so-called annual rudbeckia or gloriosa daisies, which bloom the first year from seed, probably belong to this species.

SALVIA sage

The well-known scarlet sage that is grown as an annual is really a perennial and belongs in this publication. The genus contains a large number of hardy perennial species that are easy to grow but are quite uncommon in Canadian gardens. They need light sandy soil and very good drainage.

S. azurea azure sage   *0.9-1.2 m (3-4 ft.)   Sept., Oct.*
A very beautiful perennial with gray leaves and pale blue nettle-like flowers.

Cultivars   *Sept., Oct.*
   'Angustifolia' — much narrower leaves than the species
                      0.9 m (3 ft.)

   'Grandiflora'   *1.4 m (4½ ft.)*
   Downy leaves and dense spikes of gentian blue flowers. Although reported to be tender at Vermont, it is hardy at Ottawa. Synonym: *S. pitcheri*

S. beckeri   *0.9 m (3 ft.)   June-Aug.*
A fairly hardy salvia with blue and violet flowers in well-spaced whorls on fairly short spikes. The leaves are up to 0.8 m (2½ ft.) long, ovate, and hairy beneath. This species was grown at Ottawa from 1905 to 1925.

S. glutinosa Jupiter's distaff   *0.9 m (3 ft.)   July*
A very unusual species because of its whorls of pale yellow flowers on slender sticky stems and its yellow pointed leaf tips.

S. haematodes see *Salvia pratensis*

S. jurisicii   *0.3 m (1 ft.)   June-Aug.*
An unusual low-growing mat-like species with finely cut feathery leaves and clusters of violet flowers.

S. nemerosa see *Salvia × superba*

S. nutans nodding sage   *0.6-0.9 m (2-3 ft.)   July*
A fine species with drooping violet blue flowers on hairy stems, and oblong, toothed, wrinkled leaves.

S. patens gentian sage   *0.4-0.6 m (1½-2 ft.)   Aug., Sept.*
Although it is not hardy in most areas, it is included in this publication because its roots, similar to dahlias, may be stored and planted out each year. This is one of the showiest of all blue plants. The flowers, borne on a few remote whorls, have a corolla tube over 5 cm (2 in.) long with the tip slightly opened.

170

Cultivars   *0.8 m (2½ ft.)*   *July, Aug.*
   'Alba' — white flowers
   'Cambridge Blue' — pale blue flowers

S. pitcheri see *Salvia azurea* 'Grandiflora'

S. pratensis meadow sage   *0.9 m (3 ft.)*   *June-Sept.*
A very rugged, hardy plant with well-spaced whorls of blue, purple, rose pink, and white flowers. Oblong heart-shaped basal leaves and narrowly pointed leaves on the flower stems. Synonym: *S. haematodes*

Cultivars   *0.6 m (2 ft.)*   *June*
   'Alba' — white flowers
   'Rosea' — rose pink flowers
   'Tenori' — deep blue flowers
   'Variegata' — light blue flowers, mid-lobe of lower lip white

S. sclarea clary   *0.9-1.1 m (3-3½ ft.)*   *Aug.-Oct.*
A striking but short-lived herb with pale mauve flowers and conspicuous white and rose bracts. Rather unpleasant smelling when bruised.

   'Turkestanica'   *0.9 m (3 ft.)*   *Aug.*
   White flowers with enlarged pinkish purple bracts directly beneath; unpleasant smelling.

S. × superba   *0.8 m (2½ ft.)*   *July, Aug.*
A very fine hybrid with masses of dense racemes of rich purple flowers and reddish violet calyxes that persist after the flowers have dropped. Extremely effective in bold groups in the border. Sage green aromatic leaves. Synonyms: *S. nemerosa, S. virgata* 'Nemerosa'

Cultivars
   'East Friesland' — dwarf   *0.4 m (1½ ft.)*   *July*

   'Lubeca' — identical with the species, but only about half as high   *0.4 m (1½ ft.)*   *July, Aug.*

   'May Night' — dwarf, with violet blue flowers   *0.4 m (1½ ft.)*   *June, Aug.*

   'Purple Glory' — new, with outstanding qualities   *0.8 m (2½ ft.)*   *July, Aug.*

S. uliginosa 'Bog Sage'   *1.2-1.5 m (4-5 ft.)*   *Aug.-Oct.*
Tall, graceful with a branched four-angled stem, and lanceolate leaves, hairy above, and smooth, black-dotted beneath. The rich blue flowers often persist until frost. No hardier than the so-called hardy chrysanthemums, and should be covered with straw in the winter for protection.

S. virgata Oriental sage   *0.6 m (2 ft.)*   *July, Aug.*
A well-branched species with nettle-like leaves and stems, and branching spikes bearing whorls of light blue funnel-shaped flowers.

'Nemerosa' see *Salvia × superba*

SANGUINARIA

S. canadensis bloodroot   *23 cm (9 in.)*   *Apr.*
Common in Eastern Canadian woodlands; it has not been exploited as well as it should. A very useful little perennial for sun and part shade. Its distinctive gray blue foliage and snow white flowers are extremely ornamental.

   'Multiplex'   *23 cm (9 in.)*   *Apr.*
   The double form is scarce and rather expensive, but worth the effort needed to locate it in catalogs.

SANGUISORBA   (Color photo p. 89)

A genus containing a few very attractive species with showy pinnate leaves and flowers resembling a bottlebrush. They grow in any good garden soil and flower over a long period. Many are better known under the old generic name *Poterium*.

S. canadensis American burnet   *0.9 m (3 ft.)*   *all summer*
Deeply cut grayish green foliage and white cylindrical flower heads up to 15 cm (6 in.) long. Synonym: *Poterium canadensis*

S. obtusa Japanese burnet   *0.9 m (3 ft.)*   *all summer*
Pink fluffy bottlebrushes hanging obliquely from wiry 0.9-m (3-ft.) stems arising from pale green pinnate foliage. Synonym: *Poterium obtusum*

SAPONARIA soapwort

The cultivated soapworts are best in a rock garden, where their compact habit and June flowers are much appreciated. The leaves and roots of some species lather like soap and have been used for cleaning. Easy to cultivate except in soil containing lime. The soapworts included here are very hardy.

S. × boisseri   *8 cm (3 in.)*   *May, June*
Intermediate between the two species *S. caespitosa × S. ocynoides*, with freely produced pink flowers on mat-forming plants.

S. caespitosa   *8 cm (3 in.)*   *May, June*
Very pretty; forming shiny green cushions of large pink flowers.

S. lutea   *8 cm (3 in.)*   *May, June*
A mat-forming midget with creamy white flowers and purple stamens. Not too hardy, it needs protection.

S. ocymoides rock soapwort   *23 cm (9 in.)*   *June*
A creeping plant with star-shaped rose pink flowers. A very popular plant for rock gardens and dry walls.

'Splendens'   *23 cm (9 in.)   June*
Larger and deeper rose pink flowers than the species.

S. officinalis bouncingbet   *0.9 m (3 ft.)   July-Oct.*
A ubiquitous perennial with pinkish or mauve flowers arranged in close terminal clusters on upright plants. It has become widely naturalized in North America. It soon becomes invasive in a garden.

The following double cultivars are far superior to the species:
'Albo Plena' — white flowers   *0.8 m (2½ ft.)   July-Sept.*
'Roseo-Plena' — rose flowers   *0.8 m (2½ ft.)   July-Sept.*
'Rubra Plena' — red flowers   *0.8 m (2½ ft.)   July-Sept.*

SAXIFRAGA

Most of the 300 or so species in this genus are specialists' rock plants and should not be included in a publication on herbaceous perennials. The following species, however, are easy to grow and might be useful in groups at the front of the border. They are best in light shade and average soil.

S. aizoon see *Saxifraga paniculata*

S. paniculata   *8-15 cm (3-6 in.)   May*
Rosettes of bright green foliage forming neat spreading clumps, and white or pink flowers. Synonym: *S. aizoon*

S. umbrosa London pride   *38 cm (15 in.)   May, June*
A good perennial in the milder parts of British Columbia for odd shady corners or an edging where little else will grow. Leathery rosettes of evergreen leaves and dainty sprays of pink flowers.

Cultivars   *May, June*
'Aurea Punctata'   *38 cm (15 in.)*
Pretty variegated leaves and golden spots on the flowers.

'Primuloides' — very free flowering   *15 cm (6 in.)*

'Primuloides Rubra'   *15 cm (6 in.)*
Same as the previous cultivar, but with red flowers.

SCABIOSA scabious, pincushionflower

One of the most popular and hardiest of garden perennials for borders and for cut flowers. Called pincushionflower because the globe-shaped flower heads have protruding stamens. Useful in borders and rock gardens. Best in well-drained sandy loam soil with lime or chalk added if the soil is too acid.

S. caucasica blue bonnets   *0.9 m (3 ft.)   July-Sept.*
A beautiful perennial plant that flourishes in cool moist climates, but is hard to establish in Eastern Canada. Best in light soil with the addition of leaf mold and an annual dress-

ing of well-rotted manure or organic matter. After 3 or 4 years, the plants should be divided. Some of the newer cultivars seem more rugged and are worth trying.

Cultivars
'All Blue' — good blue flowers, free flowering

'Bressingham White' — white flowers

'Clive Greaves' — rich blue flowers, an excellent constitution

'Challenger' — fine blue flowers

'Mrs. Willmott' — ivory white flowers

'Penhill Blue' — deep violet blue flowers

S. fischeri   *0.4 m (1½ ft.)   July-Oct.*
A small, more rugged plant with heads of blue flowers. Not as showy as *S. caucasica*, but much more reliable.

S. graminifolia
grassleaf scabious   *30 cm (12 in.)   May-Aug.*
Forms neat clumps from which arise lilac rose pincushion-flowers on erect stems and narrow silvery leaves.

S. lucida   *23 cm (9 in.)   May-Sept.*
Small, attractive rosy lilac flowers and finely cut silvery foliage.

S. ochroleuca   *0.6 m (2 ft.)   July-Sept.*
Small 3-cm (1-in.) globes of creamy yellow flowers on wiry 0.6-m (2-ft.) stems.

SEDUM stonecrop   (Color photo p. 89)

Most of the plants in this large genus are best used in rock gardens and as ground covers on hot sunny slopes. A few species are useful in a border; they are adaptable to almost any soil, but best in gravelly kinds. All species have fleshy leaves.

S. acre common stonecrop   *8 cm (3 in.)   July, Aug.*
A very common stonecrop with widespread habitat from Britain to Persia, Norway, and Morocco. It has become very common in North America, where its creeping mats of glorious evergreen foliage are seen in many sunny areas climbing over rocks, walls, and cliffs. Very few plants can surpass it as a ground cover in such locations. Because it is a very invasive plant, care must be taken to restrict it to the use suggested here.

S. ewersii   *23 cm (9 in.)   Aug., Sept.*
Glaucous leaves, a red-stemmed twigging much-branched rootstock, and purplish pink flower clusters produced late in the summer.

S. kamtschaticum   *15 cm (6 in.)*   *June-Sept.*
An evergreen species that forms neat mats with spoon-shaped scalloped leaves and vivid orange yellow flowers.

    'Variegatum'   *15 cm (6 in.)*   *June-Sept.*
    Leaves broadly margined cream, tinged pink; bright yellow flowers.

S. maximum purple-leaved sedum   *0.6 m (2 ft.)*   *Sept., Oct.*
A very variable species; a thick rootstock and carrot-like roots; greenish white flowers 2 cm (¾ in.) across in crowded, terminal, and lateral corymbs.

    'Atropurpureum'   *0.6 m (2 ft.)*   *Sept., Oct.*
    Much more desirable than the species, with purple-tinted mahogany leaves and vivid purple stems; soft creamy rose flowers in crowded clusters. A striking and most effective late-flowering plant for the front of the flower border.

S. sieboldii   *30 cm (12 in.)*   *Sept., Oct.*
A very refined species with low, arching, slender purplish stems and whorls of three sessile, almost completely round blue green leaves with wavy pinkish margins that turn rose in autumn. The pink flowers are borne in much-branched, densely packed umbels.

    'Medico-variegatum'   *23 cm (9 in.)*   *Sept., Oct.*
    Large blotches of yellow in the middle of the leaves. Not as hardy as the species, it is usually used as a pot plant.

S. spathulifolium   *8 cm (3 in.)*   *May, June*
This is one of the smaller creeping clump-forming species, but it is a better ground cover than most of the others. Red-tinged leaves and yellow flowers. The cultivars are much superior.

Cultivars
    'Casablanca' — neat mealy gray rosettes and yellow flowers
                            8 cm (3 in.)   *May, June*

    'Major'   *15 cm (6 in.)*   *May, June*
    Plum-colored foliage, larger yellow flowers than the species.

    'Purpureum'   *0.9 m (3 ft.)*   *June*
    Deep purple young leaves turning mealy white; yellow flowers are larger than the species.

S. spectabile showy stonecrop   *0.4 m (1½ ft.)*   *Sept., Oct.*
The most commonly grown species and perhaps the showiest. An erect plant with broad, glaucous, sessile, obovate fleshy leaves. The flowers are bright pink and extremely showy. They are 1 cm (½ in.) across and borne in large flat-topped corymbs up to 15 cm (6 in.) wide. The flowers seem to be extremely attractive to butterflies.

Cultivars   *Sept., Oct.*
    'Album' — rather dirty white flowers   *0.6 m (2 ft.)*

    'Atropurpureum' — darker flowers than the species
                            0.4 m (1½ ft.)

    'Autumn Joy' (Herbstfreude)   *0.6 m (2 ft.)*
    Massive heads of bright rosy salmon flowers, tinted bronze.

    'Brilliant' — raspberry carmine flowers
                         0.4-0.6 m (1½-2 ft.)

    'Carmen' — carmine rose flowers   *0.4-0.6 m (1½-2 ft.)*

    'Meteor' — large wine red flower clusters
                         0.4-0.6 m (1½-2 ft.)

    'Variegatum' — white and yellow markings on the leaves
                         0.4-0.6 m (1½-2 ft.)

S. spurium   *23 cm (9 in.)*   *July, Aug.*
An evergreen species forming a large mat with hairy much-branched creeping stems. Abundant pink flowers produced in dense flat terminal spines on reddish stems. An excellent ground cover for dry banks or sunny open areas.

Cultivars
    'Album' — muddy white flowers

    'Bronze Carpet' — reddish bronze foliage

    'Coccineum' — deep red flowers

    'Dragonsblood' (Schorbus ter Blut) — ruby red flowers, and scarlet leaves in the fall

    'Erdblut' — dark red flowers and brown foliage

    'Green Mantle' — a dense green carpeting plant

    'Roseum' — rose flowers

SEMPERVIVUM hen-and-chickens, houseleek

Ideal plants for growing on rocks and in crevices of soil in rock gardens. They thrive in poor soil and make good companions for the sedums. Some species are excellent edging plants. The leaves of all sempervivums are arranged in close rosettes with a flower stalk arising from the center. After it has flowered, the rosette dies, but is soon replaced by a number of young rosettes growing around the edge of the old one.

S. tectorum hen-and-chickens   *30 cm (12 in.)*   *July, Aug.*
The large green rosettes and pink and red flowers make this one of the best species for edging.

    'Calcareum'   *30 cm (12 in.)*   *July, Aug.*
    Glaucous leaves with brown purple tips.

SENECIO

S. clivorum see *Ligularia dentata*

S. japonica see *Ligularia japonica*

SIDALCEA prairie mallow   (Color photo p. 89)

This genus consists of about eight species of herbaceous perennials of western North America. The species are not particularly striking, but they have given rise to many showy cultivars, which mostly resemble small hollyhocks, but are much more graceful.

They thrive in any garden soil that has a good moisture-holding capacity. After they have flowered they should be cut back closely to give renewed root strength.

Cultivars	*June-Sept.*
'Croftway Red' – rich deep flowers	*0.9 m (3 ft.)*
'Dainty' – rose pink flowers with a white eye	*0.9 m (3 ft.)*
'Duchess' – clear rose flowers	*0.9 m (3 ft.)*
'Elsie Heugh' – clear satiny pink flowers	*1.1 m (3½ ft.)*
'Interlaken' – clear pink flowers	*0.9 m (3 ft.)*
'Loveliness' – shell pink flowers, compact habit	*0.8 m (2½ ft.)*
'Mrs. Alderson' – large clear pink flowers	*0.9 m (3 ft.)*
'Mrs. Borrodaile' – carmine crimson flowers	*0.9 m (3 ft.)*
'Mrs. Galloway' – deep rose red flowers	*0.9 m (3 ft.)*
'Nimmerdor' – closely set spikes of rosy crimson flowers	*0.9 m (3 ft.)*
'Oberon' – neat and erect, soft rose pink flowers	*0.8 m (2½ ft.)*
'Puck' – very dwarf, clear pink flowers	*0.6 m (2 ft.)*
'Rev. Page Roberts' – light rose pink flowers	*0.9 m (3 ft.)*
'Rose Queen' – rose flowers	*1.1 m (3½ ft.)*
'Sussex Beauty' – fine pink flowers	*0.9 m (3 ft.)*
'Titania' – erect and sturdy, satiny pink flowers	*0.9 m (3 ft.)*
'Wensleydale' – compact habit, large rosy red flowers	*1.2 m (4 ft.)*
'William Smith' – warm salmon pink flowers	*0.9 m (3 ft.)*

SMILACINA

S. racemosa false spikenhard   *0.6-0.9 m (2-3 ft.)   May, June*
A native woodland plant like Solomon's seal, with soft green broad leaves and panicles of fragrant creamy white flowers. Very useful for the shady part of a flower border or for naturalizing.

S. stellata
star-flowered lily-of-the-valley   *0.6 m (2 ft.)   May, June*
Similar to *S. racemosa*, but smaller.

SOLIDAGO goldenrod   (Color photo p. 89)

Goldenrods are like prophets without honor in Canada. They have become so associated with the early fall scene throughout the country that most gardeners would not use them in borders. Also, many Canadians have the mistaken idea that they are the main cause of hay fever, probably because they bloom at the same time as ragweed, the real culprit. Many species of goldenrod are attractive for the border. Many new cultivars are more refined, smaller, much more compact, and some, which are more apt to be distinctive in the border, actually bloom in July before the main species start.

Goldenrods do well in either sunny or semishaded locations in most soils.

S. brachystachys   *23 cm (9 in.)   Aug., Sept.*
Neat little bushes with short spikes of deep yellow.

S. caesia wreath goldenrod   *0.6-0.9 m (2-3 ft.)   Aug.-Oct.*
Very free flowering with small heads of flowers in the axils of leaves all the way up the stem and terminating in short thyrsus.

S. missouriensis   *0.6-0.9 m (2-3 ft.)   July-Sept.*
Smooth stems and graceful short spikes of yellow flowers.

S. petiolaris   *0.9-1.5 m (3-5 ft.)   Sept., Oct.*
Later flowering; hairy stems, narrow elliptic leaves, and large raggedy flowers.

S. uliginosa swamp goldenrod *0.6-1.2 m (2-4 ft.)   Sept., Oct.*
Very graceful, late flowering; rounded oblong leaves on smooth stems of small yellow flowers.

S. virgaurea   *0.3-1.2 m (1-4 ft.)   July-Oct.*
A European species with smooth stems and flat plate-like heads of flowers.

Cultivars	*Aug., Sept., except where noted*
'Cloth of Gold' – strong, free flowering	*0.4 m (1½ ft.)*
'Crown of Rays' – very attractive light yellow flowers	*0.8 m (2½ ft.)*

'Golden Falls' – beautiful golden sprays of flowers
0.9 m (3 ft.)

'Golden Gates' *0.8 m (2½ ft.)*
Light yellowish green foliage and shapely, slightly arched yellow flower spikes.

'Goldenmosa' – round fluffy blossoms remindful of mimosa
0.8-0.9 m (2½-3 ft.)

'Golden Plume' – bright yellow flowers *0.8 m (2½ ft.)*

'Golden Shower' – arching sprays of deep yellow flowers
0.8 m (2½ ft.)

'Golden Thumb' (Queenie) – free flowering, dwarf, yellow flowers *0.3 m (1 ft.)*

'Golden Wings' – branching sprays of deep golden yellow flowers *1.5-1.8 m (5-6 ft.)*

'Goldstrahl' – bright canary yellow flowers *0.9 m (3 ft.)*

'Laurin' – deep yellow flowers on very compact plants
0.3 m (1 ft.)

'Leda' – very erect habit, spikes of sunshine yellow flowers
0.9 m (3 ft.)

'Ledsham' – bright yellow flowers *0.3 m (1 ft.)*

'Lemore' – wide branching heads of soft primrose
0.6 m (2 ft.)

'Leraft' – distinctive, bright yellow flowers *0.8 m (2½ ft.)*

'Lesale' – daisy-like medium yellow flowers *0.9 m (3 ft.)*

'Lesden' – new, distinctive hybrid with compact habit and yellow flowers *0.9 m (3 ft.)*

'Leslie' – clear yellow flowers *0.6 m (2 ft.)*

'Mimosa' – graceful arching plumes of yellow flowers
1.1 m (3½ ft.)

'Peter Pan' – bright yellow plumes of flowers *0.9 m (3 ft.)*

'Praecox' – flowering very much earlier (June, July) than the species *1.1 m (3½ ft.)*

SOLIDASTER

A name comprising the parent generic names *Aster* and *Solidago*. An intergeneric hybrid obtained by crossing the perennial aster with a goldenrod. The plants flower at the same time as the parents and need the same cultural treatment.

S. luteus
A good garden plant with narrow aster-like leaves and small yellow flowers.

SPHAERALCEA

S. coccinea *1.8 m (6 ft.) July*
A native prairie plant with silvery gray, finely divided leaves and red flowers. It has a running woody rootstock and is useful as a ground cover in dry locations. Synonym: *Malvastrum coccineum*

SPIRAEA

S. aruncus see *Aruncus sylvester*

S. digitata see *Filipendula palmata*

S. filipendula see *F. vulgaris*

S. gigantea see *F. camtschatica*

S. hexapetala see *F. vulgaris*

S. lobata see *F. rubra*

S. palmata see *F. purpurea*

S. rubra see *F. rubra*

S. ulmaria see *F. ulmaria*

S. venusta see *F. rubra* 'Venusta'

STACHYS woundwort

Very easily grown plants belonging to the sage family with whorls of red or purple flowers arranged on spikes. They bring an Old World charm to the flower border with their odd flowers and contrasting foliage. They grow well in very poor soil in sun or part shade.

S. grandiflora *0.3-0.6 m (1-2 ft.) May, June*
Very broad ovate, obtuse leaves, which are wrinkled and very roughly hairy. The whorls of rich violet flowers are much showier in this species because each individual floret is an inch long. Synonym: *S. macrantha*

'Rosea' – rose-colored flowers
0.3-0.6 m (1-2 ft.) May, June

S. lanata see *Stachys olympica*

S. macrantha see *Stachys grandiflora*

S. olympica woolly hedge-nettle, woolly betony, lamb's-ear
0.3-0.4 m (1-1½ ft.) July, Aug.
A silvery-leaved plant with a very woolly texture and shaped like a lamb's ear, from which it gets one of its common names. It is an excellent plant for the front of the border, particularly where some distinctive leaf texture is needed.

Its crimson flowers are not very showy, and should be removed early so that the foliage effect is more pronounced. Synonym: *S. lanata*

'Silver Carpet' — dwarf, compact, with silvery foliage
15 cm (6 in.)   July, Aug.

STEIRONEMA

S. ciliata see *Lysimachia ciliata*

STOKESIA

S. laevis cornflower aster   *0.4 m (1½ ft.)   July-Sept.*
In the right soil conditions this plant produces an abundance of large lavender blue flowers similar to the annual China aster. The leaves are long, leathery, and dark green forming a compact mound. The plant requires a sunny location and an exceptionally well-drained, moist soil. In Canada it does best in southern Ontario and milder localities of British Columbia.

Cultivars   *July-Sept.*
  'Alba' — white flowers   *0.3 m (1 ft.)*
  'Blue Danube' — blue flowers   *0.3 m (1 ft.)*
  'Blue Moon' — very large silvery blue flowers   *0.3 m (1 ft.)*
  'Blue Star' — very large blue flowers   *0.3 m (1 ft.)*
  'Lutea' — yellow flowers   *0.3 m (1 ft.)*
  'Rosea' — soft pink flowers   *0.3 m (1 ft.)*
  'Silvery Moon' — good white flowers   *0.3 m (1 ft.)*
  'Superba' — rich mauve purple flowers   *0.3 m (1 ft.)*
  'Wyoming' — deep blue flowers   *38 cm (15 in.)*

SYMPHYANDRA   (Color photo p. 89)

A genus of campanula-like plants for the front of the border and the rock garden.

S. hofmannii   *0.3-0.6 m (1-2 ft.)   July, Aug.*
A species with large lilac blue flowers.

S. pendula   *46 cm (18 in.)   July, Aug.*
Beautiful cream pendulous funnel-shaped flowers produced in panicles.

S. wanneri   *15 cm (6 in.)   all summer*
Hairy stems and leaves, and nodding bell-shaped blue flowers.

SYMPHYTUM comfrey

Most of these are, at best, coarse perennials, used for naturalizing in sunny places in a wild garden. Most of the species are very invasive and grow in almost impossible areas.

S. asperum prickly comfrey   *1.2-1.8 m (4-6 ft.)   July, Aug.*
A very vigorous perennial for large borders, where it may be used to fill a gap toward the rear. The plant looks like an overgrown pulmonaria with its pink flower buds that change to blue.

S. peregrinum   *0.9-1.2 m (3-4 ft.)   July, Aug.*
A good plant for sun or shade, with much-branched stems carrying terminal trusses of single blue drooping bell-shaped flowers.

TELESONIX

T. jamesii see *Boykinia jamesii*

TEUCRIUM

T. chamaedrys   *0.6 m (2 ft.)   July-Sept.*
Although its flowers are not particularly beautiful, this *Teucrium* is useful in gardens as a formal clipped or informal unclipped hedge, a ground cover, an edging, and for planting in rock gardens. It is a dwarf shrub with shiny dark green leaves and little rosy red flowers in late summer.

THALICTRUM meadow rue   (Color photo p. 89)

These are very attractive perennials with beautiful cut leaves, somewhat reminiscent of the maidenhair fern. They have extremely delicate light airy flowers. These have no true petals, instead they have petaloid sepals, which together with numerous colorful stamens give color to the inflorescence. The thalictrums are mostly tall graceful plants that are best suited to the back of the border. They will grow in light shade and in full sun if the soil is rich and moist. All the species are very hardy except *T. dipterocarpum*.

T. adiantifolium see *Thalictrum minus*

T. aquilegifolium
columbine meadow rue   *0.6-0.9 m (2-3 ft.)   May-July*
An attractive plant with pale purple fluffy flowers in spreading loose panicles.

Cultivars
  'Album' white meadow rue — white flowers
   0.8 m (2½ ft.)   May-July

  'Atropurpureum'   *1.2 m (4 ft.)   June-Aug.*
  Large plant with fluffy purple flowers. Synonym: 'Purpureum'

  'Bees Purple' — purple rose flowers
   0.9 m (3 ft.)   June-Aug.

  'Dwarf Purple' — purple fluffy blooms
   0.8 m (2½ ft.)   June-Aug.

'Purple Cloud'   0.8 m (2½ ft.)   June-Aug.
Neater than the other cultivars, rosy purple flowers.

'Thundercloud'   0.8 m (2½ ft.)   June-Aug.
Excellent, outstanding, has grown well in Ottawa; soft fluffy reddish purple flowers.

T. delavayi   0.6-0.9 m (2-3 ft.)   July, Aug.
Similar to *T. dipterocarpum*, but much smaller and with rosy violet flowers.

T. diffusiflorum   0.6-0.9 m (2-3 ft.)   July, Aug.
Probably the most attractive species in the genus. It has large clear lilac flowers of perfect form and texture, and very finely cut gray green foliage.

T. dipterocarpum Yunnan meadow rue
A vigorous species, although not as winter-hardy as *T. aquilegifolium*. Large compound leaves and tall graceful panicles of deep lavender flowers with showy protruding yellow anthers.

In Eastern Canada this beautiful plant needs a mulch of well-rotted manure or compost applied each year and no cultivating around the plant. In late fall apply a straw mulch around the crowns for winter protection.

Cultivars   July, Aug.
'Album' — white flowers   0.6-1.5 m (2-5 ft.)

'Hewitt's Double'   0.6 m (2 ft.)
Rich mauve completely round flowers that last a long time when cut.

T. flavum yellow meadow rue 0.6-0.9 m (2-3 ft.)   July, Aug.
Quite distinctive with gray green finely cut leaves and feathery heads of soft yellow flowers.

Cultivars
'Glaucum' — soft yellow flowers and blue gray foliage
                               1.5-1.8 m (5-6 ft.)   July

'Illuminator'   1.1 m (3½ ft.)   June, July
Lemon yellow flowers with variable yellow and green foliage.

T. glaucum see *Thalictrum speciosissimum*

T. kiusianum   15 cm (6 in.)   July
An uncommon dwarf species from Japan with purplish leaves, pink lilac sepals, and striking blue stamens. Best in part shade and very leafy soil.

T. minus   0.6-0.9 m (2-3 ft.)   July
Grown only because of its delicate compact tufts of leaves similar to maidenhair fern. Synonym: *T. adiantifolium*

T. rochebrunianum
lavender mist   1.2-1.8 m (4-6 ft.)   July-Sept.
A fairly new species, said to be far superior to *T. dipterocarpum* and much hardier. Vigorous clumps produce huge masses of lavender violet flowers with light yellow stamens.

T. speciosissimum
dusty meadow rue   0.6-0.9 m (2-3 ft.)   July, Aug.
A handsome border perennial with blue gray leaves and plumy panicles of fragrant soft yellow flowers. Synonym: *T. glaucum*

THERMOPSIS false lupine   (Color photo p. 90)

Very attractive, hardy, and long-suffering plants that grow well in light sandy soil. Long terminal spikes of yellow lupine-like flowers.

T. caroliniana   0.9 m (3 ft.)   July, Aug.
Late flowering; erect racemes of sparkling yellow flowers, very effective with delphiniums.

T. montana   0.6 m (2 ft.)   June, July
More restrained than the other species, with divided leaves and erect stiff spikes of yellow flowers.

T. rhombifolia   38 cm (15 in.)   June, July
A smaller species with yellow flowers loosely arranged in 10-cm (4-in.) racemes.

THYMUS mother-of-thyme, creeping thyme
(Color photo p. 90)

Most of the species in this genus are dwarf shrubs or creeping rock garden plants, but they are often listed in catalogs and used for the front of borders. They do well in full sun in well-drained medium soil.

T. lanuginosus woolly mountain thyme   5 cm (2 in.)   June
Minute very woolly gray leaves and pinkish flowers.

T. nitidus   23 cm (9 in.)   June
A miniature shrub with fragrant silvery gray leaves and pink flowers.

T. serpyllum
mother-of-thyme   10-15 cm (4-6 in.)   June, July
Very hardy small-leaved creeping plants, which form close mats and make excellent ground covers for slopes, in crevices of rocks, and in cracks in flagstone walks. Rosy purple flowers.

Cultivars
'Album' — white flowers   5-10 cm (2-4 in.)   July

'Argenteus'   10-15 cm (4-6 in.)   June
Silver markings on the leaves, fragrant.

Trillium grandiflorum, wake robin, trillium

'Citriodorus'　　　　　*15 cm (6 in.)　　July, Aug.*
Lemon-scented thyme, yellowish-tinted foliage with distinct lemon scent.

'Coccineus' – rich red flowers on a larger plant
　　　　　　　　　　15-20 cm (6-8 in.)　　June

T. vulgaris garden thyme　　*15 cm (6 in.)　　July, Aug.*
A somewhat weedy species, with lilac flowers. It has a more loosely spreading habit than *T. serphyllum*.

TIARELLA foamflower

Eastern North American plants for woodlands or shady gardens. Attractive heart-shaped leaves and crimson-stemmed spikes of foam-like creamy flowers. Best in good rich moist soil.

T. cordifolia foamflower　　*23 cm (9 in.)　　May, June*
Simple racemes of starry creamy white flowers opening from pink-tinged buds. The young leaves are spotted green and veined with deep red, later changing to reddish bronze.

T. polyphylla　　　　　*30 cm (12 in.)　　May-July*
Very attractive leaves veined yellowish green and sprays of minute pearly white flowers.

T. trifoliata　　　　　*23 cm (9 in.)　　May, June*
A recently cultivated native of western North America with small sprays of tiny white flowers arising from mound-like plants.

T. wherryi　　　　　*38 cm (15 in.)　　May, June*
A very attractive plant with feathery spikes of creamy flowers and hairy heart-shaped leaves.

TRADESCANTIA　　(Color photo p. 90)

T. andersoniana
A series of garden hybrids with bright flowers over 3 cm (an inch) in diameter and erect growth. The following clones are included under this specific name.

Cultivars　　　　　*0.6 m (2 ft.)　　June-Sept.*
　'Alba' – white flowers

　'Blue Stone' – strong deep blue flowers

　'Coerulea' – bright blue flowers

　'Flore Pleno' – semidouble, mauve blue flowers

　'Iris' – large Oxford blue flowers

　'Iris Pritchard' – white flowers shaded pale violet, with violet feathery staminal tuft

　'James Shatton' – rose purple flowers

　'James C. Weguelin' – large porcelain blue flowers

'Leonora' – large clear blue flowers

'Osprey' – huge feathery white flowers, shaded blue centers

'Pauline' – large rosy mauve flowers

'Purewell Giant' – carmine purple flowers

'Purple Dome' – spectacular rosy purple flowers

'Red Cloud' – large reddish purple flowers

'Snow Cap' – large white flowers

'Valour' – reddish purple flowers

T. virginiana spiderwort　　*5 cm (2 in.)　　June-Sept.*
This hardy perennial species is the only one rugged enough for garden use in Canada, but the numerous cultivars included above under *T. andersoniana* far more than compensate for this. The species has linear-lanceolate green leaves 15-36 cm (6-14 in.) long, and blue to purple flowers in few to many flowered umbels. The individual flowers only last a day, but new ones replace them, and the old ones soon drop leaving the plant tidy.

TRILLIUM wake robin

Most of these very delightful woodland plants are native to North America. Their chief characteristic is that all parts of the plant are in threes: three petals, three stamens, and trifoliate leaves. Although they are best in shade, they can be grown in full sun if the soil is well supplied with organic matter and kept moist during the growing season.

T. cernuum   *0.3-0.4 m (1-1½ ft.)*   *May, June*
Medium-sized nodding flowers with broadly rhombic leaves on short stalks.

T. erectum purple trillium, ill-scented wake robin — reddish disagreeable-smelling flowers   *0.3 m (1 ft.)*   *May, June*

Cultivars   *0.3 m (1 ft.)*   *May, June*
    'Album' — white flowers
    'Ochroleucum' — creamy yellow flowers

T. grandiflorum
wake robin, trillium   *0.4 m (1½ ft.)*   *May, June*
The provincial flower emblem for Ontario. This is by far the best trillium to grow because of its very showy soft white graceful flowers of good size with very long anthers.

T. sessile   *0.3 m (1 ft.)*   *May, June*
Sweet-scented purple and sometimes greenish flowers that nestle tightly in the leaves on the same stalks.

TRITOMA see KNIPHOFIA

TROLLIUS globeflower (Color photo p. 91)

This genus contains some very fine perennials with erect stems, usually palmate leaves, and very showy global buttercup-like flowers. The beauty of the flowers is enhanced by the yellow or orange sepals, which are globe-shaped and have shorter flat nectaries. These compact-growing plants flower abundantly, especially in moist rather stiff soils.

T. altaicus Siberian globeflower   *0.3-0.6 m (1-2 ft.)*   *June*
Yellow or pale orange flowers with 15 to 20 sepals and narrow leaves like *T. europaeus*.

T. anemonifolius   *0.4 m (1½ ft.)*   *June*
Rounded leaves divided almost to the base into three shallowly three-lobed and toothed divisions. Golden yellow flowers, 5 cm (2 in.) across.

T. asiaticus   *0.4 m (1½ ft.)*   *May, June*
Bright orange yellow (not globe-shaped) flowers and bright orange red anthers. Bronzy green very finely divided leaves.

Cultivars   *0.4 m (1½ ft.)*
    'Aurantiacus' — very good orange flowers   *May, June*

    'Byrne's Giant'   *June, July*
A fine variety; citron yellow flowers, larger than the species and later flowering.

    'Fortunei' — rich yellow flowers and bright orange red anthers   *May, June*

T. chinensis Chinese globeflower   *0.3 m (1 ft.)*   *May, June*
T. europaeus and *T. chinensis* have been responsible for many good cultivars. This species has stout grooved stems and partly kidney-shaped, partly rounded leaves. The golden yellow flowers are on peduncles 30 cm (12 in.) long.

T. europaeus European globeflower   *0.6 m (2 ft.)*   *May, June*
This is the giant European globeflower. Stout and sturdy, it forms a large clump with neatly divided foliage and clear yellow, globular flowers on stems 0.3-0.6 m (1-2 ft.) tall.

Cultivars   *May, June*
    'Alabaster' — the palest yellow flowers   *0.8 m (2½ ft.)*

    'Bees' Orange' — large orange gold flowers   *0.6 m (2 ft.)*

    'Canary Bird' — pale yellow flowers   *0.6 m (2 ft.)*

    'Earliest-of-All' — medium yellow flowers   *0.6 m (2 ft.)*

    'Empire Day' — early, large, rich orange yellow flowers
                               0.6 m (2 ft.)

    'Fireglobe' — deep orange yellow flowers   *0.6 m (2 ft.)*

    'First Lancers' — fiery orange flowers   *0.9 m (3 ft.)*

    'Glory of Leiden' — yellow flowers   *0.6 m (2 ft.)*

    'Golden Monarch' — medium yellow huge flowers
                               0.6 m (2 ft.)

    'Golden Wonder' — fine deep yellow flowers   *0.6 m (2 ft.)*

    'Goldquelle' — large golden yellow flowers   *0.6 m (2 ft.)*

    'Helios' — citron yellow flowers   *0.6 m (2 ft.)*

    'Lemon Queen' — good pale yellow flowers   *0.6 m (2 ft.)*

    'Newry Giant' — very large pale yellow flowers
                               0.9 m (3 ft.)

    'Orange Princess' — orange yellow flowers   *0.8 m (2½ ft.)*

    'Pritchard's Giant' — large, medium yellow flowers
                               0.6 m (2 ft.)

    'Princess Juliana' — free flowering, clear yellow flowers
                          0.8 m (2½ ft.)

    'Salamander' — fiery orange flowers   *0.6 m (2 ft.)*

    'Superbus' — early clear lemon flowers   *0.6 m (2 ft.)*

T. laxus   *38 cm (15 in.)*   *May, June*
Unique solitary greenish yellow flowers.

1 *Trollius asiaticus*, globeflower
2 *Trollius europaeus* 'Glory of Leiden', European globeflower

T. ledebouri　　　　　　　　　0.6 m (2 ft.)　June
A fine species with deep orange cup-shaped flowers and vivid orange stamens. Flowers a little later than the other species.

'Golden Queen'　　　　0.9-1.2 m (3-4 ft.)　June
Much taller than the species, with larger butter-yellow flowers up to 10 cm (4 in.) across and similar vivid orange stamens.

T. pumilus　　　　　　　　25 cm (10 in.)　May, June
A charming little species with 3-cm (1-in.) blooms of clear yellow and rounded glossy leaves.

'Stenopetalus' — shining yellow flowers
　　　　　　　　　　　　30 cm (12 in.)　May, June

T. yunnanensis　　　　　　　0.6 m (2 ft.)　May, June
Flattened, single, 8-cm (3-in.) buttercup-like flowers and broad mottled anemone-like leaves.

UVULARIA

U. grandiflora　　　　　　　0.4 m (½ ft.)　May
A very useful North American plant with pale yellow drooping flowers, arising from slender stems. A good plant for woodlands and other shady places.

VALERIANA　　(Color photo p. 91)

V. officinalis valerian　　1.2-1.5 m 4-5 ft.　July, Aug.
An old-fashioned border plant that self-sows readily and persists with little trouble. It has large corymbs made up of myriads of small white or pinkish white fragrant flowers. The dark green much-dissected pinnate leaves are graceful during most of the season. It grows well in the shade and perhaps is best used at the back of a moist shaded border or for naturalizing in woodlands.

VERBASCUM mullein

These are very old fashioned plants, mostly biennials, known generally in Canada because of the wild roadside mullein. A few are perennial enough to be included in borders, and the hybrids are worth planting, if only because of their ability to combine well with astilbes, perennial phlox, and delphiniums. They need full sun and good well-drained soil. See also Chapter 3, Biennials.

V. bombyciferum　　　　　0.8 m (2½ ft.)　June, July
A very beautiful biennial with large bright yellow flowers and large silvery woolly leaves. Synonym: 'Broussa'

'Silver Spire'　　　　　　0.8 m (2½ ft.)　June, July
Even more white tomentum than the species, forming felt-like silvery leaves through which bright yellow flowers shine.

V. chaixii nettle-leaved mullein   *0.9 m (3 ft.)   June-Aug.*
Very heavy, white, felted crenate leaves and yellow flowers with prominent purple-haired filaments.

V. nigrum dark mullein   *0.6-0.9 m (2-3 ft.)   June-Oct.*
Long spikes of small yellow flowers bearing purple anthers. The heart-shaped stalked leaves are slightly downy beneath.

'Album' — white flowers   *0.6-0.9 m (2-3 ft.)   June-Oct.*

V. olympicum   *1.5-1.8 m (5-6 ft.)   June-Sept.*
A very large robust mullein with branching candelabra-like sprigs of bright golden flowers with whitish filament hairs. It produces large rosettes of silvery gray felted leaves, which are attractive enough to make the plant a desirable ornamental.

V. phoeniceum purple mullein *0.9-1.2 m (3-4 ft.)   May-Aug.*
A variable species with rose, pink, and purple flowers on spikes 0.8 m (2½ ft.) long. It is most likely the predominant parent of many cultivars listed below.

Cultivars	*June, July*
'C. L. Adams' — golden yellow flowers	*1.8 m (6 ft.)*
'Cotswold Queen' — terra-cotta flowers	*0.9 m (3 ft.)*
'Gainsborough' — pale yellow flowers	*0.9 m (3 ft.)*
'Golden Bush' — bushy type yellow flowers	*0.6 m (2 ft.)*
'Hartleyi' — biscuit yellow, suffused plum flowers	*1.2 m (4 ft.)*
'Mont Blanc' — white flowers	*0.9 m (3 ft.)*
'Pink Domino' — deep rose flowers	*1.1 m (3½ ft.)*
'Royal Highland' — apricot and yellow flowers	*1.8-2.4 m (6-8 ft.)*

V. thapsiforme   *0.9-1.5 m (3-5 ft.)   June-Aug.*
This species has large yellow flowers in clusters arranged to form a spike. The soft broad lance-shaped leaves, and the flower stems are densely covered with yellow hairs.

VERBENA vervain

This very attractive genus is best known in Canada as the popular garden verbena *(V. × hybrida)* used as a bedding annual. All but one of the species mentioned here are tender perennials that are hardy only in the milder parts of British Columbia, but they are very showy plants and should be grown a great deal more in that area.

V. bonariensis   *1.2-1.5 m (4-5 ft.)   July-Oct.*
Spikes of rich purple flowers on clustered heads.

V. chamaedryfolia see *Verbena peruviana*

V. corymbosa   *0.6 m (2 ft.)   July, Aug.*
A tender South American species, which forms a compact, dark green plant with dense terminal clusters of blooms formed from fragrant funnel-shaped blue flowers. It does best in a moist, sunny area.

V. hastata   *0.6-0.9 m (2-3 ft.)   July-Sept.*
A native Canadian plant with stiff nettle-like stems and leaves, and heads of small violet flowers.

'Album' — white flowers   *0.6 m (2 ft.)   July-Sept.*

V. peruviana   *15 cm (6 in.)   July-Oct.*
A creeping, mat-forming species with extremely brilliant scarlet flowers. An excellent ground cover in a warm, sunny, well-drained location. Synonym: *V. chamaedryfolia*

VERONICA speedwell   (Color photo p. 91)

The veronicas owe their popularity to the fact that blue is scarce in the perennial border and also most of them are very hardy and easy to grow. The trailing types are extremely useful for rockeries, in crevices, and on flagstone walks. Most of the speedwells are best in a sunny well-drained location, and a few will grow well in the shade.

V. cinerea   *15 cm (6 in.)   June, July*
A dwarf perennial with small spikes of rose flowers, and incurved ash gray leaves with very short hairs.

V. incana   *0.4 m (1½ ft.)   June, July*
A very beautiful vigorously hardy silvery gray plant with terminal spikes of soft blue flowers.

'Rosea' — pink flowers   *0.4 m (1½ ft.)   June, July*

V. latifolia Hungarian speedwell
A bushy perennial with dark green, narrow, toothed leaves on long slender spires of lavender blue flowers. Grows in almost any soil and withstands part shade. Synonym: *V. teucrium*

Cultivars	*June, July*
'Amethystina' — bright blue flowers	*30 cm (12 in.)*
'Crater Lake Blue' — fairly erect, deep blue flowers	*38 cm (15 in.)*
'Royal Blue' — early flowering, blue flowers	*30 cm (12 in.)*
'Shirley Blue' — intense blue flowers	*38 cm (15 in.)*

V. longifolia beach speedwell *0.6-1.2 m (2-4 ft.)   July-Sept.*
This species has either opposite paired leaves or three leaves arranged in whorls, and closely packed spikes of blue flowers.

Cultivars   *July–Sept.*

'Albiflora' – white flowers   *0.8-0.9 m (2½-3 ft.)*

'Rosea' – pink flowers   *0.8-0.9 m (2½-3 ft.)*

'Subsessilis' clump speedwell   *0.6 m (2 ft.)*
Heavily branched, close-clustered plant with light blue flowers, much better than the species.

V. pectinata   *10 cm (4 in.)*   *June, July*
A prostrate, creeping, woolly mat with racemes of white-centered blue flowers.

'Rosea' – pink flowers   *10 cm (4 in.)*   *June, July*

V. prostrata   *20 cm (8 in.)*   *June, July*
A species with mat-forming sterile stems and erect or ascending flowering stems bearing deep blue flowers in dense racemes. Synonym: *V. rupestris*

Cultivars   *June, July*
  'Loddon Blue' – bright blue flowers   *13-15 cm (5-6 in.)*
  'Mrs. Holt' – pink flowers   *15 cm (6 in.)*
  'Silver Queen' – palest blue flowers   *15 cm (6 in.)*
  'Spode Blue' – clear light blue flowers   *15 cm (6 in.)*

V. repens
A dwarf mat-forming species with prostrate slender stems bearing numerous pale blue flowers in the axils of the leaves. Quite useful for crevices in patios and flagstone walks or a ground cover for early spring flowering bulbs. Not hardy in areas colder than southern Ontario.

V. rupestris see *Veronica prostrata*

V. spicata spiked speedwell   *0.4 m (1½ ft.)*   *July, Aug.*
A very valuable plant for the front of the border. Narrow leaves and bright blue flowers in stiff dense spikes. *V. spicata* and *V. longifolia* are mainly responsible for most of the following cultivars, which are much more ornamental and desirable.

Cultivars

  'Barcarolle' – deep rosy pink flowers
        0.3 m (1 ft.)   *June-Aug.*

  'Blue Champion' – medium blue flowers
        0.8 m (2½ ft.)   *July*

  'Blue Peter'   *0.4 m (1½ ft.)*   *June-Aug.*
  Deep blue flowers in compact trusses, long serrated leaves.

  'Blue Spire' – deep violet blue flowers, erect
        0.4 m (1½ ft.)   *June-Aug.*

  'Erica' – pale orchid purple flowers
        0.3 m (1 ft.)   *June-Aug.*

'Icicle' – white flowers   *0.4 m (1½ ft.)*   *June-Aug.*

'Minuet' – gray green foliage, pink flowers
        0.3 m (1 ft.)   *June-Aug.*

'Romily Purple' – dark blue violet flowers, upright
        0.3 m (1 ft.)   *June-Aug.*

'Saraband'   *0.4 m (1½ ft.)*   *June-Aug.*
Gray leaved, erect spikes of deep violet flowers.

'True Blue' – pure blue flowers   *0.4 m (1½ ft.)*   *June*

'Well's Variety'   *0.4 m (1½ ft.)*   *June*
Bright blue flowers, green foliage.

'Wendy' – clear blue flowers, silvery foliage
        0.4 m (1½ ft.)   *June*

V. teucrium see *Veronica latifolia*

VERONICASTRUM

V. virginicum Culver's-root   *1.2 m (4 ft.)*   *Aug., Sept.*
Very closely related to the veronicas, and sometimes classified as such. Resembling a very vigorous veronica with slender stiff spires of white or pale blue flowers. Two characters, according to many botanists, identify this plant as a distinct species: the leaves are in whorls of three to six and are 15-23 cm (6-9 in.) apart along the stem; and the corolla lobes are much shorter than the tube.

VINCA periwinkle

A genus of erect or decumbent perennial herbs or subshrubs, most of them are hardy within the Ottawa area and a few species have all the attributes that make an ideal trailing ground cover plant. They are easy to grow and, although best in shade, grow in full sun if adequate moisture is supplied.

V. major   *20 cm (8 in.)*   *May-July*
Heart-shaped leaves, blue purple flowers, and larger in all parts than *V. minor*. The calyx lobes on the species are ciliate and as long as the corolla tube, whereas those of *V. minor* are one-third as long and glabrous. A very rampant grower, the flowerless shoots root only at the tips.

Cultivars
  'Rubescens' – red purple flowers; leaves narrower and hairier than the species

  'Variegata' – blue flowers, blotched creamy white leaves

V. minor   *15 cm (6 in.)*   *May-Sept.*
The most commonly grown *Vinca*. The prostrate flowerless stems root into the ground at all the nodes. The flowering stems are short and erect bearing blue purple flowers. It has a more prostrate habit than *V. major* and smaller flowers.

Cultivars
'Alba' — white flowers
'Alboplena' — double white flowers
'Argenteo-variegata' — silver variegated flowers
'Atropurpurea' — deep purple flowers
'Azurea' — sky blue flowers
'Bowles Variety' — large purple-blue flowers
'Caeruleo-plena' — double blue flowers
'Cuprea' — reddish bronze flowers
'La Grave' — very large lavender flowers
'Multiplex' — double plum purple flowers
'Punicea' — red purple flowers
'Roseoplena' — double reddish purple flowers
'Variegata' — yellow variegated flowers

VIOLA violas, violets (for pansies see Chapter 3, Biennials)

Bedding violas, sometimes called violettas, are useful in rock gardens and perennial borders, for edging beds and mixing with tulips. In very cold regions these plants, like pansies, are best treated as biennials by sowing the seeds in cold frames and covering them for the winter, or as annuals by sowing them indoors in early March. Most of the other species of viola are interesting for woodland planting. Descriptions of a few of the good showy perennials follow.

V. cornuta violet   *15-23 cm (6-9 in.)   all summer*
The flowers of this species are smaller than those of the bedding violas, and the plants are hardier. Many of the cultivars have larger flowers and a wide range of colors. Most of the cultivars last for many years in Eastern Canada, but they need protection on the Prairies.

Cultivars
'Amethyst' — pale amethyst violet flowers
'Ardross Gem' — blue and gold suffused bicolor
'Blue Carpet' — bright blue flowers
'Bullion' — lemon yellow flowers
'Jersey Gem' — very reliable deep blue flowers
'Nora Leigh' — light blue flowers

V. cucullata purple violet   *8-15 cm (3-6 in.)   Apr.-June*
A native woodland plant that grows in wet areas from Newfoundland to Georgia. This species is the provincial floral emblem for New Brunswick. Small bluish violet flowers with white and purple venation.

V. × florariensis   *15 cm (6 in.)   all summer*
A hybrid (*V. cornuta* × a form of *V. tricolor*) from Henry Correvon's garden, Floriare, near Geneva, Switzerland. Well known for a long season of bloom and its purple and white flowers.

V. gracilis   *15 cm (6 in.)   May, June*
Very dainty, with narrow foliage and delicate deep violet flowers.

Cultivars
'Black Knight' — very rich purple black flowers
'Lutea' — deep golden yellow flowers
'Major' — velvety violet purple flowers

V. jooi   *8 cm (3 in.)   May*
Very compact, with pinkish lilac fragrant flowers.

V. labradorica   *10-15 cm (4-6 in.)   all summer*
Compact, with heart-shaped hazy purple leaves. The flowers are a lilac version of the English violet.

V. odorata sweet violet   *8-15 cm (3-6 in.)   May, June*
This is the deep purple sweet-scented wild violet of England. It spreads by runners and when growing freely overruns other plants.

Cultivars
Most of the many cultivars are more tender than the species.

'Alba' — white flowers 8-15 cm (3-6 in.) high, flowering in May, June

'Double Russian' — small, double, purple flowers

'Parma' — the common greenhouse variety

'Roxina' — a 15-cm (6-in.) plant with many deep rose flowers, with pleasing fragrance in early spring

'Royal Robe' — fragrant, dark purple flowers on 20-cm (8-in.) stems

'White Wonder' (The Czar) — faint blue flowers with yellow streaks

V. papilionacea   *5-13 cm (2-5 in.)   May*
A rhizomatous stemless violet with broad oval leaves 10-13 cm
(4 or 5 in.) wide, deeply cordate at the base. Large, rich
violet flowers with white centers.

'Priceana' confederate violet   *5-13 cm (2-5 in.)   May*
Grayish blue flowers with blue veins.

V. pedata bird's-foot violet   *15 cm (6 in.)   June, July*
Deeply cut basal foliage, and small flowers with violet on the
upper petal and lilac on the lower ones. They thrive in poor
sandy, acid soil and full sun.

VISCARIA

V. alpina see *Lychnis alpina*

184

The four planting plans presented here are actual plans of borders at the Plant Research Institute's ornamental grounds. The borders can be seen any time during the growing season. One plan emphasizes spring and fall effect, one provides only summer display, and two are for all-season interest. A few changes here and there will be made to these borders as time goes on. Better and more durable cultivars will be substituted, but plants with the same or similar colors and heights will be kept.

Several plants in the plans are not included in the text of this book. These plants are needed to round out the complement of species and varieties in any perennial border, but they are not herbaceous perennials in the strictest sense. Such plants are lilies, which enhance the summer display, tulips, for color in spring, and annuals, to brighten the border all season long. Sweet peas need 1.5-m (5-ft.) twigs for support, to produce a tall effect. Otherwise, use the 'Knee High' strain of these delightful plants, which are almost self-supporting and are available in all the colors mentioned in the plans.

The plans, of course, are not meant to be duplicated by the reader. Few gardeners would have the space that is needed for these examples. The plans are meant to show how herbaceous perennials may be satisfactorily placed and grouped. From the advice presented, you can, and should, make your own plan by incorporating your particular choices of plants and in this way producing your own intimate and distinctive border.

Under each border drawing there is a list of plants. The list gives the plants in numerical order corresponding to the numbers in the spaces in the plan. The numbers on the plan refer to the numbers in the list that precede the botanical name of the plant. After the name of the plant the numeral in italic signifies the number of such plants needed in the particular space on the plan. In the list, a variety or cultivar is indicated by single quotation marks, for example, 'Fair Lady'. A common name of a true herbaceous perennial plant can readily be found in the descriptive section, Part II, of this book, wherein entries are alphabetical by scientific name.

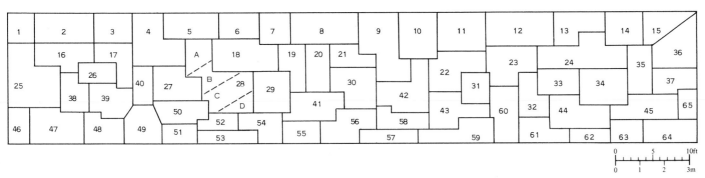

A SPRING- AND FALL-FLOWERING BORDER

1 *Oenothera tetragona 4*
2 *Aster* 'Fair Lady' *5*
3 *Helianthus atrorubens* 'Monarch' *2*
4 *Centaurea macrocephala 4*
5 *Boltonia asteroides 2*
6 *Galega officinalis* 'Her Majesty' *2*
7 *Rudbeckia laciniata* 'Golden Glow' *1*
8 *Delphinium* 'Blue Jay' *3*
9 *Lupinus polyphyllus* (Russell hybrid) 'Lilac Time' *3* and 'Josephine' *3*
10 *Lavatera thuringiaca 3*
11 *Aster* 'The Cardinal' *6*
12 *Solidago* 'Golden Wings' *3*
13 *Iris* (tall bearded type) 'Happy Birthday' *6*
14 *Eremurus himalaicus 1*
15 *Centaurea dealbata* 'John Coutts' *5*
16 *Iris sibirica* 'Caesar's Brother' *10*
17 *Aster* 'September Ruby' *4*
18 *Paeonia* 'Snow Mountain' *2*
19 *Aster* 'The Bishop' *6*
20 *Iris* (tall bearded type) 'Lady Mohr' *6*
21 *Verbena bonariensis 5*
22 *Monarda fistulosa* 'Melissa' *4*
23 *Sidalcea malvaeflora* 'Wensleydale' *5*
24 *Lythrum salicaria* 'Brightness' *4*
25 *Tulipa* (triumph type) 'Princess Beatrix' *50* and *Colchicum autumnale*, autumn-crocus *12*
26 *Helenium autumnale* 'Bruno' *4*
27 *Paeonia* 'Festiva Maxima' *1*
28 *Iris* (tall bearded type) 'Deep Black' *8*
   'Chivalry' *6*
   'Elizabeth of England' *8*
   'Lady Boscawen' *7*
29 *Cimicifuga racemosa 3*
30 *Physostegia virginiana 11*
31 *Verbascum chaexii* 'C. L. Adams' *1*
32 *Tulipa* (breeder type) 'Georges Grappe' *36*
33 *Doronicum plantagineum* 'Excelsum' *2*
34 *Helenium autumnale* 'Riverton Gem' *3*
35 *Chrysanthemum rubellum* 'Mary Stoker' *6*

36 *Aster* 'Winston Churchill' *6*
37 *Campanula glomerata* 'Dahurica' *6*
38 *Chrysanthemum*, any deep yellow cultivar *5*
39 *Trollius europaeus* 'Canary Bird' *6*
40 *Lilium*, lily 'Tarantella' *10* and
   Lilium 'Imperial Silver' *10*
41 *Rudbeckia fulgida* 'Deamii' *10*
42 *Oenothera tetragona 12*
43 *Narcissus* 'Snow Princess' *24*
44 *Iris sibirica* 'Ottawa' *18*
45 *Mertensia virginica 11*
46 *Aster* 'Little Pink Beauty' *9*
47 *Hyacinthus* 'Carnegie' *12*
48 *Chrysanthemum*, any yellow cultivar *7*
49 *Dicentra spectabilis 5*
50 *Tulipa* (single, early type) 'Pink Beauty' *24*
51 *Phlox subulata* 'Lilacina' *7*
52 *Iberis sempervirens 8*
53 *Campanula carpatica* 'Ditton Blue' *6*
54 *Chrysanthemum*, any red cultivar *10*
55 *Sedum spectabile 3*
56 *Euphorbia epithymoides 6*
57 *Crocus* 'Snow Bunting' *24* and *Arabis caucasica 6*
58 *Tulipa* (double, late type) 'Clara Carder' *24*
59 *Crocus biflorus 36* and *Aster* 'Lady-in-Blue' *9*
60 *Salvia uliginosa 6*
61 *Pulmonaria saccharata 10*
62 *Hyacinthus azureus 18*
63 *Crocus* 'Peter Pan' *30*
64 *Scilla sibirica 48*
65 *Galanthus nivalis 24*

Note: Numbers 47, 62, 63, 64, 65, can also be used as summer bedding plants.

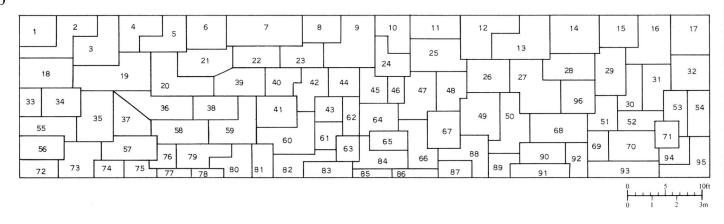

A scale bar: 0 5 10ft / 0 1 2 3m

A BORDER OF INTEREST IN SPRING, SUMMER, AND FALL

1 *Althaea rosea* 'Chater Yellow' *2*
2 *Aster ericoides* 'Ringdove' *7*
3 *Ligularia dentata* 'Desdemona' *3*
4 *Delphinium* 'Guinevere' *3*
5 *Coreopsis grandiflora* 'Mayfield Giant' *3*
6 *Lythrum virgatum* 'Morden Pink' *2*
7 *Helenium autumnale* 'Gypsy' *3*
8 *Aster* 'Davey's True Blue' *3*
9 *Solidago cutleri* 'Mimosa' *4*
10 *Thalictrum aquilegifolium* 'Lavender Mist' *3*
11 *Aster* 'Crimson Brocade' *5*
12 *Aster* 'Aquilla' *6*
13 *Dahlia* 'Jane Lausche' *3*
14 *Helenium autumnale* 'Riverton Gem' *3*
15 *Boltonia asteroides* *2*
16 *Chrysanthemum coccineum* *4*
17 *Aster amellus* 'Moorheim Gem' *5*
18 *Lythrum virgatum* *3*
19 *Aster* 'Peace' *10*
20 *Cimicifuga japonica* *3*
21 *Aster* 'Picture' *5*
22 *Helenium autumnale* 'Spatrot' *2*
23 *Nepeta tartarica* *2*
24 *Helenium autumnale* *2*
25 *Phlox paniculata* 'Caroline Van Den Berg' *4*
26 *Achillea* 'Coronation Gold' *3*
27 *Verbascum phoeniceum* 'Royal Highland' *3*
28 *Aster* 'Orchid Pink' *5*
29 *Heliopsis* 'Gigantea' *2*
30 *Physostegia virginiana* *3*
31 *Campanula persicifolia* 'Telham Beauty' *4*
32 *Lilium martagon* 'Achievement' *15*
33 *Phlox paniculata* 'G. V. Llewellyn' *5*
34 *Aster linosyris* *5*
35 *Lathyrus odoratus*, any mauve cultivar *12*
36 *Heliopsis* 'Gold Greenheart' *3*
37 *Trollius europaeus* 'First Lancers' *4*

38 *Solidaster luteus* *3*
39 *Rudbeckia fulgida* 'Goldquelle' *4*
40 *Monarda fistulosa* 'Cambridge Scarlet' *3*
41 *Aconitum henryi* 'Spark's Variety' *3*
42 *Hemerocallis* 'Cibola' *4*
43 *Aconitum napellus* 'Bressingham Spire' *3*
44 *Physostegia virginiana* *4*
45 *Aster* 'Harrington's Pink' *5*
46 *Achillea millefolium* 'Cerise Queen' *2*
47 *Physostegia virginiana* *11*
48 *Gladiolus* 'Greenland' *12*
49 *Monarda fistulosa* 'Blue Stocking' *5*
50 *Aster ericoides* 'Golden Spray' *5*
51 *Achillea ptarmica* 'Angel's Breath' *5*
52 *Trollius laxus* 'Golden Queen' *6*
53 *Tulipa kaufmanniana* 'Yellow Dawn' *18*
54 *Colchicum autumnale* *6*
55 *Sanguisorba obtusa* *6*
56 *Aster* 'Peter Harrison' *9*
57 *Iris sibirica* 'Perry's Blue' *10*
58 *Tulipa* (breeder type) 'Georges Grappe' *30*
59 *Platycodon grandiflorum* var. *mariesii* *9*
60 *Achillea ptarmica* 'Perry's Giant White' *12*
61 *Aster amellus* 'Rudolph Gothe' *6*
62 *Paeonia* 'Philipe Rivoire' *2*
63 *Narcissus* 'Cheerfulness' *18*
64 *Iris* (tall bearded type) 'Violet Harmony' *8*
65 *Tulipa multiflora* 'Rose Mist' *18*
66 *Hosta lancifolia* *4*
67 *Aster* 'Lil Fardell' *3*
68 *Physostegia virginiana* 'Bouquet Rose' *9*
69 *Chrysanthemum*, any bronze cultivar *5*
70 *Polemonium caeruleum* 'Blue Pearl' *5*
71 *Rudbeckia laciniata* 'Golden Glow' *3*
72 *Nepeta faassenii* *4*
73 *Chelone obliqua* *5*
74 *Viola wittrockiana* 'Swiss Giant' *12*
75 *Arabis caucasica* *5*
76 *Coreopsis verticillata* 'Golden Showers' *6*

77 *Crocus* 'Vanguard' *18*
78 *Sedum middendorffianum 3*
79 *Erigeron speciosus* 'Mrs. F. H. Beale' *6*
80 *Oenothera tetragona 7*
81 *Chrysanthemum*, a bronze cultivar *6*
82 *Gaillardia × grandiflora* 'Monarch Strain' *6*
83 *Campanula glomerata* 'Nana Alba' *4*
84 *Mertensia virginica 8*
85 *Phlox subulata* 'Rosea' *3*
86 *Tulipa kaufmanniana 12*

87 *Arabis alpina* 'Rosea' *5*
88 *Aster* 'Purple Feather' *10*
89 *Gaillardia grandiflora* 'Goblin' *4*
90 *Gaillardia grandiflora* 'Ipswich Beauty' *5*
91 *Chrysanthemum*, any bronze cultivar *6*
92 *Aster* 'Blue Radiance' *5*
93 *Chrysanthemum*, any copper cultivar *7*
94 *Crocus* 'E. P. Bowles' *18*
95 *Geranium prostratum 6*
96 *Lilium* 'Black Dragon' *16*

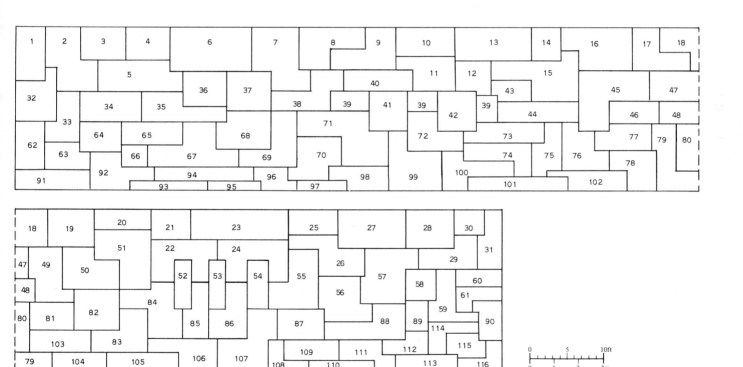

A LONG BORDER OF PLANTS THAT FLOWER MAINLY IN SUMMER

1 *Physostegia virginiana* 'Bouquet Rose' *8*
2 *Achillea filipendulina* 'Cloth of Gold' *6*
3 *Althaea rosea* 'Chater Yellow' *2*
4 *Aconitum henryi* 'Spark's Variety' *2*
5 *Delphinium* 'King Arthur' *5*
6 *Cephalaria gigantea 5*
7 *Heliopsis helianthoides 4*
8 *Thalictrum aquilegifolium* 'Album' *3*
9 *Solidago virgaurea* 'Golden Wings' *3*
10 *Aster* 'Ada Ballard' *4*
11 *Gladiolus* 'Dusty Miller' *24*
12 *Monarda fistulosa* 'Souris' *3*

13 *Thermopsis villosa 3*
14 *Delphinium* 'Astolat' *1*
15 *Aster* 'Adoration' *6*
16 *Lythrum virgatum 5*
17 *Valeriana officinalis 4*
18 *Delphinium* 'Galahad' *5*
19 *Physostegia virginiana* 'Vivid' *10*
20 *Solidago virgaurea* 'Golden Wings' *3*
21 *Veronica longifolia 3*
22 *Helenium autumnale* 'Gypsy' *5*
23 *Aster* 'Lil Fardell' *5*
24 *Lythrum virgatum* 'Morden Pink' *5*
25 *Eryngium amethystinum 3*
26 *Phlox paniculata* 'Fairy's Petticoat' *4*

27 *Lavatera thuringiaca 2*
28 *Phlox paniculata* 'Henderson's Late White' *4*
29 *Physostegia virginiana* 'Summer Snow' *6*
30 *Rudbeckia laciniata* 'Golden Glow' *1*
31 *Dahlia* 'Sam Herst' *3*
32 *Centaurea dealbata* 'Sternbergii' *5*
33 *Aquilegia* 'Mrs. Scott Elliott' *6*
34 *Phlox paniculata* 'Our Gracie' *3*
35 *Clematis recta* 'Mandschurica' *3*
36 *Scabiosa caucasica 2*
37 *Lythrum virgatum* 'Brightness' *4*
38 *Lilium* 'Croesus' *30*
39 *Coreopsis grandiflora* 'Golden Showers' *6*
40 *Dahlia* 'Honey Bear' *3*
41 *Lathyrus odoratus,* any mauve cultivar *25*
42 *Lathyrus odoratus,* any red cultivar *25*
43 *Phlox* 'Prince George' *2*
44 *Hemerocallis* 'Sovereign' *4*
45 *Lilium* 'Harlequin' *30*
46 *Hemerocallis* 'Bess Ross' *3*
47 *Paeonia* 'Mrs. Livingstone Farad' *2*
48 *Aster* 'Countess of Dudley' *5*
49 *Delphinium* 'Blue Bees' *4*
50 *Tulipa* (Darwin type) 'Aristocrat' *50*
51 *Hemerocallis* 'Merry Christmas' *6*
52 *Lilium,* Pink Glory strain *15*
53 *Lilium,* 'Verona' *15*
54 *Lilium* 'T. A. Havameyer' *15*
55 *Sanguisorba obtusa 4*
56 *Limonium latifolium 4*
57 *Liatris pycnostachya 5*
58 *Delphinium* 'Guinevere' *2*
59 *Oenothera tetragona 5*
60 *Hemerocallis* 'Pink Bowknot' *3*
61 *Achillea millefolium* 'Fire King' *6*
62 *Aster* 'Blue Bird' *6*
63 *Asperula longifolia 6*
64 *Cimicifuga americana 2*
65 *Achillea* 'The Pearl' *4*
66 *Lilium* 'Croesus' *10*
67 *Iris* (bearded) 'Golden Majesty' *17*
68 *Physostegia virginiana* 'Summer Glow' *9*
69 *Iris* (tall bearded type) 'Pierre Menard' *10*
70 *Iris sibirica* 'Gatineau' *11*
71 *Oenothera tetragona 16*
72 *Bergenia cordifolia 3*
73 *Physostegia virginiana* 'Summer Show' *5*
74 *Campanula persicifolia 4*
75 *Iris* (tall bearded type) 'Copper Halo' *10*
76 *Gladiolus* 'Peter Pears' *24*
77 *Phlox paniculata* 'Firefly' *3*
78 *Achillea ptarmica* 'Angels' Breath' *6*
79 *Oenothera tetragona 8*

80 *Achillea taygetea 5*
81 *Iris* (tall bearded type) 'Pluie d'Or' *8*
82 *Ipomoea rubro-caerulea praecox 16*
83 *Iris* (tall bearded type) 'Crinkled Ivory' *13*
84 *Thalictrum aquilegifolium* 'Thunder Cloud' *4*
85 *Phlox paniculata* 'Flamboyant' *4*
86 *Hemerocallis* 'Midwest Majesty' *3*
87 *Aruncus sylvester 4*
88 *Chrysanthemum maximum* 'Wirral Pride' *5*
89 *Iris* (tall bearded type) 'Sable Night' *8*
90 *Lilium* 'Golden Chalice' *24*
91 *Nepeta* 'Blue Beauty' *5*
92 *Polemonium caeruleum* 'Blue Pearl' *6*
93 *Arabis* × *arendsii 6*
94 *Phlox subulata* 'Autumn Rose' *8*
95 *Thymus serpyllum* 'Annie Hall' *5*
96 *Dianthus* 'John Ball' *6*
97 *Phlox subulata* 'Rosea' *4*
98 *Heuchera sanguinea 6*
99 *Erigeron speciosus* 'Mrs. F. H. Beale' *8*
100 *Iris* (dwarf bearded type) 'Fairy' *11*
101 *Iberis sempervirens 15*
102 *Incarvillea grandiflora 6*
103 *Achillea ptarmica* 'Perry's Giant White' *4*
104 *Arabis caucasica 5*
105 *Iberis sempervirens 4*
106 *Gypsophila paniculata* 'Rosenschleier' *3*
107 *Rudbeckia laciniata* 'Goldsturm' *4*
108 *Asperula longifolia 3*
109 *Alyssum saxatile 4*
110 *Linum perenne 3*
111 *Iris sibirica* 'Dragon Fly' *10*
112 *Tradescantia virginica* 'Purple Dome' *3*
113 *Dictamnus albus 3*
114 *Erigeron* 'Vanity' *4*
115 *Colchicum autumnale 9*
116 *Astilbe simplicifolia* 'Granat' *4*

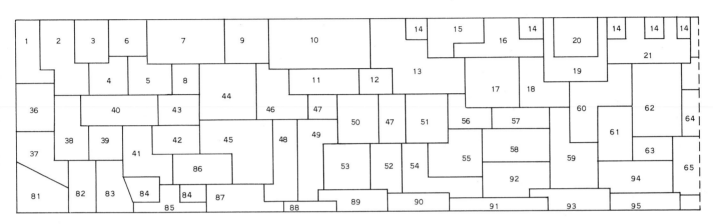

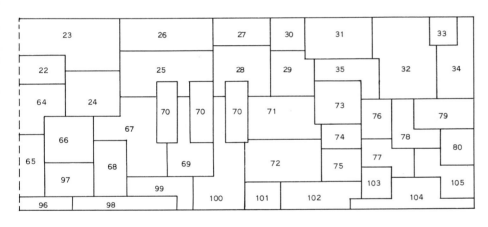

A LONG BORDER CONTAINING PLANTS THAT FLOWER IN SPRING, SUMMER, AND FALL

1 *Aster* 'Purple Feather' *6*

2 *Lythrum salicaria* 'Morden Gleam' *4*

3 *Aster* 'Lil Fardell' *2*

4 *Thermopsis villosa 2*

5 *Aster* 'Aquila' *2*

6 *Valeriana officinalis 2*

7 *Boltonia asteroides 3*

8 *Physostegia virginiana* 'Summer Glow' *3*

9 *Gladiolus* 'Red Pepper' *24*

10 *Helenium autumnalis* 'Riverton Beauty' *3*

11 *Lilium* 'Burgundy Strain' *16*

12 *Coreopsis grandiflora* 'Mayfield Giant' *1*

13 *Solidago virguarea* 'Golden Gates' *6*

14 *Lychnis chalcedonica 5*

15 *Aster* 'Mistress Quickly' *3*

16 *Aster* 'Harrington's Pink' *4*

17 *Cimicifuga americana 3*

18 *Phlox paniculata* 'Henderson's Late White' *3*

19 *Tulipa forsteriana* 'Red Emperor' *20*
Tulipa 'Eichleri' *20* and
Chionodoxa luciliae 50

20 *Trollius ledebouri* 'Golden Queen' *2*

21 *Helenium autumnale 3*

22 *Physostegia virginiana* 'Vivid' *5*

23 *Anemone hupehensis 7*

24 *Chrysanthemum* 'Prairie Moon' *3*

25 *Thalictrum adiantifolium 4*

26 *Dahlia* 'Jane Lausche' *4*

27 *Delphinium* 'Guinevere' *3*

28 *Boltonia latisquama 3*

29 *Gladiolus* 'Focus' *15*

30 *Phlox* 'Starfire' *2*

31 *Hibiscus moscheutos 1*

32 *Coreopsis* 'Mayfield Giant' *3*

33 *Cephalaria tatarica 1*

34 *Clematis* 'Robert Brydon' *3*

35 *Heliopsis scabra 2*

36 *Chrysanthemum* 'Ahnasti' *4*

37 *Sedum spectabile 5*

38 *Hosta lancifolia 3*

39 *Althaea*, any red *4*
40 *Paeonia* 'Kansas' *2*
41 *Colchicum autumnale 12* and
   Tulipa (double, late type) 'Mount Tacoma' *24*
42 *Paeonia* 'M. Jules Elie' *1*
43 *Coreopsis* 'Golden Showers' *4*
44 *Hemerocallis* 'Crimson Pirate' *4*
45 *Echinacea* 'White King' *4*
46 *Monarda* 'Cambridge Scarlet' *3*
47 *Tulipa* 'Darwin Sunkist' *28*
48 *Oenothera tetragona 6*
49 *Iris sibirica* 'Rimouski' *12*
50 *Delphinium* 'Black Night' *3*
51 *Delphinium* 'Astolat' *3*
52 *Tulipa* (double, early type) 'Murillo' *24*
53 *Campanula persicifolia 5*
54 *Achillea* 'The Pearl' *5*
55 *Thalictrum* 'Golden' *3*
56 *Aster linosyris 3*
57 *Phlox* 'July Glow' *3*
58 *Lilium* 'Joan Evans' *24*
59 *Monarda* 'Sunset' *4*
60 *Achillea* 'Cloth of Gold' *4*
61 *Hemerocallis* 'Revolute' *2*
62 *Hemerocallis* 'Sovereign' *4*
63 *Chrysanthemum arcticum 2*
64 *Lilium* 'Harmony' *25*
65 *Chrysanthemum* 'Mois' *3* and
   Tulipa (lily-flowered type) 'Marietta' *24*
66 *Ipomoea* hybrid *6* and
   Eranthis cilicica 50
67 *Helenium* 'Fountain' *4*
68 *Aquilegia* 'Crimson Star' *7*
69 *Lythrum roseum 4*
70 *Echinops* 'Taplow Blue' *8*
71 *Paeonia* 'Red Charm' *2*
72 *Phlox* 'Eventide' *6*
73 *Paeonia* 'Edulis Superba' *1*
74 *Coreopsis* 'Golden Showers' *2*
75 *Iris* (tall bearded type) 'Happy Birthday' *11*
76 *Lathyrus odoratus*, any lilac cultivar *12*
77 *Chrysanthemum*, an orange-red cultivar *5*
78 *Oenothera tetragona 4*
79 *Centaurea montana 4*
80 *Aster* 'Peter Harrison' *3*
81 *Nepeta faassenii* 'Blue Beauty' *4*
82 *Asperula longifolia 5*
83 *Polemonium caeruleum* 'Blue Pearl' *4*
84 *Narcissus* 'Flower Record' *24*
85 *Iberis sempervirens 9*
86 *Salvia farinacea 5*
87 *Astilbe pumila 8*
88 *Phlox subulata* 'Sunningdale Red' *3*
89 *Iris* (dwarf bearded type) 'Endymion' *12*
90 *Pulmonaria saccharata 3*
91 *Asperula longifolia 4*
92 *Physostegia virginiana* 'Summer Snow' *9*
93 *Eryngium planum* 'Blue Dwarf' *6*
94 *Gaillardia grandiflora* 'Monarch Strain' *4*
95 *Chrysanthemum*, any bronze cultivar *4*
96 *Arabis suendermannii 6*
97 *Iris* (tall bearded type) 'Rose Taffeta' *12*
98 *Phlox subulata 6*
99 *Campanula persicifolia* 'Alba' *4*
100 *Sanguisorba obtusa 6*
101 *Iberis sempervirens* 'Snowflake' *2*
102 *Tulipa* (triumph type) 'Blizzard' *24*
103 *Phlox subulata* 'Alba' *2*
104 *Aster* 'Lilac Time' *10*
105 *Chrysanthemum*, a yellow cultivar *5*

INDEX OF COMMON AND SCIENTIFIC NAMES OF PLANTS

(Page numbers of color photographs are printed in italics)

Achillea 37, 92
  ageratifolia 64, 92
    'Aizoon' 64, 92
  argentia 64, 92
  clavennae 92
  filipendulina 64, 92
    × *clypeolata* 'Coronation Gold' *72*, 92
    'Gold Plate' 64, 92
    'Parker's Variety' 92
    × *lewisii* 'King Edward' 64, 92
  millefolium 92
    'Cerise Queen' 92
    'Fire King' 92
    'Roseum' 92
  ptarmica 92
    'Angels' Breath' 92
    'Boule de Neige' 92
    'Lilac Queen' 92
    'Old Ivory' 92
    'The Pearl' 92
  × *taygetea* 92
  tomentosa 64, 92
    'Moonlight' 92
aconite 45
Aconitum 48, 93
  autumnale see *Aconitum henryi* 93
  bicolor 93
  carmichaelii 93
    'Wilsonii' 93
  fischeri see *Aconitum carmichaelli* 93
  henryi 93
    'Spark's Variety' 93
  lycoctonum see *Aconitum vulparia* 93
  napellus 93
    'Blue Sceptre' 93
    'Bressingham Spire' 93
    'Kelmscot' *72*, 93
    'Newry Blue' 93
  vulparia 93
  wilsoni see *Aconitum carmichaelii* 'Wilsonii' 93
adonis
  spring 93
Adonis 93
  amurensis 93
    'Plena' 93
    var. *vernalis* 93

ageratum
  hardy 128
agrimony
  double hemp 128
  hemp 128
Agrostemma see *Lychnis* 93, 151
  coronaria see *Lychnis coronaria* 93, 151
Ajuga 93
  genevensis 93
    'Brocklebankii' 93
    'Pink Spires' 93
  pyramidalis 93
    'Metallica-crispa' 93
    'Tottenham Blue' 93
  reptans 94
    'Alba' 94
    'Atro Purpurea' 94
    'Gaiety' 94
    'Variegata' 94
alkanet 95
  tufted 95
Alstroemeria 94
  aurantiaca 94
Althaea 94
  rosea 29, 94
    strains: Chater 29
        Imperator 29
        Powderpuff 29
        Triumph 29
alumroot *81*, 136
alyssum 42
Alyssum 37, 94
  repens *72*, 94
  saxatile *72*, 94
    'Citrinum' 94
    'Compactum' 94
    'Dudley Neville' 94
    'Flore Pleno' 94
    'Silver Queen' 94
Amsonia 94
  montana 94
  tabernaemontana 94
anchusa 42
Anchusa 95, 132
  azurea 95
    'Dropmore' 95
    'Feltham Pride' 95
    'Loddon Royalist' 95
    'Morning Glory' 95
    'Opal' 95
    'Pride of Dover' 95
  caespitosa 95
  myosotidiflora see *Brunnera macrophylla* 95, 107
anemone 95
  Chinese *73*, 95

grape-leaved 96
Japanese 95
rue 96
snowdrop *73*, 96
wood 95
Anemone 37, *73*, 95, 168
  acutiloba see *Hepatica acutiloba* 95, 136
  hupehensis 64, *73*, 95
    var. *japonica* 64, 95
      'Alba' 95
  × *hybrida* 95
    'Kriemhilde' 95
    'Lorelei' 95
    'Louise Uhink' 95
    'Margarita' 95
    'Marie Manchart' 95
    'Montrose' 95
    'Queen Charlotte' 95
    'September Charm' 95
    'September Sprite' 95
    'Whirlwind' 95
  japonica see *Anemone hupehensis* var. *japonica* *73*, 95
  × *lesseri* 95
  nemorosa 95
  patens 95
  pulsatilla *73*, 96
    'Alba' 96
    'Mallandieri' 96
    'Mrs. Van der Elst' 96
    'Red Clock' 96
  sylvestris *73*, 96
  vitifolia 96
Anemonella 96
  thalictroides 96
angel's-hair 100
Anthemis 37, 96
  aizoon see *Achillea ageratifolia* 'Aizoon' 92, 96
  biebersteiniana 96
  cupaniana 96
  macedonica 96
  nobilis 96
    'Treneague' 96
  sancti-johannis 96, 97
  tinctoria 97
    'Beauty of Grallagh' 97
    'Golden Dawn' 97
    'Grallagh Gold' 97
    'Kelwayi' 97
    'Moonlight' 96, 97
    'Mrs. E. G. Buxton' 97
    'Perry's Variety' 97
    'Thora Perry' 97
Antirrhinum 48
Aquilegia 37, 48, *74*, 97
  alpina 97

'Hensol Harebell'   97
caerulea   97, 98
canadensis   97, 98
chrysantha   97, 98
discolor   98
ecalcarata   98
einseleana   98
flabellata   98
  'Nana'   98
  'Nana Alba'   98
glandulosa   98
× *helenae*   98
jonesii   98
jucunda see *Aquilegia*
  glandulosa   98
longissima   98
strains: Clematiflora Hybrida   98
  Copper Queen   98
  Crimson Star   98
  McKana's Giants   97, 98
  Mrs. Scott-Elliot's   98
  Rose Queen   98
  Snow Queen   98
vulgaris   98
Arabis   37, 73, 98
albida see *Arabis caucasica*
  98, 99
albida 'Rosea' see *Arabis* ×
  arendsii   98, 99
alpina   98
  'Rosea'   73, 98
× *arendsii*   99
  'Pink Charm'   73, 99
  'Rosabella'   99
aubrietioides   99
blepharophylla   99
  'Spring Charm'   99
caucasica   99
  'Albida'   99
  'Coccinea'   99
  'Flore Pleno'   99
  'Sulphurea'   99
  'Variegata'   99
kellereri   99
procurrens   99
sturii see *Arabis procurrens*   99
Armeria   74, 99
caespitosa see *Armeria juniperifolia*
  99
juniperifolia   99
  'Alba'   99
  'Beechwood'   99
  'Bevan'   100
maritima   74, 100
  'Alba'   100
  'Alpina'   100
  'Laucheana'   100
  'Laucheana Six Hills'   100
  'Vindictive'   100

pseud-armeria   100
  'Bees' Ruby'   100
  'Bloodstone'   100
  'Glory of Holland'   100
Artemisia   24, 74, 100
albula   100
  'Lambrook Silver'   100
  'Silver Queen'   100
discolor   100
frigida   100
lactiflora   24, 100
lanata   100
maritima 'Nutans'   see *Artemisia*
  nutans   100
nutans   100
palmeri   74, 100
purshiana   100
schmidtiana 'Nana'   100
  'Silver Mount'   74, 100
stelleriana   101
Arum dracunculus see *Dracunculus*
  vulgaris   101, 124
Aruncus   101
  sylvester   74, 101
  'Kneiffi'   101
Asclepias   101
  tuberosa   101
Asperula   101
  odorata   101
asphodel   101
  white   101
Asphodeline   101
  lutea   101
Asphodelus   101
  cerasiferus   101
aster   40, 44, 74
  alpine   102
  blue wood   102
  cornflower   175
  golden   115
  heath   102
  Italian   102
  New England   102
  New York   103
  smooth   102
Aster   37, 74, 101, 174
acris   101
  'Nanus'   101
  'Roseus'   101
× *alpellus*
  'Summer Greeting'   102
  'Triumph'   102
alpinus   102
  'Beechwood'   102
  'Joy'   102
  'Wargrave'   102
amellus   102

'Brilliant'   102
'King George'   102
'Lac de Geneve'   102
'Lady Hindlip'   102
'Mauve Beauty'   102
'Moorheim Gem'   102
'Nocturne'   102
'Perry's Favorite'   102
'Sonia'   102
'Sonnewendi'   102
× *cordi-belgii* 'Pioneer'   102
cordifolius   102
  'Silver Spray'   102
  'Sweet Lavender'   102
ericoides   102
  'Blue Star'   102
  'Brimstone'   102
  'Chastity'   102
  'Delight'   102
  'Esther'   102
  'Golden Spray'   102
  'Ringdove'   102
× *frikartii*   102
laevis   102
linosyris   102
novae-angliae   102, 103
  'Barr's Pink'   102
  'Crimson Beauty'   102
  'Harrington's Pink'   102
  'Incomparabilis'   102
  'Lil Fardell'   102
  'Lye End Beauty'   102
  'Mount Rainier'   102
  'Red Star'   102
  'September Ruby'   102
  'Survivor'   102
  'Treasure'   102
novi-belgii   103
  'Ada Ballard'   103
  'Alpenglow'   104
  'Appleblossom'   103
  'Audrey'   104
  'Beechwood Beacon'   104
  'Beechwood Challenger'   103
  'Blondie'   103
  'Blue Bonnet'   103
  'Blue Bouquet'   104
  'Blue Radiance'   104
  'Bonanza'   104
  'Buxton's Blue'   104
  'Canterbury Carpet'   104
  'Chartwell'   103
  'Chequers'   104
  'Crimson Brocade'   103
  'Davey's True Blue'   103
  'Dawn'   103
  'Destiny'   103
  'Elegance'   103
  'Erica'   104

'Ernest Ballard' 103
'Eventide' 104
'Fair Lady' 103
'Fairy' 104
'Fellowship' 103
'Flamingo' 103
'Gayborder Charm' 104
'Gayborder Royal' 103
'Gayborder Splendour' 104
'Gayborder Violet' 103
'Glorious' 103
'Goldflame' *74,* 103
'Guy Ballard' 104
'Harrison's Blue' 103
'Hilda Ballard' 103
'Janet McMullen' 103
'Jenny' 104
'Lady-in-blue' 104
'Lavender Gown' 104
'Lavender Midget' 104
'Little Blue Baby' 104
'Little Boy Blue' 104
'Little Pink Lady' 104
'Little Pink Pyramid' 104
'Little Red Boy' 104
'Mabel Reeves' 103
'Margaret Rose' 104
'Mauve Ballard' 103
'Midget' 104
'Orchid Pink' 103
'Peter Harrison' 103, 104
'Peter Pan' 104
'Picture' 103
'Pink Bouquet' 104
'Pink Lace' 104
'Plenty' 103
'Purple Feather' 104
'Purple Prelude' 104
'Queen Mary' 103
'Red Sunset' 103
'Romany' 104
'Rose Serenade' 104
'Royal Amethyst' 104
'Royal Gem' cultivars 104
'Royal Opal' 103, 104
'Royal Pearl' 103, 104
'Royal Sapphire' 104
'Snowball' 104
'The Archbishop' 103
'The Bishop' 103
'The Cardinal' 103
'Twinkle' 103
thomsonii 102
yunnanensis 'Napsbury' 104
astilbe *75,* 104
Astilbe 37, *75,* 104
'Amethyst' 104
'Avalanche' 104
'Bonn' 104

'Cattleya' 104
'Coblence' 104
'Cologne' 104
'Deutschland' 104
'Dusseldorf' 104
'Elna' 104
'Erica' 104
'Fanal' 104
'Fire' 104
'Granat' 104
'Intermezzo' 104
'Irene Rotsieper' 104, 105
'Jo Orphorst' 104
'Ostrich Plume' 104
'Peach Blossom' 105
'Pink Pearl' 105
'Red Sentinel' 105
'Rhineland' 105
'Tamarix' 105
'Venus' 105
'Vesuvius' 105
'W. D. Willen' 105
'White Queen' 105
'William Reeves' 105

Astrantia 105
carniolica 'Major' 105
'Maxima' 105
'Rubra' 105
'Shaggy' 105
major see *Astrantia carniolica*
'Major' 105

Aubrieta 37, *75,* 105
deltoidea 105
'Argenteo-variegata' 105
'Aureo-variegata' 105
'Bonfire' 105
'Borsch's White' 105
'Carnival' 105
'Crimson Bedder' 105
'Dr. Mules' 105
'Gloriosa' 105
'Gurgedyke' *75,* 105
'Lilac Time' 105
'Mary Poppins' 105
'Maurice Pritchard' 105
'Oakington Lavender' 105
'Studland' 105
'Wanda' 105

auricula 166
avens 131
Bulgarian 131
mountain 124
scarlet 132

baby's-breath 24, 132
bachelor's-button
hardy 111
white 168

yellow 168
balloonflower *87,* 162
balm
bee 155
Baptisia 105
australis 105
'Old Orchard' hybrids 105
bracteata 105
leucantha 106
tinctoria 106
barrenwort 125
beardtongue *86,* 159
bear's-tail
Cretan 110
bee balm 155
Belamcanda 106
chinensis 106
flabellata 106
bellflower *75,* 108
broad-leaved 109
Carpathian 108
Chinese *87,* 162
clustered 108
Dalmatian 109
giant 157
milky *75,* 109
peach-leaved 109
spotted 110
spurred 108

Bellis *75,* 106
perennis *75,* 106, 127
'Dresden China' 106
'Monstrosa' 106
'Rob Roy' 106
bells
blue, of Scotland 110
Canterbury 28, 43, *70*
coral *81,* 136
Coventry 110
golden 107
bergamot 155
wild *85,* 155
Bergenia 48, *75,* 106
ciliata 106
cordifolia *75,* 106
crassifolia 106
ligulata see *Bergenia ciliata* 106
stracheyi 106
betony
woolly 174
bills
mosquito 122
bishop's-hat 125
bladder cherry 162
blanketflower *80,* 129

blazingstar *83,* 147
  dotted 147
bleedingheart *79,* 121, 122
  plume *79,* 121
  western 121
bloodroot 36, 170
blue
  loddon 136
bluebells 154
  Himalayan 154
  mountain 154
  prairie 154
  Siberian 154
  Virginian *85,* 154
blue bonnets 171
blue-eyed Mary 157
bluet
  mountain *75,* 111
Bocconia 106
  cordata see *Macleaya cordata* 106
Boltonia 106
  asteroides 106
    'Snowbank' 106
  latisquama 107
    'Nana' 107
bouncingbet 171
Boykinia 107
  aconitifolia 107
  jamesii 107
  rotundifolia 107
bronzeleaf
  feathered 168
brown-eyed Susan 169
Brunnera 38, 107, 123
  macrophylla 107
bugbane 115
  Kamchatka 116
bugle 93
  carpet 94
  Geneva 93
bugloss 95
  Italian 95
  Siberian 107
bunchberry 118
Buphthalmum 107
  salicifolium 107
  speciosum 107
burnet
  American 170
  Japanese *89,* 170
buttercup 168
butterflyweed 101
calamint 107

alpine 107
Calamintha 107
  alpina 107
  grandiflora 107
  nepeta 107
  nepetoides see *Calamintha nepeta*
    107
Caltha 107
  leptosepala 107
  palustris 108
    'Alba' 108
    'Holubyi' 108
    'Plena' 108
Campanula 36, *75,* 108
  alliarifolia 108
  carpatica 108
    'Alba' 108
    'Blue Carpet' 108
    'Ditton Blue' 108
    'Harvest Blue' 108
    'Queen of Sommerville' 108
    'Riverslea' 108
    var. *turbinata* 108
      'Pallida' 108
      'White Star' 108
  cochlearifolia 108
    'Alba' 108
    'Miranda' 108
    'Miss Willmott' 108
  collina 108
  elatines var. *garganica* 108, 109
    'Hirsuta' 108
    'W. H. Paine' 108
  glomerata 108
    'Acaulis' 108
    'Dahurica' 108
    'Joan Elliot' 109
    'Nana Alba' 109
    'Nana Lilacina' 109
    'Purple Pixie' 109
    'Superba' 109
  istriaca see *Campanula elatines* var.
    garganica 'Hirsuta' 108, 109
  lactiflora *75,* 109
    'London Anna' 109
    'Pouffe' 109
    'Pritchard's Variety' 109
  latifolia 109
    'Alba' 109
    'Brantwood' 109
    'Gloaming' 109
    'Macrantha' 109
  latiloba see *Campanula*
    persicifolia 109
  medium 28
  muralis see *Campanula*
    portenschlagiana 109

  persicifolia 109
    'Beechwood' 109
    'Blue Bells' 109
    'Blue Gardenia' 109
    'Cantab' 109
    'Fleur de Neige' 109
    'Moerheimi' (or 'Summer Skies')
      109
    'Mount Hood' 109
    'Mrs. H. Harrison' 109
    'Pride of Exmouth' 109
    'Snowdrift' 109
    'Telham Beauty' 109
    'Wirral Belle' 109
  portenschlagiana 109
  poscharskyana 110
    'Lisduggan' 110
    'Stella' 110
  punctata 110
  pusilla see *Campanula cochlearifolia*
    108, 110
  rotundifolia 110
    'Olympica' 110
    'Purple Gem' 110
  trachelium 110
    'Bernice' 110
campion *84,* 151
  Arctic 151
  rose 151
candytuft 138
  perennial 139
  white 123
cardinalflower 150
  blue 150
carnation 120
cartwheelflower 136
Catananche 110
  caerulea 110
    'Major' 110
    'Perry's White' 110
catchfly
  German *84,* 152
catmint 156
catnip 156
  Tartarian 156
Cautleya 110
  robusta 110
Celmisia 110
  spectabilis 110
Celsia 110
  acaulis 110
  arcturus 110
  cretica 111
centaurea
  Persian 111
Centaurea 37, *75,* 111

196

dealbata 111
  'John Coutts' 111
  'Sternbergii' 111
glastifolia 111
macrocephala 111
montana 75, 111
  'Alba' 111
  'Parham's' 111
  'Rosea' 111
  'Violetta' 111
Centranthus 111
 ruber 111
  'Albus' 111
  'Atrococcineus' 111
Cephalaria 111
 alpina 111
 gigantea 76, 111
 tatarica see *Cephalaria gigantea* 111
Cerastium 37, 111
 alpinum 111
  'Lanatum' 111
 biebersteinii 111
 tomentosum 111
Ceratostigma 111
 plumbaginoides 111
 willmottiana 111
chalkplant 132
chamomile 96
Cheiranthus 111
 cheiri 32
Chelone 111
 digitalis see *Penstemon digitalis*
  111, 159
 glabra 112
 lyonii 112
cherry
 bladder 162
chickweed 111
chrysanthemum 40, 42, 44, 45, 77, 112
 Korean 115
 Japanese 112
Chrysanthemum 37, 46, 112
 arcticum 112, 114
 balsamita 112
 coccineum 112
  'Agnes M. Kelway' 112
  'Allurement' 112
  'Apollo' 112
  'Avalanche' 112
  'Brenda' 112
  'Buckeye' 112
  'Carl Vogt' 112
  'Comet' 112
  'Crimson Giant' 112
  'Eileen May Robinson' 112
  'Evenglow' 112
  'Harold Robinson' 112

'Helen' 112
'H. M. Stanley' 112
'Inferno' 112
'J. M. Tweedy' 112
'Jubilee Gem' 112
'Kelway Glorious' 112
'Madeleine' 112
'Marjorie Deed' 112
'Mrs. D. D. Bliss' 112
'Mrs. E. C. Beckwith' 112
'Pink Bouquet' 112
'Queen Mary' 113
'Scarlet Glow' 112
'Senator' 113
'Vanessa' 113
'Venus' 113
'White Madeleine' 76, 113
coreanum see *Chrysanthemum*
 sibiricum 113, 115
corymbosum 113
mawi 112
maximum 113
 'Aglaia' 113, 114
 'Beaute Nivelloise' 113
 'Cobham Gold' 76, 113
 'Edgebrook Grant' 113
 'Esther Read' 76, 113
 'Everest' 113
 'Horace Read' 113
 'H. Siebert' 113
 'Ian Murray' 113
 'Jennifer Read' 113
 'King Edward VII' 113
 'Majestic' 113
 'Marconi' 113
 'Mark Riegel' 113
 'Sedgewick' 113
 'Stone Mountain' 113
 'Thomas E. Killeen' 76, 113
 'Wirral Pride' 113
 'Wirral Snowball' 114
 'Wirral Supreme' 76, 114
morifolium 64, 112
 'Acaena' 115
 'Ahnasti' 115
 'Astoria' 115
 'Autumn Song' 114
 'Beckethon' 115
 'Bonnie Brandon' 115
 'Brown Eyes' 114
 'Campaigner' 114
 'Christopher Columbus' 114
 'Coquette' 77, 114
 'Golden Carpet' 77, 115
 'Joan Brandon' 115
 'Julie Brandon' 115
 'Knock-out' 114
 'Larry' 114
 'Moeis' 115

'Muted Sunshine' 115
'Nootka' 115
'Ontario Nugget' 115
'Pink Haze' 77, 114
'Pink Sentinel' 114
'Powder River' 114
'Prairie Sun' 115
'Purple Pirate' 114
'Purple Star' 114
'Purple Waters' 114
'Redskin' 77, 114
'Roll Call' 77, 114
'Ruby Mound' 77, 114
'Salmon Minn Pink' 114
'Stadium Queen' 114
'White Cushion' 114
'Zonta' 114
nipponicum 113, 114, 115
rubellum 114, 115
 'Clara Curtis' 115
 'Duchess of Edinburgh' 115
 'Mary Stoker' 115
 'Paul Boissier' 115
sibiricum 113, 114, 115
uliginosum 112, 115
weyitchii 115
Chrysogonum 115
 virginianum 115
Chrysopsis 115
 falcata 115
 mariana 115
 villosa 115
Cimicifuga 115
 americana 115
 cordifolia see *Cimicifuga americana*
  115
 dahurica 115
 foetida 'Intermedia' see *Cimicifuga*
  simplex 116
 japonica var. *acerina* 116
 racemosa 78, 116
 simplex 116
  'Armleuchter' 116
  'White Pearl' 116
cinquefoil 164
 Kashmirian 164
clary 170
clematis
 bush 116
 fragrant tube 116
Clematis 116
 heracleifolia var. *davidiana* 116
  'Cote d'Azure' 116
  'Crépuscule' 116
 integrifolia 116
  'Coerulea' 116
  'Hendersoni' 116
 recta 116

'Grandiflora' 116
'Mandshurica' 116
'Purpurea' 116
Codonopsis 116
clematidea 116, 117
convolvulacea 116
ovata 117
columbine *74, 97*
alpine 97
Altai 98
dwarf fan 98
European 98
fan 98
golden 97
Jones 98
longspur 98
Rocky Mountain 97
white fan 98
wild 97
comfrey 175
prickly 175
coneflower 169
purple *79,* 124
sweet 169
Convallaria 117
majalis 36, 117
'Aureo-variegata' 117
'Flore Plena' see 'Prolificans' 117
'Fortin's Giant' 117
'Fortunei' 117
'Prolificans' 117
'Rosea' 117
coreopsis
thread-leaved 118
Coreopsis 37, 47, 117
auriculata 117
'Nana' 117
'Superba' 117
grandiflora *78,* 117
'Badengold' 117
'Golden Plume' *78,* 117
'Goldfink' 117
'Mayfield Giant' 117
'Newgold' 117
'Rubythroat' 117
'Sunburst' 117
'Sunchild' 117
lanceolata 117
rosea 118
'Nana' 118
verticillata 118
'Golden Showers' *78,* 118
'Grandiflora' 118
Cornus 118
canadensis *78,* 118
corydalis
Chinese 118

Corydalis 118
bulbosa 118
cava 118
cheilanthifolia 118
lutea 118
nobilis 118, 119
wilsoni 119
costmary 112
cowslip 167
Himalayan 166
Jerusalem 167
cranesbill *81,* 130
Armenian 131
bloody 131
lilac 131
meadow 131
creeping Jenny *84,* 152
cress
false rock- 105
rock- *73,* 98
crocus
prairie 95
cross
Maltese *84,* 151
crown-of-thorns 128
Culver's-root 181
cupid's-dart 110

daisy
Arctic 112
Caucasian 113
dwarf pink 112
English *75,* 106, 127
gloriosa *88,* 169
Japanese oxeye 115
Michaelmas 101, 102
moon 112, 115
oxeye 113
painted 112
Shasta *13,* 40, *76,* 113
Danes'-blood 108
day-lily 16, *66*
dwarf yellow 135
lemon 135
tawny 135
dead-nettle 145
giant 146
spotted 145
delphinium 17, 36, 40, 43, 44, 45, *66*
belladonna 119
butterfly 119
common 119
European 17
'Anne Page' 18
'Anona' 18
'Arcadia' 18

'Audrey Mott' 18
'Blackmore's Glorious' 18
'Blue Lagoon' 18
'Blue Riband' 18
'Blue Rosette' 18
'Bridesmaid' 18
'Cambria' 18
'Charles F. Langdon' 18
'C. H. Middleton' 18
'Dame Myra Curtis' 18
'Duchess of Portland' 18
'Etonian' 18
'Eve Gower' 18
'Father Thames' 18
'Fleur Celeste' 18
'Frederick Grisewood' 18
'F. W. Smith' 18
'George Bishop' 18
'Janice' 18
'Jennifer Langdon' 18
'Julia Langdon' 18
'Lady Eleanor' 18
'Lady Guinevere' 18
'Mayflower' 18
'Mrs. Frank Bishop' 18
'Nell Gwynn' 18
'Silver Moon' 18
'Sir Neville Pearson' 18
'Swanlake' 18
'Sylvia' 18
'Watkin Samuel' 18
'W. B. Cranfield' 18
'Wrexham Glory' 18
Pacific Giant Strain 16, 17, 119
Delphinium 46, 47, 48, 119
ajacis 119
× *Belladonna* 119
'Blue Bees' 119
'Capri' 119
'Cliveden Beauty' 119
'Coelestinum' 119
'Lamartine' 119
'Naples' 119
'Pink Sensation' 119
'Wendy' 119
elatum 119
formosum 'Coelestinum' 119
grandiflorum 119
'Blue Butterfly' 119
'Blue Gem' 119
'Blue Mirror' 119
'Connecticut Yankee' 119
'White Butterfly' 119
menziesii 119
nudicaule 119
tatsienense 119
desertcandle
Himalayan 126

Dianella 119
  intermedia 119
  tasmanica 119
dianthus 45

Dianthus 37, 48, 120
  × *allwoodii* 120
    'Little Joe' 120
  arenarius 120
  × *arvernensis* 120
  barbatus 32, 120
    'Harlequin' 32
    'Indian Carpet' 32
    'Pheasant Eye' 32
    'Scarlet Beauty' 32
  caesius see *Dianthus gratianopo-litanus* 120
  caryophyllus 120
  deltoides 120
    'Alba' 120
    'Brilliant' 120
    'Erectus' 120
    'Flashing Light' 120
    'Huntsman' 120
    'Wisley' *78*, 120
  gratianopolitanus 120
    'Flore Pleno' 120
    'Rose Queen' 120
  knappi 120
  plumarius 120
    'Candystripe' 121
    'Caprice' *78*, 121
    'Cheyenne' 121
    'Cyclops' 121
    'Dinah' 121
    'Dubonnet' 121
    'Emile Pare' 121
    'Emperor' 121
    'Evangeline' 121
    'Excelsior' 121
    'Her Majesty' 121
    'Highland Queen' 121
    'Inchmery' 121
    'John Ball' 121
    'Mrs. Sinkins' 121
    'Old Spice' 121
    'Pink Princess' 121
    'Shadow Valley' 121
    'Sweet Memory' 121
  superbus 120, 121
    'Blue Loveliness' 121
    'White Loveliness' 121

Dicentra 45, 121, 132
  chrysantha 121
  cucullaria 118, 121
  eximia *79*, 121
    'Alba' 121
  formosa 121
    'Adrian Bloom' 121

    'Bountiful' *79*, 121
    'Summer Beauty' 121
    'Sweetheart' 122
    ssp. *oregana* 122
  oregana see *Dicentra formosa* ssp. *oregana* 122
  spectabilis *79*, 121, 122
    'Alba' 122
dictamnus 42
Dictamnus 122
  albus *79*, 122
    'Caucasicus' see *Dictamnus albus*
    'Gigantius' 122
    'Gigantius' 122
    'Purpureus' 122
    'Ruber' see *Dictamnus albus* 'Purpureus' 122
  fraxinella 36
Dierama 122
  pendulum 122
  pulcherrima 122
    'Nanum' 122
Digitalis 47, 122
  purpurea 29, 122
    'Foxy' 29, *70*
Dodecatheon 122
  hendersonii 122
  meadia 122
  pauciflorum see *Dodecatheon pulchel-lum* 122, 123
  pulchellum 122, 123
dogwood 157
Doronicum 37, 123
  caucasicum 123
    'Madame Mason' 123
  cordatum 123
  pardalianches 123
    'Bunch of Gold' 123
    'Spring Beauty' 123
  plantagineum 123
    'Excelsum' Harper Crewe variety 123
Dracocephalum 123
  altaiense see *Dracocephalum grandi-florum* 123
  argunense 123, 124
  forrestii 123
  grandiflorum 124
  hemsleyanum 124, 156
  ruyschiana 124
    'Speciosum' see *Dracocephalum argunense* 123, 124
  wilsonii 124, 156
Dracunculus 124
  vulgaris 124
dragonhead 123
  false *87*, 162

dragonplant 124
dropwood 24, 129
Dryas 124
  drummondii *79*, 124
  integrifolia 124
  octopetala 124
    'Minor' 124
  × *suendermannii* 124
Dutchman's-breeches 118, 121

eardrops
  golden 121
Echinacea 37, 38, 124
  angustifolia 124
  purpurea *79*, 124
    'Bright Star' (Leuchtstern) 124
    'Robert Bloom' 124
    'The King' 125
    'White Luster' 124, 125
Echinops 125
  exaltatus *79*, 125
    'Veitch's Blue' 125
  humilis 125
    'Blue Cloud' 125
    'Nivalis' 125
    'Taplow Blue' 125
  ritro see *Echinops exaltatus* 125
  sphaerocephalus see *Echinops exaltatus* 125
elecampane 139
Epimedium 125
  alpinum 'Rubrum' see *Epimedium coccineum* 125
  coccineum 125
  grandiflorum 125
    'Rose Queen' 125
    'Violaceum' 125
  macranthum see *Epimedium grandi-florum* 125
  niveum see *Epimedium* × *youngianum* 'Niveum' 126
  perralderianum 125
  pinnatum 125
    var. *colchicum* 126
  × *rubrum* see *Epimedium coccineum* 125, 126
  × *versicolor* 126
  × *warleyense* 126
  × *youngianum* 126
    'Niveum' 125, 126
Eremurus 64, 126
  bungei see *Eremurus stenophyllus* 126
  elwesii 126
    'Albus' 126
  himalaicus 126
  × *himbrob* 126

olgae 126
robustus 126
× *Shelford* 126
stenophyllus 126
   'Highdown Gold' 127
   'Isobel' 127
   'Magnificus' 127
   'Moonlight' 127
   'Rosalind' 127
   'Sir Arthur Hazelrigg' 127
   'White Beauty' 127
erigeron 44

Erigeron 64, 127
aurantiacus 127
caespitosus 80, 127
leiomerus 127
macranthus 127
mucronatus 127
speciosus 127
   'Azure Beauty' 127
   'Double Beauty' 127
   'Mrs. E. H. Beale' 127
   'Quakeress' 80, 127
   'Summertime' 127
   'Wuppertal' 127
miscellaneous cultivars
   'Amity' 80, 127
   'B. Ladhams' 127
   'Charity' 127
   'Darkest-of-All' 127
   'Dignity' 127
   'Felicity' 127
   'Festivity' 127
   'Foerster's Siebling' 127
   'Gaiety' 127
   'Lilofee' 80, 127
   'Mesa-Grande' 127
   'Prosperity' 127
   'Serenity' 127
   'Vanity' 127
   'Violetta' 127

Eryngium 127
   'Donard Variety' 80, 127
agavifolium 127
alpinum 'Improved Form' 127
amethystinum 127
bourgatii 127
bromeliifolium 128
dichotomum 128
oliverianum 128
pandanifolium 128
planum 128
tripartitum 128
   'Violetta' 128

Erysimum 128
asperum 128
hieracifolium 32

Eupatorium 128, 150

ageratoides 128
cannabinum 'Plenum' 128
coelestinum 128
purpureum 128
Euphorbia 48, 128
corollata 128
cyparissias 128
epithymoides 80, 128
griffithi 80, 128
meloformis 128
polychroma see *Euphorbia*
   epithymoides 128
pulcherrima 128
sikkimensis 128
splendens 128
wulfenii 128
evening-primrose 86, 156
showy 156

fair-maid-of-France 168
feather
Kansas 147
figwort
Cape 162
Filipendula 129
camtschatica 129
   'Rosea' 129
multijuga 129
palmata 129
purpurea see also *Filipendula*
   palmata 129
   'Alba' 129
   'Purpurascens' 129
rubra 129
   'Venusta' 129
ulmaria 129
   'Aurea' 129
   'Flore Pleno' 129
vulgaris 129
   'Flore Pleno' 24, 129
flax 149
yellow 83, 149
fleabane 80, 127
fleeceflower 164
Himalayan 164
mountain 164
flower-of-Jove 151
foamflower 177
forget-me-not 28, 107, 156
true 156
wood 156

Forsythia 107
foxglove 29, 43, 44, 45, 70, 122
fumitory 118
yellow wall 118
funkia 137

Funkia 129
subcordata see *Hosta plantaginea*
   129, 138

gaillardia 45
Gaillardia 37, 38, 129
aristata 80, 129
× *grandiflora* 129
   'Baby Cole' 129
   'Burgundy' 129
   'Goblin' 129
   'Ipswich Beauty' 129
   'Mandarin' 129
   'Portola' hybrids 129
   'Wirral Flame' 129
Galega 130
officinalis 130
   'Alba' 130
   'Carnea' 130
   'Hartlandii' 130
   'Her Majesty' 130
   'Lady Wilson' 130
orientalis 130
gasplant 36, 79, 122
gaura 130
Gaura 130
lindheimeri 130
gayfeather 83, 147
dwarf 147
grassleaf 147
spike 147
gentian
fringed 36
white milkweed 130
willow 130
yellow 130
Gentiana 130
asclepiadea 130
   'Alba' 130
crinita 36
lutea 130
septemfida 130
geranium
mint 112
Geranium 130
anemonifolium 130
armenum see also *Geranium*
   psilostemon 130, 131
cinereum 130
   'Album' 130
   var. *subcaulescens* 130
dalmaticum 81, 130
endressi 130
   'A. T. Johnson' 130
   'Rose Clair' 130
   'Wargrave Pink' 130

200

farreri see *Geranium napuligerum*
130, 131
grandiflorum see *Geranium meeboldii*
130, 131
ibericum 130
'Album' 130
'Johnson's Blue' 130
var. *platypetalum* 130
macrorrhizum 131
meeboldii 131
'Alpinum' 131
napuligerum 131
nodosum 131
phaeum 131
platypetalum see *Geranium ibericum*
var. *platypetalum* 130, 131
pratense 131
'Album' 131
'Album Plenum' 131
'Coeruleum Plenum' 131
'Purpureum Plenum' 131
'Silver Queen' 131
psilostemon see also *Geranium*
armenum 130, 131
renardii 131
sanguineum 131
'Album' 131
'Lancastriense' 131
'Prostratum' 131
sylvaticum 131
'Mayflower' 131
wallichianum 131
'Buxton's Blue' 131
wlassovianum 131

Geum 131
× *borisii* 131
bulgaricum 131
chiloense 132
'Dolly North' 132
'Fire Opal' 132
'Georgenberg' 132
'Golden West' 132
'Lady Stratheden' 132
'Mrs. Bradshaw' 132
'Orangeman' 132
'Prince of Orange' 132
'Princess Juliana' 132
'Red Wings' 132
'Rubin' 132
× *heldreichii* 132
'Superbum' 132
rossii 132
ghostplant 100

Gillenia 132
trifoliata 132

Glaucidium 132
plamatum 132

Glaucium 132

flavum 132
globeflower *91,* 178
Chinese 178
European 178
Siberian 178
goat's-beard *74,* 101
goat's-rue 130
gold-dust 94
goldenrod *89,* 173, 174
swamp 173
wreath 173
goldenstar 115
golden tuft *72,* 94
goldilocks 102
gooseberry
Cape 162
groundsel
golden *83,* 147
Gunnera 132
manicata 132
gypsophila 40
Gypsophila 132
bodgeri see *Gypsophila paniculata*
'Compacta Plena' 132, 133
manginii 132
monstrosa 132
paniculata 24, 132
'Bristol Fairy' 132
'Compacta Plena' 132, 133
'Flamingo' 133
'Perfecta' 133
'Pink Star' 133
'Rosenschlier' *81,* 133

Haplopappus 133
coronopifolius see *Haplopappus*
glutinosus 133
glutinosus 133
lyalli 133
harebell 110
hawkweed 137
shaggy 137
hedge-nettle
woolly 174
helenium 45
Helenium 133
autumnale 133
'Allgold' 133
'Bruno' 133
'Butterpat' 133
'Chipperfield Orange' 133
'Coppelia' 133
'Copper Spray' 133
'Crimson Beauty' 133
'Fountain' *81,* 133

'Golden Fox' 133
'Golden Youth' 133
'Goldlackzwerg' 133
'Gypsy' 133
'July Sun' 133
'Karneol' 133
'Madame Canivet' 133
'Mahogany' 133
'Moerheim Beauty' 133
'Red Indian' 134
'Riverton Beauty' 134
'Riverton Gem' *81,* 134
'Spatrot' 134
'The Bishop' 134
'Waltraud' 134
'Wyndley' 134
hoopesii 134
helianthus 45

Helianthus 37, 134
atrorubens 134
decapetalus 134
'Badirector Linne' *81,* 134
'Capenock Star' 134
'Capenock Star Supreme' 134
'Flore Pleno' 134
'Loddon Gold' 134
'Morning Sun' 134
'Multiflorus' *81,* 134
'Soleil d'Or' 134
'Triomphe de Grand' 134
laetiflorus 134
'Mrs. Mellish' 134
orgyalis see *Helianthus salicifolius*
134
salicifolius 134
scaberrimus see *Helianthus*
laetiflorus 134
sparsifolius see *Helianthus*
atrorubens 134

Heliopsis 37, 134
helianthoides 134
scabra 134
'Ballerina' 134
'Gigantea' 134
'Incomparabilis' 134
'Golden Plume' 134
'Golden Rays' 135
'Goldgefieder' 134
'Gold Greenheart' 135
'Light of Loddon' 135
'Patula' 135
'Summer Sun' 135
hellebore 135
black *81,* 135
Corsican 135
green 135
stinking 135

Helleborus 135
  corsicus 135
  foetidus 135
  niger 36, *81,* 135
    'Altifolius' 135
    'Keesen Variety' 135
    'Potter's Wheel' 135
  orientalis 135
    'Albion Otto' 135
  viridis 135
Hemerocallis 48, 135
    'Alan' 17
    'Bess Ross' 17
    'Blithe Spirit' 17
    'Burning Daylight' 17
    'Capri' 17
    'Cartwheels' 17
    'Cathedral Towers' 17
    'Cheery Pink' 17
    'Cibola' 17
    'Colonel Joe' 17
    'Colonial Dame' 17
    'Crimson Glory' 17
    'Crimson Pirate' 17
    'Dorcas' 17
    'Evelyn Claar' 17
    'Frances Fay' 17
    'Gladys Kendall' 17
    'Golden Hours' 17
    'Jewel Russel' 17
    'Luxury Lace' 17
    'Lyric' 17
    'Mabel Fuller' 17
    'May Hall' 17
    'Midwest Star' 17
    'Moon Ruffles' 17
    'Nantahalla' 17
    'Nashville' 17
    'Neyron Rose' 17
    'Painted Lady' 17
    'Potentate' 17
    'Salmon Sheen' 17
    'Springside' 17
    'Swansdown' 17
    'Taylor Russell' 17
    'Theron' 17
    'War Eagle' 17
  flava see *Hemerocallis*
    lilioasphodelus 135
  fulva 135
    var. *kwanso* 135
  gracilis see *Hemerocallis minor* 135
  lilioasphodelus 135
  minor 135
  multiflora 135
hen-and-chickens 172
Hepatica 37, 135
  acutiloba 136
    'Alba Plena' 136

    'Candidissima' 136
    'Flore Plena' 136
    'Purpurea' 136
Heracleum 136
  mantegazzianum 136
Hesperis 136
  matronalis 136
    'Bloom's' 136
    'Bressingham Blaze' 136
    'Carmen' 136
    'Damask' 136
    'Firebird' 136
    'Freedom' 136
    'Gaiety' *81,* 136
    'Garnet' 136
    'Hartsman' 136
    'Jubilee' 136
    'Mary Rose' 136
    'Oakington Jewel' 136
    'Pearl Drops' 136
    'Pluie de Feu' 136
    'Red Spangles' 136
    'Rhapsody' 136
    'Scarlet Sentinel' 136
    'Scintillation' 136
    'Snowflakes' 136
    'Sparkler' 136
    'Splendour' 136
    'Sunset' 136
Heuchera 47, 136
  angulosa 136
  brizoides × *Tiarella cordifolia* 136
  × *Heucherella tiarelloides* 136
  sanguinea 136
  transilvanica 137
  triloba 137
Hibiscus 137
  moscheutos 137
    'Annie Hemming' 137
    'Bessie Ross' 137
    'Poinsettia' 137
    'Satan' 137
    'Southern Belle' *82,* 137
    'Super Rose' *82,* 137
    'The Clown' 137
  palustris see *Hibiscus moscheutos* 137
Hieracium 137
  bombycinum 137
  villosum 137
holly
  sea *80,* 127
hollyhock 29, 30, 45, 48, *70,* 94, 137
honesty 150
  perennial 150
Hosta *82,* 137
  albo-marginata 137

  caerulea see *Hosta ventricosa* 137, 138
  crispula 138
  decorata 138
    'Marginata Alba' 138
  elata 138
  fortunei 138
    'Gigantea' see *Hosta elata* 138
    'Marginata Alba' see *Hosta crispula* 138
  glauca see *Hosta sieboldiana* 138
  lancifolia *82,* 138
    var. *albo-marginata* see *Hosta albo-marginata* 137, 138
    'Fortis' see *Hosta undulata* 'Irromena' 138
    'Tardiflora' see *Hosta tardiflora* 138
  media picta see *Hosta undulata* 138
  plantaginea 129, 138
    'Grandiflora' 138
  sieboldiana 138
  tardiflora 138
  undulata 138
    'Irromena' 138
    'Univittata' 138
  variegata see *Hosta undulata* 138
  ventricosa 138
houseleek 172

Iberis 138
  corifolia see *Iberis saxatilis*
    'Corifolia' 138, 139
  gibraltarica 138
  saxatilus 139
    'Corifolia' 139
  sempervirens 139
    'Christmas Snow' 139
    'Little Gem' 139
    'Purity' 139
    'Snowflake' 139
Incarvillea 139
  compacta 139
  delavayi *82,* 139
    'Bees Pink' 139
  grandiflora 139
    'Brevipes' 139
indigo
  blue false 105
  false 105
inula
  Caucasian 140
Inula 139
  ensifolia 139
  grandiflora 139
  grandulosa see *Inula orientalis* 139, 140
  helenium 139

hookeri 139
magnifica 139
Oculus-Christii 139
orientalis 139, 140
   'Laciniata' 140
   var. *superba* 140
royleana 140
iris 19, 36, 40, 45, *67, 82,* 140
  crested 144
  dwarf bearded 140, 141
  intermediate bearded 141
    'Alien' 142
    'Allah' 142
    'Cloud Fluff' 142
    'Easter Bunny' 142
    'First Lilac' 142
    'Kiss Me Kate' 142
    'Lilli Hoog' 142
    'Lillipinkput' 142
    'Pink Debut' 142
    'Pink Fancy' 142
    'Sugar' 142
    'Yellow Dresden' 142
    'Zwanenburg' 142

  Japanese *82,* 142
    'Blue Coat' 142
    'Diamond Night' 142
    'Imperial Palace' 142
    'Imperial Robe' 142
    'Ivory Glow' 142
    'Ocean Mist' 142
    'Pink Frost' *82,* 142
    'Red Titan' 142
    'Rose Tower' 142
    'Royal Pageant' 143
    'Snowy Hills' 143
  other beardless 144
  roof, of Japan 144
  Siberian *82,*143
    'Blue Herald' 143
    'Cool Spring' 143
    'Eric the Red' 143
    'Gatineau' 143
    'Mandy Morse' *82,* 143
    'Mountain Lake' 143
    'Royal Ensign' 143
    'Snow Crest' *82,* 143
    'Tunkhannock' 143
    'Tycoon' 143
    'White Dove' 143
    'White Swirl' 143
  spuria 143
  tall bearded 19, 140
  vesper 144
  yellowband 144
  yellow flag 144
  Yunnan 144
Iris *67, 82,* 140

arenaria 141
aurea 144
bulleyana 144
chamaeiris 141
cristata 144
delavayi 144
dichotoma 144
douglasiana 144
  'Margot Holmes' 144
fimbriata see *Iris japonica* 144
flavescens 144
flavissima 'Arenaria' see *Iris arenaria* 141
forrestii 144
graminea 143, 144
innominata *82,* 144
japonica 144
lacustris 144
mellita 141
minuta 144
ochroleuca 143, 144
  'Queen Victoria' 144
orientalis 144
  'Nana' 144
pseudacorus 144
pumila 141
  'Ablaze' 141
  'Angel Eyes' 141
  'April Morn' 141
  'Atomic Blue' 141
  'Baria' 141
  'Blue Mascot' 141
  'Brassie' 141
  'Bright White' 141
  'Butterball' 141
  'Cherry Spot' 141
  'Cup and Saucer' 141
  'Dale Dennis' 141
  'Dark Fairy' 141
  'Dear Love' 141
  'Easter Holiday' 141
  'Flaxen' 141
  'Golden Fair' 141
  'Green Spot' 141
  'Honey Bear' 141
  'Knotty Pine' 141
  'Lemon Flare' 140, 141
  'Pam' 140, 141
  'Promise' 141
  'Red Gem' 141
  'Sparkling Eyes' 141
  'Veri-Gay' 141
sanguinea see *Iris orientalis* 144
setosa 144
tectorum 144
versicolor 144
wilsonii 144
Jacob's ladder 163

creeping 163
  dwarf 163
Jenny
  creeping *84,* 152
Joe-Pye weed 128
Jupiter's distaff 169

Kansas feather 147
king
  silver 100
Kirengsshoma 145
  palmata 145
knapweed 111
  globe 111
Kniphofia 145, 178
  aloides see *Kniphofia uvaria* 145
  caulescens 145
  foliosa 145
  galpinii 145
  tubergeni 145
  uvaria 145
    'Alcazar' 145
    'Bees' Lemon' 145
    'Bees' Sunset' 145
    'Buttercup' 145
    'Coral Sea' 145
    'Earliest of All' 145
    'Maid of Orleans' 145
    'Mount Etna' 145
    'Royal Standard' 145
    'Springtime' 145
    'Summer Sunshine' 145
    'White Fairy' 145
knotweed 164
  alpine 164
  Himalayan 164
  Japanese 164

ladder
  creeping Jacob's 163
  dwarf Jacob's 163
  Jacob's 163
lamb's-ear 174
Lamium 145
  galeobdolon 'Florentium' 145
  maculatum 145
    'Aureum' 145
  orvala 146
  veronicaefolium 146
lantern
  Chinese 162
larkspur
  perennial 119

Lathyrus 146
  grandiflora 146
  latifolius 146

'Pink Beauty' 146
'Roseus' 146
'Snow Queen' 146
'White Pearl' 146
rotundifolia 146
Lavandula 146
officinalis see Lavandula spica 146
spica 146
'Alba' 146
'Gruppenhall' 146
'Hidcote Blue' 146
'Jean Davies' 146
'Munstead' 146
'Twinkle Purple' 146
vera see Lavandula spica 146
lavatera
tree 146
Lavatera 146
cachemirica 146
olbia 146
'Rosea' 146
thuringiaca 83, 146
lavender
common 146
common sea- 148
sea- 148
leadwort 111
leopardplant 148
leopard's-bane 123
great 123
liatris 45
Liatris 147
callilepis see Liatris spicata 147
graminifolia 147
punctata 147
pycnostachya 147
scariosa 147
'Alba' 147
'Nana' 147
'September Glory' 147
'Silver Tips' 83, 147
'White Spire' 147
spicata 147
'Alba' 147
'Kobold' 147
Ligularia 147
clivorum see Ligularia dentata 147, 173
dentata 147
'Desdemona' 83, 147
'Gregynog Gold' 147
'Othello' 147
× hessei 148
hodgsonii 148
intermedia 148
japonica 148, 173
stenocephala 'Globosa' 148

tussilaginea 148
'Argentea' 148
'Aureo Maculata' 148
veitchiana 148
wilsoniana 83, 148
Lilium 48
convallarium see Convallaria majalis 117
lily 36
blackberry- 106
day- 135
dwarf yellow 135
lemon 135
tawny 135
foxtail 126
Madonna 41, 43
-of-the-valley 36, 44, 117
star-flowered 173
plantain- 137, 138
blue 138
fragrant 138
lance-leaved 82, 138
wavey-leaved 138
torch- 145
Limonium 148
bellidifolium 148
cosyrense 148
dumosum see Limonium tataricum 148
'Angustifolium' 148
elatum 148
eximium 148
'Album' 148
felicularis 148
gmelini 148
latifolium 148
'Blue Cloud' 148
'Chilwell Beauty' 148
'Elegance' 148
'Grandiflora' 148
'Grittleton Variety' 148
'Violetta' 148
minimum 148
tataricum 148
'Angustifolium' 148
'Nanum' 148
Linaria 149
dalmatica 149
purpurea 149
'Canon J. Went' 149
triornithophora 149
Lindelofia 149
longiflora 149
Linum 149
alpinum 149
austriacum 149
'Album' 149
flavum 83, 149
'Cloth of Gold' 149

'Compactum' 149
narbonense 149
'Heavenly Blue' 149
'June Perfield' 149
'Peto's Variety' 149
'Six Hill's Variety' 149
perenne 149
'Album' 149
var. lewisii 149
salsoloides 149
'Alpinum' 149
liriope
grape-hyacinth 149
Liriope 149
muscari 149
spicata 149
'Majestic' 149
Lithospermum 150
canescens 150
intermedium 150
lobelia cultivars 150
'Gloire Ste. Anne' 150
'Hustoman' 150
'Purple Emperor' 150
'Queen Victoria' 150
'The Bishop' 150
Lobelia 150
cardinalis 150
fulgens 150
siphilitica 150
× speciosa 83, 150
London pride 171
loosestrife 84, 152
dotted-leaved 152
gooseneck 152
purple 84, 152
wand 84, 152
yellow 152
Lunaria 150
annua 150
biennis see Lunaria annua 150
rediviva 150
lungwort 167
lupine 36, 44, 150
false 90, 176
Russell strains 83, 151
Lupinus 36, 150
polyphyllus 150
'Betty Astel' 151
'Blue Jacket' 151
'Blushing Bride' 151
'Celandine' 151
'Charmaine' 151
'Commando' 151
'Elsie Waters' 151
'Fireglow' 151
'Flaming June' 151

204

'Fred Yale' 151
'George Russell' 151
'Guardsmen' 151
'Happy Days' 151
'Josephine' 151
'Lady Fayne' 151
'Lilac Time' 151
'Monkgate' 151
'Mrs. Micklethwaite' 151
'Mrs. Noel Terry' 151
'Patricia of York' 151
'Radiant' 151
'Rapture' 151
'Susan of York' 151
'Sweetheart' 151
'Thundercloud' 151
Lychnis 151
  alpina 151, 183
  × *arkwrightii* 151
  chalcedonica 84, 151
    'Alba' 151
    'Alba Plena' 151
    'Rubra Plena' 151
    'Salmonea' 151
  coronaria 93, 151
    'Abbotswood Rose' 151
    'Alba' 151
    'Atrosanguinea' 151
    'Flore Pleno' 151
  flos-jovis 151
    'Hort's Variety' 152
  × *haageana* 84, 152
  viscaria 152
    'Splendens' 152
    'Splendens Flore-Pleno' 84, 152
Lysimachia 152
  brachystachys 152
  ciliata 84, 152, 175
  clethroides 152
  ephemerum 152
  fortunei 152
  nummularia 152
    'Aurea' 152
  punctata 152
  vulgaris 152
Lythrum 152
  salicaria 152
    'Brightness' 152
    'Lady Sackville' 152
    'Morden's Gleam' 84, 152
    'Robert' 152
    'The Beacon' 152
  virgatum 152
    'Dropmore Purple' 152
    'Morden Pink' 84, 152
    'Rose Queen' 152
    'The Rocket' 152

Macleaya 153
  cordata 106, 153
  microcarpa 153
    'Coral Plume' 153
madwort 72, 94
  rock 94
mallow 45
  Cashmerian 146
  musk 153
  prairie 89, 173
  tree- 83, 146
Maltese cross 84, 151
Malva 153
  alcea 153
    'Fastigiata' 153
  moschata 153
    'Alba' 153
Malvastrum 153
  coccineum see *Sphaeralcea coccinea*
    153, 174
marguerite
  golden 96, 97
marigold
  common marsh 108
  marsh 107
Mary
  blue-eyed 157
masterwort 105
mayapple
  American 163
mayflower
  Himalayan 163
meadow rue 89, 175
  columbine 89, 175
  dusty 176
  white 175
  yellow 176
  Yunnan 176
meadowsweet 129
Meconopsis 153
  baileyii see *Meconopsis betonicifolia*
    153
  betonicifolia 153
    'Pratense' 153
  cambrica 153
    'Flore Plena' 153
  quintuplinervia 154
Mertensia 123, 154
  ciliata 154
  echioides 154
  lanceolata 154
  longiflora 154
  paniculata 154
  primuloides see *Mertensia echioides*
    154
  sibirica 154

virginica 154
milfoil 92
milkweed 101
Mimulus 154
  × *burnetii* 154
  cardinalis 154
    'Cerise Queen' 154
    'Grandiflorus' 154
    'Rose Queen' 154
  cupreus 154
    'Bee's Dazzler' 154
    'Bonfire' 154
    'Chelsea Pensioner' 154
    'Fireflame' 154
    'Leopard' 154
    'Plymtre' 154
    'Queen's Prize' 154
    'Red Emperor' 154
    'Whitecroft Scarlet' 154
  langsdorfii see *Mimulus luteus*
    'Guttatus' 154, 155
  lewisii 154
    'Alba' 154
  luteus 154, 155
    'Alpinus' 155
    'A. T. Johnson' 155
    'Duplex' 155
    'Guttatus' 154, 155
    'Rivularis' 155
    'Youngeanus' 155
  moschatus 155
  primuloides 155
  ringens 155
mist
  lavender 176
mistflower 128
Moltkia 155
  × *intermedia* 155
monarda 13, 45
Monarda 155
  didyma 155
  fistulosa 155

    'Adam' 155
    'Blue Stocking' 155
    'Cambridge Scarlet' 155
    'Croftway Pink' 155
    'Granite Pink' 155
    'Granite Purple' 155
    'Melissa' 155
    'Pillar Box' 155
    'Prairie Glow' 155
    'Prairie Night' 155
    'Salmon Queen' 155
    'Snow Maiden' 155
    'Souris' 155
    'Sunset' 155
moneywort 84, 152

golden 152
monkeyflower 154
  red 154
  scarlet 154
monk's-hood 93
  common *72, 93*
Morina 155
  longifolia 155
mosquito bills 122
mother-of-thyme *90,* 176
mourning widow 131
mugwort
  white 24, 100
mullein 30, 179
  Cretan 111
  dark 180
  nettle-leaved 180
  purple 180
musk
  monkey 155
muskflower 155
Myosotis 107, 156
  alpestris 28, 156
    'Pink Beauty' 156
    'Ruth Fisher' 156
  palustris see *Myosotis scorpioides* 156
  scorpioides 28, 156
    var. *sempervirens* 156
  sylvatica 28, 156

navelseed 157
navelwort 157
nepeta 24
Nepeta 156
  cataria 156
  × *faassenii* 156
  grandiflora 156
  hemsleyana see *Dracocephalum hemsleyanum* 156
  mussinii see *Nepeta* × *faassenii* 156
  tartarica 156
    'Blue Beauty' 156
    'Six Hills Giant' 156
  ucranica see *Nepeta grandiflora* 156
  wilsonii see *Dracocephalum wilsonii* 124, 156

nettle
  dead- 145
  giant dead- 146
  spotted dead- 145
  woolly hedge- 174

obedientplant *87,* 162
oenothera 45
Oenothera 156

caespitosa 156
  'Eximea' 156
fruticosa var. *youngii* see *Oenothera tetragona* 156, 157
missouriensis 156
speciosa 156
  var. *childsii* 157
tetragona 157
  var. *fraseri* see *Oenothera tetragona* ssp. *glauca* 157
  ssp. *glauca* 157
    'Highlight' 157
    'Illumination' 157
    'Riparia' 157
    'W. Cuthbertson' 157
    'Yellow River' 157

old woman 101
Omphalodes 157
  cappadocica 157
    'Anthea Bloom' 157
  verna 157
    'Alba' 157
Ostrowskia 157
  magnifica 157
Oswego tea 155
oxeye 107
  heart-leaved 107
  willowleaf 107
oxlip 166
Ozark sundrops 156

Pachysandra 157
  procumbens 157
  terminalis 157
    'Variegata' 157
Paeonia 46, 69, 158
  anomala 158
  emodi 158
  lactiflora 22, 158
    'America' 22
    'Charles Haines' 22
    'Chocolate Soldier' 22
    'Claire de Lune' 22
    'Convoy' 22
    'Crusader' 22
    'Diana Parks' 22
    'Festiva Maxima' 22
    'Hansina Brand' 22
    'Hattie Lafuze' 22
    'Illini Belle' 22
    'Isani Gidui' 22
    'Joyce Ellen' 22
    'Jules Elie' 22
    'Kansas' 22
    'Laura Magnuson' 22

'Mrs. G. E. Hemerik' 22
'Mrs. Livingstone Farrand' 22
'Mrs. Wilder Bancroft' 22
'Oriental Gold' 22
'Philip Rivoire' 22
'President Lincoln' 22
'Red Charm' 22
'Sara Bernhardt' 22
'Sea Shell' 22
'Snow Mountain' 22
'The Bride' 22
'White Cap' 22
mlokosewitschi 158
officinalis 158
  'Albo Plena' 158
  'Rosea Plena' 158
  'Rubra Plena' 158
russi 158
× *smouthii* 158
tenuifolia 158
  'Plena' 158
veitchii 158
  'Woodwardii' 158
wittmanniana 158
pansy *13,* 31, 43, *71*
  'Apricot Queen' 32
  'Coronation Gold' 32
  'Gipsy Queen' 32
  'Goldie' 32, *71*
  'Orange King' 32
Papaver 46, 158
  nudicaule 30, 158
    Akabana Scarlet strain 30
    Coonara strain 30
    Kelmscott strain 30
    Pencilstalk Giant strain 30
    Red Cardinal strain 30
    Sutton's Lemon Yellow strain 30
    Sutton's Orange strain 30
    Sutton's Pink Shades strain 30
    Tangerine strain 30
  orientale 158
    'Barr's White' 21
    'Beauty of Livermore' 21
    'Carousel' 21
    'Cerise Beauty' 21
    'Coral Cup' 21
    'Crimson Pompon' 21
    'Curtis' Giant Flame' 21
    'Field Marshall van der Glotz' 21
    'Helen Elizabeth' 21
    'Juliette' 21, 36
    'Lavender Glory' 21
    'Mary Finan' 21
    'Mrs. Perry' 21
    'Olympia' 21
    'Pink Lassie' 21
    'Pinnacle' 21
    'Salmon Glow' 21

206

'Salome' 21
'Springtime' 21
'Valencia' 21
parsnip
giant 136
pasqueflower 96
pea
everlasting 146
perennial 146
Peltiphyllum 158
peltatum 158
penstemon
blue 159
shell-leaf 159
Utah 160
Penstemon 159
acuminatus 159
alpinus 159
flathead late hybrids 159
barbatus 159
'Coccineus' 159
'Rosy Elf' 159
'Torreyi' 159
cobaea 159
'Ozark' 159
crandallii 159
'Glabescens' 159
diffusus see *Penstemon serrulatus*
159, 160
digitalis 111, 159
glaber 159
'Roseus' 159
× *gloxinoides* 159
'Alice Hindly' 159
'Cherry Red' 159
'Firebird' 159
'Garnet' 159
'Majestic' 159
'Newberry' 159
'Ruby' 159
'White Bedder' 159
grandiflorus 159
Fate hybrid strain 159
Seeba hybrid strain 159
heterophyllus 159
laevigatus var. *digitalis* see *Penste-*
mon digitalis 159
murrayanus 160
ovatus 160
palmeri 160
serrulatus 160
unilateralis 160
utahensis 160
unaffiliated cultivars 160
'Prairie Dusk' 160
'Prairie Fire' 160
Six Hills hybrid 160

peony 22, 23, 40, 41, 42,
43, 44, 45, *69,* 158
common red 158
fernleaf 158
lemon 158
Russo's 158
periwinkle 181, 182
Phlomis 160
cashmiriana 160
fruticosa 160
russelliana 160
samia 160
tuberosa 160
viscosa see *Phlomis russelliana* 160
phlox 24, 25, 40, 42, 44
creeping 161
hairy 160
Hood's 161
meadow 161
mountain 161
periwinkle 160
summer 161
trailing 161
wild blue 160
Phlox 24, 25, 38, 160
adsurgens 160
amoena 160
'Rosea' 160
'Variegata' 160
carolina see *Phlox glaberrima* 160
divaricata 122, 160
var. *canadensis* 160
var. *laphamii* 160
glaberrima 160
'Buckeye' 160
hoodii 161
maculata 161
nivalis 161
'Camla' 161
'Camla Alba' 161
'Dixie Brilliant' 161
'Gladwynne' 161
'Sylvestris' 161
ovata 161
var. *pulchra* 161
paniculata 24, 36, 38, 44, 161
'Annie Laurie' 25
'Barnell' 24
'Brigadier' 25
'Bruno von Zeppelin' *25*
'B. Symons Jeune' 24
'Caroline van den Berg' 24
'Daily Sketch' 36
'Dodo Hanbury Forbes' 25
'Elizabeth Arden' 24
'Fairy's Petticoat' *25*
'Firefly' 25
'Flamboyant' 24
'Gaiety' *25*

'Gleneagle Glory' 24
'Gnome' 24
'Graf Zeppelin' 25
'Hampton Court' 25
'Harewood' 25
'Henderson's Late White' 25
'Iceberg' 25
'Joan' 25
'July Glow' *25*
'Leo Schlageter' 25
'Lord Lambourne' 25
'Marlborough' 24
'Miles Copijn' 25
'Olive Symons Jeune' 25
'Pastorale' 25
'San Antonio' 25
'Spatrot' 25
'Stirling' 24
'Tenor' 25
'Vintage Wine' 24
'Windsor' 25
pilosa 161
stolonifera 161
'Blue Ridge' 161
'Lavender Lady' 161
subulata 37, 38, 161
'Alexander's Beauty' 161
'Alexander's Surprise' 161
'Autumn Rose' *86,* 161
'Betty' 161
'Blue Hills' 161
'Bonita' 161
'Brightness' *86,* 161
'Brilliant' 161
'Coreale' 161
'Eventide' 161
'Exquisite' 161
'Fairy' 161
'G. F. Wilson' 161
'Samson' 161
'Scarlet Flame' 161
'Sensation' 161
'Starglow' 161
'Star of Heaven' *86,* 162
'Temiscaming' 162
'The Bride' 162
'Vivid' 162
Phygelius 162
capensis 162
Physalis 162
alkekengi 162
'Gigantea' 162
'Monstrosa' 162
'Nana' 162
'Orbiculari' 162
franchetii see *Physalis alkekengi*
162
physic
Indian 132

Physotegia   101, 162
   virginiana   87,   162
     'Alba'   162
     'Bouquet Rose'   87, 162
     'Grandiflora'   162
     'Nana'   162
     'Rosy Spire'   162
     'Speciosa'   162
     'Speciosa Rosea'   162
     'Speciosa Rubra'   162
     'Summer Glow'   162
     'Summer Snow'   24, 162
     'Summer Spire'   162
     'Vivid'   162
pigeonberry   118
pincushionflower   171
pink   120
   Cheddar   120
   cottage   78, 120
   grass   78, 120
   maiden   78, 120
   sea   99
plantain-lily   137, 138
   blue   138
   fragrant   138
   lance-leaved   138
   wavey-leaved   138
Platycodon   162
   grandiflorum   162
     'Album'   162
     'Bristol Bell'   162
     'Bristol Bluebird'   162
     'Bristol Blush'   162
     'Bristol Bride'   162
     'Clonmere Rose'   162
     var. *japonicum*   163
     var. *mariesii*   87, 163
     'Mother of Pearl'   163
     'New Alpine'   163
plume-poppy   85, 153
Podophyllum   163
   emodi   163
   peltatum see *Podophyllum*
     emodi   163
poinsettia   128
poker
   red-hot-   145

Polemonium   163
   boreale   163
     'Superbum'   163
   caeruleum   163
     'Blue Pearl'   163
     'Lacteum'   163
     'Sapphire'   163
   carneum   163
   confertum see *Polemonium*
     viscosum   163

   lanatum see *Polemonium*
     boreale   163
   pauciflorum   163
   pulcherrimum   163
   reptans   163
   richardsonii see *Polemonium*
     boreale   163
   viscosum   163
Polygonatum   163
   commutatum   163
   giganteum see *Polygonatum*
     commutatum   163
   multiflorum   36, 87, 163
     'Flore Pleno'   163
     'Striatum'   163
Polygonum   164
   affine   164
     'Darjeeling Red'   164
     'Lowndes Variety'   164
   alpinum   164
   amplexicaule   164
     'Album'   164
     'Atrosanguineum'   164
     'Firetail'   164
   bistorta 'Superbum'   164
   campanulatum   164
     'Roseum'   164
     var. *lichiangensis*   164
   cuspidatum   164
     'Compactum'   87, 164
     'Compactum Femina'   164
   sachalinense   164
   sieboldii see *Polygonum*
     cuspidatum   164
   vaccinifolium   164
poppy
   harebell   154
   Himalayan blue   85, 153
   horn   132
   Iceland   30, 158
   Matilija   168
   Oriental   21, 36, 41, 68, 158
   plume-   85, 153
   tree   168
   Welsh   153
Potentilla   87, 164
   argyrophylla   164
     'Atrosanguinea' see *Potentilla*
     atrosanguinea   165
   atrosanguinea   165
   concolor   165
   emarginata   165
   fragiformis see *Potentilla*
     emarginata   165
   × *hopwoodiana*   165
   nepalensis   165
     'Master Floris'   165
     'Miss Willmott' see *Potentilla*
     willmottiae   165

     'Roxana'   165
   recta   165
     'Sulphurea'   165
     'Warrensii'   165
   willmottiae   165
   unclassified cultivars   165
     'Arc en Ciel'   165
     'California'   165
     'Etna'   165
     'Flambeau'   165
     'Flamenco'   165
     'Gibson's Scarlet'   165
     'Glory of Nancy'   165
     'Jupiter'   165
     'M. Rouillard'   87, 165
     'Star of the North'   165
     'White Beauty'   165
     'Wm. Rollison'   165
Poterium see *Sanguisorba*   170
   canadensis see *Sanguisorba*
     canadensis   165, 170
   obtusa see *Sanguisorba*
     obtusa   165, 170
prairie bluebell   154
prairie crocus   95
prairie mallow   173
primrose   123, 165
   Caucasian   166
   common   88, 167
   evening-   86, 156
   polyanthus   167
   showy evening-   156
   Siebold   167
   silverdust   167
   Tibetan   165
   wind-   156

Primula   37, 165
   acaulis see *Primula vulgaris*   165, 167
   alpicola   165
     'Alba'   165
     'Violacea'   165
   anisodora   165
   aurantiaca   166
   auricula   166
     'Blue Velvet'   166
     'Celtic King'   166
     'Dusty Miller'   166
     'Jean Walker'   166
     'Red Dusty Miller'   166
     'W. A. Cook'   166
   beesiana   166
   bulleyana   166
   burmanica   166
   chionantha   166
   cockburniana   166
   cortusoides   166
   denticulata   166

208

var. *cachemiriana* 166
  'Alba' *88,* 166
  'Bengal Rose' 166
  'Pritchard's Ruby' 166
  'Purple Beauty' 166
  'Red Emperor' 166
  'Stormonth's Red' 166
elatior 166
florindae 166
helodoxa 166
japonica 166
  'Miller's Crimson' 166
  'Postford White' 166
  'Rose Dubarry' 166
× *juliae* 166
  'Alba' 167
  'Betty' 167
  'Dorothy' 167
  'E. R. James' 167
  'Gloria' 167
  'Gold Jewel' *88,* 167
  'Jewel' 167
  'Kinlough Beauty' 167
  'Mrs. McGillivray' 167
  'Old Port' 167
  'Our Pat' 167
  'Pam' 167
  'Snow Cushion' 167
  'Snow White' 167
  'Wanda' 167
× *polyantha* *88,* 167
  Pacific Giant strain *88,* 167
pulverulenta 167
  'Arleen Aroon' 167
  'Lady Thursby' 167
  'Mrs. R. V. Berkeley' 167
  'Red Hugh' 167
sieboldii 167
veris 167
vulgaris *88,* 165, 167
puccoon 150

Pulmonaria 167
angustifolia 167
  'Johnston's Blue' 167
  'Mawson's Variety' 167
  'Munstead Blue' 167
  'Salmon Glory' 167
montana 167
officinalis 167
rubra see *Pulmonaria montana* 167
saccharata *88,* 167
  'Bowles Red' 167
  'Mrs. Moon' 167
  'Pink Dawn' 167
Pulsatilla see *Anemone* 95, 168
pyrethrum 40, *76,* 112

queen-of-the-meadow 129

queen-of-the-prairie 129

ragwort
  giant 148
Ranunculus 168
aconitifolius 168
  'Flore Pleno' 168
  'Plantanifolius' 168
acris 'Flore-pleno' 168
gramineus 168
repens 'Pleniflorus' 168
red-hot-poker 145
Rheum 168
australis 168
palmatum 168
  'Tanguticum' 168
rhubarb 168
  Himalaya 168
  sorrel 168
rock-cress *73,* 98
  false 105
  wall 99
rocket
  sweet 136
rockfoil
  giant *75,* 106

Rodgersia 168
pinnata 168
  'Alba' 168
  'Elegans' 168
  'Rubra' 168
podophylla 168
tabularis 168

Romneya 168
coulteri 168
trichocalyx 168

Roscoea 168
cautleoides 169
humeana 169
rose
  Christmas 36, *81,* 135
  Lenten 135

Rudbeckia 37, 38, 47
fulgida var. *speciosa* 169
  'Goldquelle' 169
  'Goldsturm' *88,* 169
laciniata 169
  'Golden Glow' 169
  'Herbstonne' 169
newmannii see *Rudbeckia fulgida*
  var. *speciosa* 169
nitida 169
subtomentosa 169
triloba *88,* 169

rue
  columbine meadow *89,* 175
  dusty meadow 176
  meadow *89,* 175
  white meadow 175
  yellow meadow 176
  Yunnan meadow 176

sage 169
  azure 169
  Bethlehem *88,* 167
  gentian 169
  Jerusalem 160
  meadow 170
  nodding 169
  Oriental 170
Salvia 169
azurea 169
  'Angustifolia' 169
  'Grandiflora' 169, 170
beckeri 169
glutinosa 169
haematodes see *Salvia pratensis*
  169, 170
jurisicii 169
nemerosa see *Salvia* × *superba*
  169, 170
nutans 169
patens 169
  'Alba' 170
  'Cambridge Blue' 170
pitcheri see *Salvia azurea*
  'Grandiflora' 169, 170
pratensis 169, 170
  'Alba' 170
  'Rosea' 170
  'Tenori' 170
  'Variegata' 170
sclarea 169, 170
  'Turkestanica' 170
× *superba* 170
  'East Friesland' 170
  'Lubeca' 170
  'May Night' 170
  'Purple Glory' 170
uliginosa 'Bog Sage' 170
virgata 170
  'Nemerosa' see *Salvia* × *superba*
  170

Sanguinaria 170
canadensis 36, 170
  'Multiplex' 170
Sanguisorba 170
canadensis 165, 170
obtusa *89,* 165, 170
Saponaria 170
× *boisseri* 170
caespitosa 170

lutea 170
ocymoides 170
  'Splendens' 171
officinalis 171
  'Albo Plena' 171
  'Roseo-Plena' 171
  'Rubra Plena' 171
Saxifraga 37, 171
 aizoon see *Saxifraga paniculata* 171
 paniculata 171
 umbrosa 171
  'Aurea Punctata' 171
  'Primuloides' 171
  'Primuloides Rubra' 171
Scabiosa 37, 171
 caucasica 171
  'All Blue' 171
  'Bressingham White' 171
  'Challenger' 171
  'Clive Greaves' 171
  'Mrs. Willmott' 171
  'Penhill Blue' 171
 fischeri 171
 graminifolia 171
 lucida 171
 ochroleuca 171
scabious 171
 giant *76*, 111
 grassleaf 171
sea holly *80*, 127
sea-lavender 148
 common 148
seal
 Solomon's 36, 87, 163
sea pink 99
sedum
 purple-leaved 172
Sedum 37, 171
 acre 171
 ewersii 171
 kamtschaticum 172
  'Variegatum' 172
 maximum 172
  'Atropurpureum' 172
 sieboldii 172
  'Medico-variegatum' 172
 spathulifolium 172
  'Casablanca' 172
  'Major' 172
  'Purpureum' 172
 spectabile 172
  'Album' 172
  'Atropurpureum' 172
  'Autumn Joy' 172
  'Brilliant' *89*, 172
  'Carmen' 172
  'Meteor' 172

  'Variegatum' 172
 spurium 172
  'Album' 172
  'Bronze Carpet' 172
  'Coccineum' 172
  'Dragonsblood' 172
  'Erdblut' 172
  'Green Mantle' 172
  'Roseum' 172
sempervivum 45
Sempervivum 172
 tectorum 172
  'Calcareum' 172
Senecio see *Ligularia* 147, 173
 clivorum see *Ligularia dentata* 147, 173
 japonica see *Ligularia japonica* 148, 173
shootingstar 122
Sidalcea 173
  'Croftway Red' 173
  'Dainty' 173
  'Duchess' 173
  'Elsie Heugh' 173
  'Interlaken' 173
  'Loveliness' 173
  'Mrs. Alderson' 173
  'Mrs. Borrodaile' 173
  'Mrs. Galloway' 173
  'Nimmerdor' 173
  'Oberon' 173
  'Puck' 173
  'Rev. Page Roberts' 173
  'Rose Queen' *89*, 173
  'Sussex Beauty' 173
  'Titania' 173
  'Wensleydale' 173
  'William Smith' 173
silver king 100
Smilacina 173
 racemosa 173
 stellata 173
snakeroot 115
 black 116
sneezeweed *81*, 133
sneezewort 92
snowball 92
snow-in-summer 111
soapwort 170
 rock 170
Solomon's seal 36, *87*, 163
Solidago 173, 174
 brachystachys 173
 caesia 173
 missouriensis 173
 petiolaris 173

 uliginosa 173
 virgaurea 173
  'Cloth of Gold' 173
  'Crown of Rays' 173
  'Golden Falls' 174
  'Golden Gates' *89*, 174
  'Goldenmosa' 174
  'Golden Plume' 174
  'Golden Shower' 174
  'Golden Thumb' 174
  'Golden Wings' *89*, 174
  'Goldstrahl' 174
  'Laurin' 174
  'Leda' 174
  'Ledsham' 174
  'Lemore' 174
  'Leraft' 174
  'Lesale' 174
  'Lesden' 174
  'Leslie' 174
  'Mimosa' 174
  'Peter Pan' 174
  'Praecox' 174
Solidaster 174
 luteus 174
speedwell *91*, 180
 beach *91*, 180
 clump 181
 Hungarian 180
 spiked 181
Sphaeralcea 174
 coccinea 153, 174
spiderwort *90*, 177
spikenhard
 false 173
Spiraea 174
 aruncus see *Aruncus sylvester* 101, 174
 digitata see *Filipendula palmata* 129, 174
 filipendula see *Filipendula vulgaris* 129, 174
 gigantea see *Filipendula camtschatica* 129, 174
 hexapetala see *Filpendula vulgaris* 129, 174
 lobata see *Filipendula rubra* 129, 174
 palmata see *Filipendula purpurea* 129, 174
 rubra see *Filipendula rubra* 129, 174
 ulmaria see *Filipendula ulmaria* 129, 174
 venusta see *Filipendula rubra*
  'Venusta' 129, 174
spurge *80*, 128
 cushion *80*, 128
 Cypress 128
 flowering 128

210

Japanese *86,* 157
Stachys 37, 174
  grandiflora 174
    'Rosea' 174
  lanata see *Stachys olympica* 174, 175
  mancrantha see *Stachys grandiflora* 174
  olympica 174
    'Silver Carpet' 175
starwort
  false 106
Statice 38
Steironema 175
  ciliata see *Lysimachia ciliata* 152, 175
Stokesia 175
  laevis 175
    'Alba' 175
    'Blue Danube' 175
    'Blue Moon' 175
    'Blue Star' 175
    'Lutea' 175
    'Rosea' 175
    'Silvery Moon' 175
    'Superba' 175
    'Wyoming' 175
stonecrop 171
  common 171
  showy *89,* 172
sun
  autumn 169
sundrops
  Ozark 156
sunflower 134
  orange 134
  prairie 134
  purple disk 134
  thinleaf *81,* 134
Susan
  brown-eyed 169
sweet William 29, 32, *71,* 120
Symphyandra 175
  hofmannii *89,* 175
  pendula 175
  wanneri 175
Symphytum 175
  asperum 175
  peregrinum 175

tea
  Oswego 155
Telesonix 175
  jamesii see *Boykinia jamesii* 107, 175
Teucrium 175
  chamaedrys 175
Thalictrum 96, 175

adiantifolium see *Thalictrum minus* 175, 176
aquilegifolium 89, 175
  'Album' 175
  'Atropurpureum' 175
  'Bees Purple' 175
  'Dwarf Purple' 175
  'Purple Cloud' 176
  'Thundercloud' *89,* 176
delavayi 176
diffusiflorum 176
dipterocarpum 176
  'Album' 176
  'Hewitt's Double' 176
flavum 176
  'Glaucum' 176
  'Illuminator' 176
glaucum see *Thalictrum speciosissimum* 176
kiusianum 176
minus 175, 176
rochebrunianum 176
speciosissimum 176
Thermopsis 176
  caroliniana *90,* 176
  montana 176
  rhombifolia 176
thistle
  globe 125
  small globe *79,* 125
thrift *74,* 99
  common *74,* 100
  great 100
thyme
  creeping 176
  garden 177
  mother-of- *90,* 176
  woolly mountain 176
Thymus 37, 176
  lanuginosus 176
  nitidus 176
  serpyllum *90,* 176
    'Album' 176
    'Argenteus' 176
    'Citriodorus' 177
    'Coccineus' 177
  vulgaris 177

Tiarella 177
  cordifolia 177
  polyphylla 177
  trifoliata 177
  wherryi 177
tickseed *78,* 117
  swamp 118
toadflax
  Dalmatian 149
  purple 149

torch-lily 145
Tradescantia 177
  andersoniana 177
    'Alba' 177
    'Blue Stone' *90,* 177
    'Coerulea' 177
    'Flore Pleno' 177
    'Iris' 177
    'Iris Pritchard' *90,* 177
    'James Shatton' 177
    'James C. Weguelin' *90,* 177
    'Leonora' 177
    'Osprey' 177
    'Pauline' 177
    'Purewell Giant' 177
    'Purple Dome' 177
    'Red Cloud' *90,* 177
    'Snow Cap' *90,* 177
    'Valour' 177
  virginiana *90,* 177
tree-mallow *83,* 146
trillium 178
  purple 178
Trillium 177
  cernuum 178
  erectum 178
    'Album' 178
    'Ochroleucum' 178
  grandiflorum 178
  sessile 178
Tritoma see *Kniphofia* 145, 178
Trollius 37, 178
  altaicus 178
  anemonifolius 178
  asiaticus 178, 179
    'Aurantiacus' 178
    'Byrne's Giant' 178
    'Fortunei' 178
  chinensis 178
  europaeus 178
    'Alabaster' 178
    'Bees' Orange' 178
    'Canary Bird' 178
    'Earliest-of-All' 178
    'Empire Day' 178
    'Fireglobe' 178
    'First Lancers' 178
    'Glory of Leiden' 178, 179
    'Golden Monarch' 178
    'Golden Wonder' 178
    'Goldquelle' 178
    'Helios' 178
    'Lemon Queen' 178
    'Newry Giant' 178
    'Orange Princess' 178
    'Pritchard's Giant' 178
    'Princess Juliana' 178
    'Salamander' 178

211

'Superbus' 178
laxus *91,* 178
ledebouri 179
  'Golden Queen' *91,* 179
pumilus 179
  'Stenopetalus' 179
yunnanensis 179
trumpetflower 139
  Chinese *82,* 139
  dwarf 139
turtlehead 111, 112
  red 112

umbrellaplant 158
Uvularia 179
  grandiflora 179

valerian *91,* 179
  red 111
Valeriana 179
  officinalis *91,* 179
veil
  rosy *81,* 133
Verbascum 30, 38, 179
  bombyciferum 179
    'Silver Spire' 179
  chaixii 180
  hapsus 30
  nigrum 180
    'Album' 180
  olympicum 30, 180
  phoeniceum 180
    'C. L. Adams' 30, 180
    'Cotswold Beauty' 30
    'Cotswold Gem' 30
    'Cotswold Queen' 30, 180
    'Gainsborough' 30, 180
    'Golden Bush' 180
    'Harkness Hybrid' 31
    'Hartleyi' 180
    'Miss Willmott' 31
    'Mont Blanc' 180
    'Pink Domino' 31, 180
    'Royal Highland' 180
  thapsiforme 180

Verbena 180
  bonariensis 180
  chamaedryfolia see *Verbena peru-*
    viana 180
  corymbosa 180
  hastata 180
    'Album' 180
  peruviana 180
veronica 40
Veronica 37, 180
  cinerea 180

incana *91,* 180
  'Rosea' 180
latifolia 180, 181
  'Amethystina' 180
  'Crater Lake Blue' 180
  'Royal Blue' 180
  'Shirley Blue' 180
longifolia *91,* 180
  'Albiflora' 181
  'Rosea' 181
  'Subsessilis' 181
pectinata 181
  'Rosea' 181
prostrata 181
  'Loddon Blue' 181
  'Mrs. Holt' 181
  'Silver Queen' 181
  'Spode Blue' 181
repens 181
rupestris see *Veronica prostrata* 181
spicata 181
  'Barcarolle' 181
  'Blue Champion' 181
  'Blue Peter' 181
  'Blue Spire' 181
  'Erica' 181
  'Icicle' 24, 181
  'Minuet' 181
  'Romily Purple' 181
  'Saraband' 181
  'True Blue' 181
  'Well's Variety' 181
  'Wendy' 181
teucrium see *Veronica latifolia*
  180, 181

Veronicastrum 181
  virginicum 181

vervain 180

Vinca 181
  major 181
    'Rubescens' 181
    'Variegata' 181
  minor 181, 182
    'Alba' 182
    'Alboplena' 182
    'Argenteo-variegata' 182
    'Atropurpurea' 182
    'Azurea' 182
    'Bowles Variety' 182
    'Caeruleo-plena' 182
    'Cuprea' 182
    'La Grave' 182
    'Multiplex' 182
    'Punicea' 182
    'Roseoplena' 182
    'Variegata' 182
viola 31, 45, 182
  bedding 32

'Archer Grant' 32
'Better Times' 32
'Bridal Morn' 32
'Eileen' 32
'Enchantress' 32
'Gladys Findlay' 32
'Iden Gem' 32
'Lily' 32
'Maggie Mott' 31, 32
'Mauve Queen' 32
'Mrs. Morrison' 32
'Mt. Spokane' 32
'Queen Elizabeth' 32
'Windsor' 32
giant Trimardeau 32
  'Alpenglow' 32
  'Berna' 31, 32
  'Delft Blue' 32
  'Jet Black' 32
  'Rheingold' 32
  'Ullswater' 31, 32
  'Wine Red' 32
strains 32

Viola 31, 182
  altaica 31
  cornuta 31, 32, 182
    'Amethyst' 182
    'Ardross Gem' 182
    'Blue Carpet' 32, 182
    'Bullion' 182
    'Buttercup' 32 ·
    'Chantreyland' 32
    'Duchess' 32
    'Jackanapes' 32
    'Jersey Gem' 32, 182
    'Lorna' 32
    'Nora Leigh' 182
    'Picotte' 32
    'Violetta' 32
  cucullata 182
  × *florariensis* 182
  gracilis 31, 182
    'Black Knight' 182
    'Lutea' 182
    'Major' 182
  jooi 182
  labradorica 182
  lutea 31
  odorata 182
    'Alba' 182
    'Double Russian' 182
    'Parma' 182
    'Roxina' 182
    'Royal Robe' 182
    'White Wonder' 182
  papilionacea 183
    'Priceana' 183
  pedata 183
  tricolor 31

visseriana   31
williamsii   31
  'Maggie Mott'   31
wittrockiana   31
violet   44, 48, 182
  bird's-foot   183
  confederate   183
  purple   182
  sweet   182
Viscaria alpina see *Lychnis alpina*
  151, 183

wake robin   177, 178
  ill-scented   178
wallflower
  English   *13,* 32
    'Blood Red'   33
    'Carmine King'   33
    'Cloth of Gold'   33
    'Eastern Queen'   33
    'Ellen Willmott'   33
    'Fire King'   33
    'Harper Crew'   33
    'Vulcan'   33
    'White Dame'   33
  Siberian   32, 33, 43, *70*
    'Golden Bedder'   33
    'Lemon Queen'   33
wandflower   122
waxbells
  yellow   145
weed
  Joe-Pye   128
whorlflower
  Himalayan   155
windflower   *73,* 95
wolf's-bane   93
wonder of Staffa   102
woodruff   36
wormwood   *74,* 100
  beach   101
  fringed   100
woundwort   174

yarrow   *72,* 92
  common   92
  fernleaf   92
  woolly   92